Fundamentals of Organic Reaction Mechanisms

Fundamentals of Organic Reaction Mechanisms

J. Milton Harris

Department of Chemistry
The University of Alabama in Huntsville
Huntsville, Alabama

Carl C. Wamser

Department of Chemistry
California State University
Fullerton, California

John Wiley & Sons, Inc.   New York   London   Sydney   Toronto

Library of Congress Cataloging in Publication Data:

Harris, Joe Milton, 1940-
    Fundamentals of organic reaction mechanisms.

    Includes bibliographies and index.
    1. Chemistry, Physical organic. 2. Chemical
reaction, Conditions and laws of. I. Wamser,
Carl C., 1944-   joint author. II. Title.
QD476.H34   547′.1′39   75-40275
ISBN 0-471-35400-7

Printed in the United States of America

10 9 8 7 6 5 4 3

Preface

Goals

This book is designed for a course that could be called either physical organic chemistry or organic reaction mechanisms. Such a course is frequently offered as a third-semester course in organic chemistry at the senior undergraduate level or as a full-year course at the beginning graduate level. Recognizing that courses of this type are highly varied at different institutions, this text is designed to be readily adjustable to the particular course and instructor. The text covers both theoretical and mechanistic aspects of physical organic chemistry in a broad, selective fashion without any attempt at comprehensive coverage. The extensive references to the original literature, plus a bibliography of suggested readings and reviews, allows ready supplementation of the material in the text. Conversely, many of the topics could be omitted at the instructor's discretion without any serious loss of continuity.

Organization

In view of the growing importance of theoretical aspects of organic reaction mechanisms, the first two chapters have been devoted to molecular orbital theory and its applications. The first chapter emphasizes those theoretical approaches that can be readily applied without extensive calculations: the Hückel method and the perturbational molecular orbital method of Dewar. The utilization of these theoretical methods is expanded and illustrated in the next chapter by presentation of theoretical treatments of aromaticity, nucleophilic aliphatic substitution, electrophilic aromatic substitution, and pericyclic reactions.

The remaining chapters present a detailed examination of organic reaction mechanisms and are intended to familiarize the student with the basic principles underlying organic reactions and the most useful approaches to mechanistic studies. Following an introductory mechanistic chapter, the remaining chapters each deal with a specific class of reactive intermediates: carbocations, carbanions, free radicals, carbenes, and excited states (photochemistry). Within this organizational scheme, all the most important mechanistic types are presented; for example, nucleophilic aliphatic substitution, electrophilic aromatic substitution, addition, and elimination reactions are included under carbocations. The most fundamental and generally useful concepts and techniques (such as linear free energy relationships, kinetics, and isotopic labeling) are introduced in Chapter 3 and further developed throughout the text as illustrative situations occur. The extensive problem sections following each chapter are designed to fortify and correlate the material presented within the chapter, as well as to introduce some related material that could not be specifically included in the text. Brief answers to the problems are included at the end of the book.

The authors are indebted to Professors Roger W. Binkley, William A. Pryor, Peter J. Stang, and S.D. Worley, who read through the manuscript and provided numerous valuable comments and criticisms. We are also particularly grateful to our wives, who did most of the typing and provided much needed support and patience.

J. MILTON HARRIS
CARL C. WAMSER

Fullerton, California

Contents

1

Theoretical Descriptions of Organic Molecules

The earliest theories of chemical bonding were presented by Lewis and Kossel and were based on the Bohr model of atomic structure.[1,2] Bohr imagined that an atom consisted of a positive nucleus surrounded by shells of electrons, first one shell with a limit of two electrons, then one with a limit of eight electrons, and so on. The most stable atoms result when the outermost shell is full, and it is this tendency to fill shells by sharing electrons (a covalent bond) or by donation or acceptance of electrons (an ionic bond) that leads to bonding. For example, the lithium fluoride molecule can be visualized as resulting from the donation of one electron from a lithium atom to a fluorine atom and the consequent electrostatic attraction of the ions thus formed (thus an ionic bond); each ion has filled electron shells (lithium—two electrons, fluorine—ten electrons). Similarly, the hydrogen molecule can be formed by sharing the single electrons of two hydrogen atoms in order to give each hydrogen atom two electrons and complete shells. In this covalent bond, an attraction between atoms exists because of attraction between one nucleus and an electron about another nucleus.

The Bohr theory of the atom was not sufficient for treatment of large atoms or molecules. Furthermore, the Kossel–Lewis theory of chemical bonding was inadequate for the prediction of such phenomena as molecular geometry. For example, water might be expected to have bond angles of 180° rather than the observed 105°. Also, this theory gave no rationalization of hindered rotation and structure about multiple bonds such as the double bond of ethylene.

The modern theory of chemical bonding is based on the quantum mechanics of Schrödinger, Heisenberg, Dirac, Jordan, Born, and others.[1,2] According to quantum mechanics the motion of an electron about a nucleus is described by a wave equation, the solution of which gives a volume about the nucleus of defined shape (an atomic orbital) in which there is certain probability the electron will be located. The orbitals each have a capacity of two electrons of opposite spins and a definite energy such that the shells of the Bohr atom are reproduced. Lowest in energy is the spherical 1s orbital, then the spherical 2s orbital and the three mutually perpendicular, dumbbell-shaped, 2p orbitals, **1**, and so on. Again, atoms are most stable when

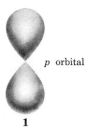

p orbital

1

valence orbitals are filled. Covalent bonding results when orbitals "overlap" and a pair of electrons of opposite spins is shared between atoms. Since all atomic orbitals except the s orbitals have a certain orientation, orbital overlap frequently has a directional character that leads to a preferred shape or geometry for the resulting molecule. For example, the oxygen atom has eight electrons about it, two in a 1s orbital, two in a 2s orbital, and four in 2p orbitals (or $1s^2 2s^2 2p_x^2 2p_y^1 2p_z^1$). Overlap with two hydrogen atoms to form water will then occur with the two unfilled and perpendicular 2p orbitals, **2**.

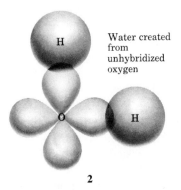

H

Water created from unhybridized oxygen

O   H

2

From this approach, water would be predicted to have bond angles of 90°. The observed bond angles of 105° have been rationalized on the basis that the hydrogen atoms repel each other (a better explanation follows).

In many cases, atoms are best described in terms of hybrid orbitals. For example, a 2s orbital and three 2p orbitals can be mathematically mixed, or hybridized, to yield four sp^3 hybrid orbitals directed toward the corners of a tetrahedron, **3** and **4**. Oxygen, of course, has a $1s^2 2s^2 2p^4$ electronic arrangement. If the six electrons in the second shell are placed in four sp^3 hybrid

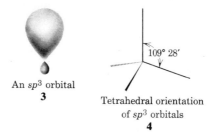

An sp^3 orbital

3

109° 28′

Tetrahedral orientation
of sp^3 orbitals

4

orbitals, two electrons would be unpaired, and overlap with the single electrons of two hydrogens would give a water molecule with bond angles of 109°28′; the actual bond angles of 105° for this molecule have been explained on the basis that lone pair–lone pair repulsion forces the lone pairs apart and the hydrogens together. Similarly, carbon has a $1s^2 2s^2 2p^2$ electronic arrangement. If the four electrons originally in the $2s$ and $2p$ orbitals are placed in the sp^3 orbitals, four bonds can be formed instead of two from the original electronic arrangement. This description in terms of hybrid orbitals is supported by the observation that carbon is virtually always tetravalent and does frequently form four tetrahedrally arranged bonds.

It is also possible to hybridize one $2s$ and only two $2p$ orbitals or one $2s$ and only one $2p$ orbital to give three sp^2 hybrid orbitals or two sp hybrid orbitals, respectively, **5** and **6**; the remaining unhybridized $2p$ orbitals, one

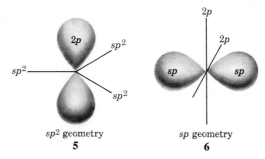

sp^2 geometry

5

sp geometry

6

in the first case and two in the second, are unchanged.

Bonds can be formed between atoms with $2p$ orbitals in two different ways: by "end-on" overlap, **7**, and by "side-by-side" overlap, **8**, of the

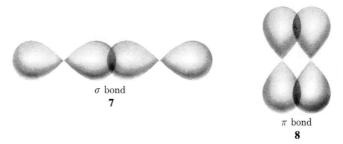

σ bond

7

π bond

8

orbitals. The bond in **7** is cylindrically symmetrical and as such is designated a σ bond. Overlap of hybrid orbitals can also occur in this end-on fashion to give σ bonds. The bond in **8** is not cylindrically symmetrical and is designated as a π bond. Union of two sp^2-hybridized carbons leads to a σ bond and a π bond, **9**. Since the π bond is not cylindrically symmetrical, rotation about the axis of the σ bond would break the π bond in **9**. It is this factor that accounts for the ethylenes being flat molecules capable of existing in *cis* and *trans* forms. Also, sp^2 hybrids are arranged trigonally and bond angles of 120° are formed between substituents on the carbons. Acetylenes can be formed by one σ and two π type overlaps of sp-hybridized carbon, **10**.

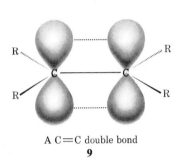

A C═C double bond
9

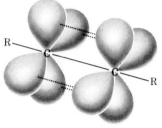

A C≡C triple bond
10

1.1 Molecular Orbital Theory

Overlap of atomic orbitals produces molecular orbitals that describe the location of electrons in a molecular environment under the influence of more than one nucleus. A most effective description of chemical bonding and molecular reactivity results from theoretical treatment of these molecular orbitals.[3-5] Ultimately, we hope that we will be able to predict the structure and properties of any organic species by use of molecular orbital calculations. Obviously, if it were possible to calculate the geometries and energies of reacting molecules at any point along a reaction coordinate (not just transition states or intermediates), the study of reaction mechanisms would be greatly simplified. We could essentially obtain a "movie" of atoms moving together and products splitting off with all of these movements on an absolute energy scale. This goal has not been reached, but much progress and many notable successes have been achieved. Calculations being used today are useful for prediction of such properties of stable molecules as total energies, bond orders, and bond lengths between atoms, bond angles, and charge densities at various atoms. These data can be used to predict relative reactivities and points of attack. Many calculations of the properties of reactive intermediates and transition states have also been performed.

The difficulties associated with the use of theoretical calculations for the study of organic reactions lies in part with the large number of atoms involved in even the most simple reactions. For example, nucleophilic substitution reactions of alkyl halides,

$$OH^- + CH_3Cl \rightarrow CH_3OH + Cl^- \qquad (1)$$

have been the subject of intense investigation for half a century, and are probably as simple and amenable to study as any organic reaction. Yet we must recall that this reaction involves not just the encounter of two molecules and constituent atoms, but also the interaction of these molecules with several solvent molecules. Thus, a complete theoretical treatment of the reaction of eq. 1 must include treatment of desolvation of the reactants, formation of some sort of encounter complex between the reactants and formation of a solvent shell about this species, and finally formation of two solvated product molecules.

As we have stated, progress is being made, and theoretical calculations provide one of our most powerful tools for the study of reaction mechanisms. In the following sections, we give an introduction to some of the more important types of theoretical techniques commonly used by organic chemists today. These techniques range from simple ones that can be performed in a matter of seconds to complicated ones that require the use of large computers. The emphasis, however, will be on those methods that can be applied easily with only the use of pencil and paper.

Qualitative Methods for Generating Molecular Orbitals

Molecular orbital theory results from an extension of the quantum mechanical treatment of atoms to molecules. Just as an electron in the field of a nucleus can be described by an atomic orbital, an electron in the field of the several nuclei constituting a molecule can be described by a molecular orbital covering the entire molecule. For example, the hydrogen molecule has two molecular orbitals (MOs), a bonding MO **11** and an antibonding MO **12**. The shapes of the MOs simply indicate a high probability that an electron will be within the designated confines. When the orbital signs of adjacent nuclei are the same (as in **11**), an attractive or bonding interaction is indicated, and when these signs change between adjacent nuclei (as in **12**), a repulsive or antibonding interaction is indicated. The

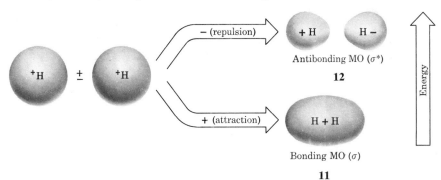

change of orbital sign between nuclei is referred to as a node. The foregoing MOs are cylindrically symmetrical and, as noted earlier, are designated σ orbitals. The asterisk of MO **12** is used to denote an antibonding MO.

As shown previously, the MOs of the hydrogen molecule can be derived by joining the atomic orbitals (AOs) of two hydrogen atoms. The general case for the combination of two AOs can be represented as in Figure 1.1a for degenerate (equal energies) orbitals, or as in Figure 1.1b for nondegenerate orbitals.[2] Each solid, horizontal line represents an energy level for an orbital.

Using these principles, it is quite simple to derive relative energy levels of molecular orbitals for molecules more complicated than H_2. For example,

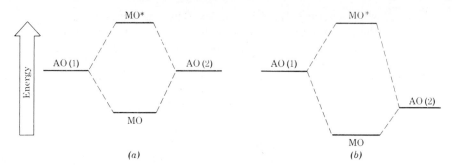

Fig. 1.1 *Bonding and antibonding MOs from the combination of (a) degenerate AOs and (b) nondegenerate AOs.*

the energy levels for diatomic molecules of the second period can be derived by combining the two atoms as shown in Figure 1.2. The $2p_z$ orbitals are considered to be oriented such that their interaction gives σ bonds, while the $2p_y$ and $2p_x$ orbitals give π bonds. Diagrams of this type are extremely useful in explaining observations such as the paramagnetism and ground-state triplet (two unpaired electrons) electron configuration of the oxygen molecule (see problem 4). This pairing method can also be used for more complicated systems such as those containing d valence orbitals or more than two atoms.

Organic chemists are in large part interested in MOs that result from combination of AOs to give π bonds. For example, the familiar delocalized MO of benzene results from overlap of the six unhybridized $2p$ orbitals of six sp^2 hybridized carbon atoms, **13**. The usual, quite effective, treatment is to consider the σ bonds as a framework for the π system (the Hückel approximation). In other words, σ bonds are considered as "localized" to two atoms while π MOs are "delocalized" over the entire carbon framework. From this point, unless stated otherwise, we will use the term MO to represent MOs of π systems.

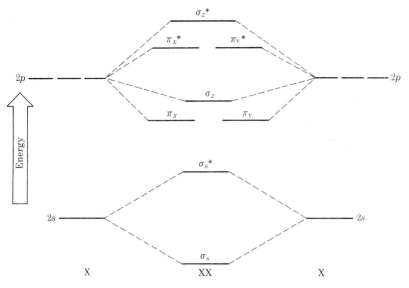

Fig. 1.2 *Energy levels for diatomic molecules (XX) of period two.*

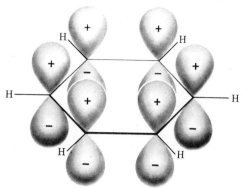

The benzene delocalized π system
13

Molecular orbitals of this type can also be obtained by the simple pairing method used earlier for diatomic molecules, since Figure 1.1 applies equally as well to the combination of AOs to produce MOs or to the combination of MOs to give new, larger MOs. Thus, to obtain the relative energies and the symmetries (arrangements of positive and negative lobes) of the MOs of 1,3-butadiene, we simply "join" two ethylenes, Figures 1.3 through 1.5. Ethylene contains π and π^* orbitals, since $2p$ orbitals can be joined in a

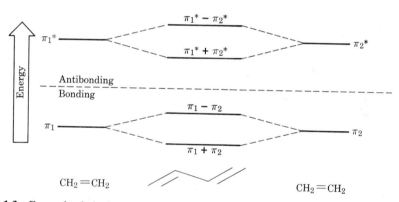

Fig. 1.3 *Energy levels for butadiene.*

bonding or an antibonding manner. Actually, the four AOs in each MO do not contribute equally to the MO; in the following section on Hückel molecular orbital theory, we will examine one method for calculating the contribution of each AO to an MO. In certain instances, however, the orbital symmetries alone will prove to be useful.

Any π system consisting of n carbons will have n MOs. Thus, butadiene has four MOs, two bonding and two antibonding, as seen in Figure 1.5, each of which is denoted by ψ_i where $i = 1$ through 4. Four electrons can be placed in the two bonding orbitals to produce two π bonds which is, of course, consistent with the structure of butadiene.

The dotted lines in Figure 1.4 represent an antibonding interaction or node between positive and negative lobes of adjacent AOs. The energy of an MO is directly related to the number of nodes in the MO. Thus, to derive

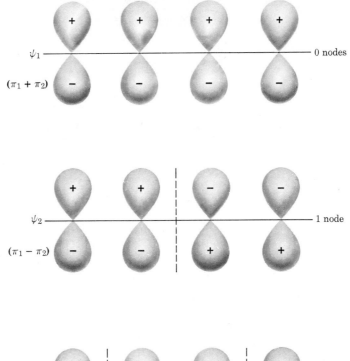

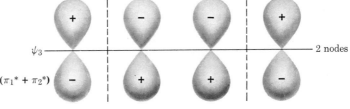

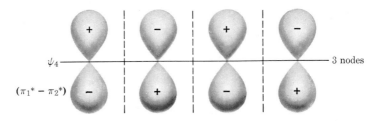

Fig. 1.4   *Approximate molecular orbitals for butadiene.*

the symmetry of an MO, it is necessary only to assign no nodes to the orbital of lowest energy, one node to the orbital of next highest energy, and so on. The symmetry of the MOs for butadiene as represented in Figure 1.4 could have been derived more rapidly by sketching the four sets of p orbitals and choosing signs for the lobes such that there was one set with no node, one with one, one with two, and one with three. The only restriction on this

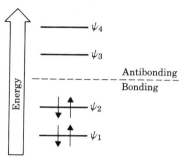

Fig. 1.5 *Energy levels for butadiene.*

technique is that the wave function must be either antisymmetric or symmetric with respect to any existing molecular symmetry element. For example, butadiene has a twofold axis of symmetry, and all of its MOs should be either symmetric or antisymmetric with respect to this axis. The hypothetical molecular orbital **14** fails in this regard, since it is symmetric to rotation if atoms 2 and 3 are viewed, but antisymmetric if atoms 1 and 4 are viewed.

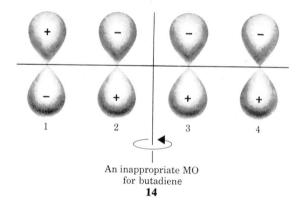

An inappropriate MO
for butadiene
14

1.2 Hückel Molecular Orbital Theory

The Hückel molecular orbital (HMO) theory was introduced in 1931, and its use, along with valence bond theory, was primarily responsible for quantum theory becoming a tool for the study of organic chemistry.[6,7] As noted earlier, the molecular orbital method results from extension of the quantum mechanical treatment of atoms to molecules. Exact solution of the quantum mechanical equations is extremely difficult and presently impossible for all except the smallest atoms. Thus, complete solution of the equations and complete description of organic molecules containing large numbers of atoms is seen to be even more difficult. Hückel avoided this problem by treating certain insolvable portions of the quantum mechanical equations as parameters and by expressing the energies of molecular orbitals in terms of these parameters.

Molecular orbital theory, as applied by present-day researchers, has gone far beyond the HMO method in complexity and accuracy of results; however, the HMO method does provide good qualitative predictions of molecular properties and continues in active use. Additionally, derivation and use of the HMO method gives an excellent illustration of the quantum mechanical method and the type and form of the resulting predictions.

Derivation of the Variation Equations[3-5, 8-10]

The best theoretical treatment of atoms is based on quantum mechanics and the Schrödinger equation,

$$H\psi = E\psi \tag{2}$$

where H is the Hamiltonian operator, ψ is the wavefunction or eigenfunction describing the atomic orbital, and E is the energy of the particular eigenfunction.

The form of eq. 2 is a very common one in quantum mechanics and may be represented generally as

$$(\alpha) f(x) = n f(x) \tag{3}$$

or

$$(\text{operator})(\text{eigenfunction}) = (\text{eigenvalue})(\text{eigenfunction}) \tag{4}$$

The result is that operation of the operator α on the eigenfunction $f(x)$ results in some multiple of the eigenfunction. For example, e^{cx} is an eigenfunction of the operator d/dx since

$$\frac{d}{dx}(e^{cx}) = ce^{cx} = \text{multiple of } e^{cx} \tag{5}$$

and x is not an eigenfunction of d/dx since

$$\frac{d}{dx}(x) = 1 \neq \text{multiple of } x \tag{6}$$

One form of quantum mechanics (Dirac) is based on this operator formalism of which the Schrödinger equation is just one result. According to this particular mode of quantum mechanics:

1. Any measurable property of a system can be symbolized by an operator.
2. States of a system are represented by eigenfunctions.
3. Operation of an operator on an eigenfunction gives an eigenvalue that is the result of a measurement of the property corresponding to the operator.

We are interested in energy, and the Hamiltonian operator is the energy operator. Thus, if the eigenfunction of some molecular state is represented by ψ, operation on ψ with H yields the energy of this state:

$$H\psi = E\psi \tag{7}$$

where H is the Hamiltonian and E is the energy. The form of H is known exactly. Therefore, to solve any molecular system, we simply need ψ. Unfortunately, the mathematical techniques necessary to solve eq. 7 for any but the most simple systems are not available. Some of the mathematical problems will be considered in the next section, but, for the moment, suffice it to say that very excellent results can be obtained by use of approximate eigenfunctions and Hamiltonian operators. As it turns out, the HMO treatment does not require detailed knowledge of the mathematical form of H or ψ.

Now that we have the Schrödinger equation, it remains to obtain eigenfunctions. This is done by utilizing the LCAO-MO approximation (linear combination of atomic orbitals to give molecular orbitals). Simply stated, the MO is given by a linear combination of AOs of the constituent atoms of the molecule

$$\psi_\mu = \sum_i a_{\mu i}\phi_i$$

$$\mu = 1, 2, 3, \ldots, n \qquad i = 1, 2, 3, \ldots, n$$

(8)

where $a_{\mu i}$ is a coefficient for each atomic orbital ϕ_i, μ denotes the MO under consideration, i is the atom, and n is the number of atoms in the molecule. For example, hydrogen will have two MOs, $\mu = 1$ or 2, each containing some combination of multiples of each ϕ_i, where $i = 1$ or 2:

$$H_1 \text{---} H_2$$

$$\psi_1 = a_{11}\phi_1 + a_{12}\phi_2$$

(9)

$$\psi_2 = a_{21}\phi_1 + a_{22}\phi_2$$

(10)

The next job is to find the best set of values for the coefficients $a_{\mu i}$ to make ψ as good an approximation as possible to the real eigenfunction. This is best done by use of the *Variation Principle:* any wave function other than the correct one will yield a value for the energy of an orbital that is numerically greater than the true value. To express this principle mathematically (proceed to eqs. 16, 20, 21, and finally 30 in order to skip the derivation), we need to rearrange the Schrödinger equation. Multiplying by ψ yields

$$\psi H \psi = \psi E \psi$$

(11)

One of the laws for the manipulation of operators states that

$$f(x)\,\alpha\,f(x) \neq \alpha\,(f(x))^2$$

(12)

Since H is an operator and E is not

$$\psi H \psi = E\psi^2$$

(13)

Integrating over all space (or all coordinates τ)

$$\int \psi H \psi \, d\tau = E \int \psi^2 \, d\tau$$

(14)

$$E = \frac{\int \psi H \psi \, d\tau}{\int \psi^2 \, d\tau}$$

(15)

According to the Variation Principle, the E obtained from eq. 15 by use of an approximate ψ will be greater than the true value of E, designated as

E_0. Thus, a mathematical statement of the Variation Principle is

$$E = \frac{\int \psi H \psi \, d\tau}{\int \psi^2 \, d\tau} \geq E_0 \tag{16}$$

In order to make ψ as good an approximation as possible to the true eigenfunction (ψ_0), the parameters a_i $\left(\text{from } \psi = \sum_i a_i \phi_i\right)$ should be chosen such that E is made as small as possible. This is done quite simply by minimizing E with respect to each of the possible coefficients,

$$\frac{\partial E}{\partial a_i} = 0 \tag{17}$$

In order to obtain values of the coefficients a_i and the energy of each MO, E_μ, it is advantageous to carry out the operation indicated by eq. 17 on the general case as expressed by eq. 8. Thus, substituting eq. 8 into eq. 16 and omitting MO subscript indexes for ease of handling gives

$$E = \frac{\int \left(\sum_i a_i \phi_i\right) H \left(\sum_j a_j \phi_j\right) d\tau}{\int \left(\sum_i a_i \phi_i\right) \left(\sum_j a_j \phi_j\right) d\tau} \tag{18}$$

$$E = \frac{\sum_i \sum_j a_i a_j \int \phi_i H \phi_j \, d\tau}{\sum_i \sum_j a_i a_j \int \phi_i \phi_j \, d\tau} \tag{19}$$

For convenience, the following substitution is made;

$$H_{ij} = \int \phi_i H \phi_j \, d\tau \tag{20}$$

$$S_{ij} = \int \phi_i \phi_j \, d\tau \tag{21}$$

Then,

$$E = \frac{\sum_i \sum_j a_i a_j H_{ij}}{\sum_i \sum_j a_i a_j S_{ij}} = \frac{x}{y} \tag{22}$$

Now, applying eq. 17,

$$\frac{\partial E}{\partial a_i} = \frac{y(\partial x/\partial a_i) - x(\partial y/\partial a_i)}{y^2} = 0 \Bigg\} \quad i = 1, 2, \ldots, n \tag{23}$$

$$0 = \frac{1}{y} \frac{\partial x}{\partial a_i} - \frac{x}{y^2} \frac{\partial y}{\partial a_i} \tag{24}$$

$$0 = \frac{\partial x}{\partial a_i} - \frac{x}{y} \frac{\partial y}{\partial a_i} \tag{25}$$

And substituting eq. 22 yields

$$\frac{\partial x}{\partial a_i} - E \frac{\partial y}{\partial a_i} = 0 \tag{26}$$

Now differentiation of x and y (as given in eq. 22) with respect to a_i gives

$$\frac{\partial x}{\partial a_i} = \sum_j 2a_j H_{ij} \tag{27}$$

$$\frac{\partial y}{\partial a_i} = \sum_j 2a_j S_{ij} \tag{28}$$

and eq 26 becomes

$$\sum_j 2a_j H_{ij} - E \sum_j 2a_j S_{ij} = 0 \qquad i, j = 1, 2, \ldots, n \tag{29}$$

$$\boxed{\sum_j a_j (H_{ij} - ES_{ij}) = 0 \qquad i, j = 1, 2, \ldots, n} \tag{30}$$

which is the variation equation. This set of simultaneous equations is the final result of application of the Variation Principle and will allow calculation of the coefficients in eq. 8, if the energy values were chosen properly. To illustrate the form of the variation equations, consider the case of the allyl π system in which atoms i and j vary from 1 to 3 and i remains constant for each individual equation:

$$a_1(H_{11} - ES_{11}) + a_2(H_{12} - ES_{12}) + a_3(H_{13} - ES_{13}) = 0$$

$$a_1(H_{21} - ES_{21}) + a_2(H_{22} - ES_{22}) + a_3(H_{23} - ES_{23}) = 0 \tag{31}$$

$$a_1(H_{31} - ES_{31}) + a_2(H_{32} - ES_{32}) + a_3(H_{33} - ES_{33}) = 0$$

The allyl π system has three MOs ($\mu = 1, 2, 3$) of different energies (Figure 1.6). Thus, there are three possible values of E that permit solution of the above set of simultaneous equations. Each of these energy values would lead to a set of a_1, a_2, and a_3 values ($a_{\mu i}$).

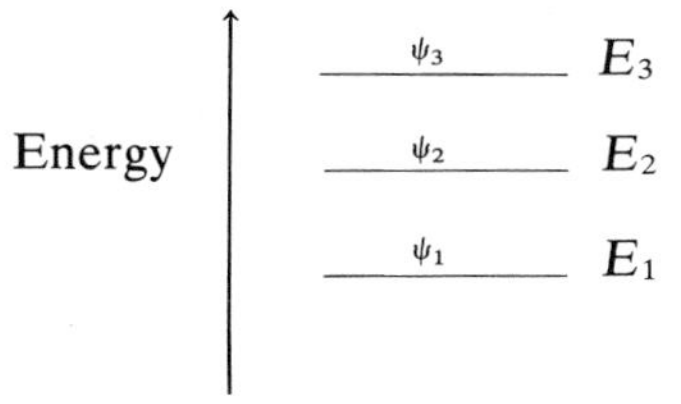

Energy

$$\psi_3 \qquad E_3 \qquad \psi_3 = a_{31}\phi_1 + a_{32}\phi_2 + a_{33}\phi_3$$

$$\psi_2 \qquad E_2 \qquad \psi_2 = a_{21}\phi_1 + a_{22}\phi_2 + a_{23}\phi_3$$

$$\psi_1 \qquad E_1 \qquad \psi_1 = a_{11}\phi_1 + a_{12}\phi_2 + a_{13}\phi_3$$

$$\phi_\mu = \sum_i a_{\mu i}\phi_i \quad (\text{eq. 8})$$

Fig. 1.6 *Energy levels and molecular orbitals for the allyl system.*

Simultaneous equations of the type of eq. 31 can be solved by separating the $(H_{ij} - ES_{ij})$ portion into a determinant (the so-called secular determinant, eq. 32) and solving this determinant for values of E. The E values can then be substituted into eq. 30 to determine $a_{\mu i}$. All of this assumes, of course, that the integrals H_{ij} and S_{ij} can be solved, which they cannot. When the exact form of the Hamiltonian operator is considered, these integrals become very complex (p. 36). It is at this point that the HMO method separates from more advanced methods; instead of attempting to solve the integrals by necessarily approximate methods, the integrals are treated as parameters, and energy is expressed in terms of these parameters.

$$|H_{ij} - ES_{ij}| = \text{Secular determinant} \tag{32}$$

$$|H_{ij} - ES_{ij}| = 0 \tag{33}$$

$$|H_{ij} - ES_{ij}| = \begin{vmatrix} H_{11}-ES_{11} & H_{12}-ES_{12} & \cdots & H_{1n}-ES_{1n} \\ H_{21}-ES_{21} & H_{22}-ES_{22} & \cdots & H_{2n}-ES_{2n} \\ \cdot & \cdot & & \cdot \\ \cdot & \cdot & & \cdot \\ \cdot & \cdot & & \cdot \\ H_{n1}-ES_{n1} & H_{n2}-ES_{n2} & \cdots & H_{nn}-ES_{nn} \end{vmatrix} = 0 \tag{34}$$

The Hückel Method

Instead of solving the secular determinant and numerically evaluating the integrals H_{ij} and S_{ij}, a very difficult and sometimes impossible task, we will simply treat the integrals as parameters and make the assumptions that constitute the Hückel molecular orbital method as it is applied to conjugated hydrocarbons. These assumptions are:

1. $H_{ii} = \alpha$ $\hspace{6.5cm}$ (35)
2. $H_{ij} = \beta$ if i and j are bonded; if they are not bonded, $H_{ij} = 0$ $\hspace{1cm}$ (36)
3. $S_{ii} = 1$, $S_{ij} = 0$ for $i \neq j$ $\hspace{5cm}$ (37)
4. σ bonds are localized and can be considered as a rigid framework for the π electrons.

The integral α is usually referred to as the Coulomb integral and represents the energy of an electron in a carbon $2p$ orbital. As long as we consider an all-carbon system, the assumption that all H_{ii} are equal is a good assumption; if heteroatoms are introduced, then different values for α must be used.[4]

The integral β is usually referred to as the resonance integral and represents the energy of interaction of two atomic orbitals on adjacent atoms i and j. The assumption that all βs are equal is rather drastic, since the interactions of two AOs in conjugated hydrocarbons can vary significantly. For example, in 1,3-butadiene $\beta_{12} \neq \beta_{23}$, because of the large variation in the bond lengths between atoms 1 and 2 and atoms 2 and 3, **15**.

15

The integral S_{ij} is referred to as the overlap integral and, for the case of $i = j$, is equal to unity (eq. 38). Atomic orbitals are defined such that eq. 38 holds (the so-called normalization condition that ensures unit probability of finding an electron having the wavefunction ϕ somewhere in all space), so

$$\int \phi_i^2 \, d\tau = 1 \tag{38}$$

the stipulation that $S_{ii} = 1$ is justified. For the case of i adjacent to j, S_{ij} can be calculated to be about 0.25 for normal bond distances between atoms i and j;[3] thus, the assumption that $S_{ij} = 0$ when $i \neq j$ is certainly not correct.

However, in view of the other assumptions already made, the reasonable results that are obtained, and the tremendous simplification of mathematical procedures, the third approximation does not seem so damaging.

The final assumption that the σ bonds are localized is probably a very good one and seems to be justified by a large body of chemical data.

At this point, the reader may justifiably feel that the HMO method uses so many dubious approximations that it is probably worthless; however, such is not the case, since results are obtained that allow a great deal of insight into the behavior of conjugated hydrocarbons. That the method does work is in large degree due to a fortuitous canceling of effects and to the types of molecules to which it is applied.[8] In essence, we can say that the HMO method works in spite of the approximations made, and the method is made mathematically simple as a result of these approximations.

Now that we have the HMO assumptions, it is informative to set up the secular determinant for the simple molecule 1,3-butadiene and see how solution of the variation equations has been simplified. Examination of eq. 34 shows that the determinant for 1,3-butadiene is

$$
\begin{vmatrix}
H_{11}-ES_{11} & H_{12}-ES_{12} & H_{13}-ES_{13} & H_{14}-ES_{14} \\
H_{21}-ES_{21} & H_{22}-ES_{22} & H_{23}-ES_{23} & H_{24}-ES_{24} \\
H_{31}-ES_{31} & H_{32}-ES_{32} & H_{33}-ES_{33} & H_{34}-ES_{34} \\
H_{41}-ES_{41} & H_{42}-ES_{42} & H_{43}-ES_{43} & H_{44}-ES_{44}
\end{vmatrix}=0
$$

Application of Hückel's assumptions 1 through 3 gives:

$$
\begin{vmatrix}
\alpha-E & \beta & 0 & 0 \\
\beta & \alpha-E & \beta & 0 \\
0 & \beta & \alpha-E & \beta \\
0 & 0 & \beta & \alpha-E
\end{vmatrix}=0
$$

Dividing by β and substituting

$$
x=\frac{\alpha-E}{\beta}
$$

yields

$$
\begin{vmatrix}
x & 1 & 0 & 0 \\
1 & x & 1 & 0 \\
0 & 1 & x & 1 \\
0 & 0 & 1 & x
\end{vmatrix}=0
$$

which can be readily solved (see Appendices 1 and 2 for techniques of solving determinants and determining roots of equations) to give

$$
x^4-3x^2+1=0
$$

$$
x=\pm1.62,\ \pm0.62
$$

Substituting for x gives values for the energies of the four MOs of 1,3-butadiene (Figure 1.7). Four electrons have been placed in the two orbitals of lowest energy, since 1,3-butadiene has four electrons. In all of our calculations, α will be treated as an energy zero, and orbitals with an energy of α will be referred to as nonbonding molecular orbitals (NBMOs). The

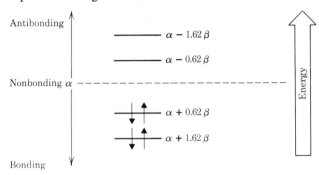

Fig. 1.7 *Energies for the four MOs of butadiene.*

other orbitals will be designated bonding or antibonding depending on their energy with respect to α.

The total energy can be calculated by using

$$E_\pi = \sum_\mu n_\mu E_\mu \tag{39}$$

where n_μ is the number of electrons in MO μ and E_μ is the energy of this MO. Thus, for butadiene

$$E_\pi = 2(\alpha + 1.62\beta) + 2(\alpha + 0.62\beta)$$
$$E = 4\alpha + 4.48\beta$$

The gain in stabilization, as a result of conjugation or delocalization of a system (delocalization energy—DE), can be defined as

$$DE = E_\pi - E_{loc} \tag{40}$$

where E_{loc} is the energy of the system if all bonds are localized. E_{loc} can be calculated by assuming that all H_{ij} values are zero except those for atoms that are joined by a double bond. For butadiene then, H_{23} would be treated as zero, and a new determinant would be solved to yield E_{loc}. Actually, this method gives the energy of a localized double bond as equal to that of ethylene ($2\alpha + 2\beta$). Thus, an easier method for calculation of E_{loc} is given by

$$E_{loc} = n_\pi(\alpha + \beta) + n_i\alpha \tag{41}$$

where n_π is the number of electrons in double bonds (four for butadiene), and n_i is the number of unpaired electrons (one for the allyl radical). As stated in its definition, α is the energy of an isolated electron in a $2p$ orbital. Solution of eq. 40 gives a delocalization energy of 0.48β for butadiene, implying that the conjugated molecule is more stable, by this amount, than the model in which the π bonds are isolated.

Delocalization energies are a poor measure of the actual importance of conjugation in a molecule. In Chapter 2, a more chemically significant measure of electron delocalization (again using E_π) is given.

Now that it is possible to determine the energies of the molecular orbitals of any conjugated hydrocarbon (limited only by the time required to solve the determinant), the next step in the HMO method is to obtain the coefficients for the AOs in each MO:

$$\psi_\mu = \sum_i a_{\mu i}\phi_i \tag{8}$$

The coefficients can be determined by substituting the known values of E into the variation equation (eq. 30). Expansion of eq. 30 and substitution of $x = (\alpha - E)/\beta$ gives

$$
\begin{aligned}
a_1 x + a_2 + \cdots + a_n &= 0 \\
a_1 + a_2 x + \cdots + a_n &= 0 \\
&\;\;\vdots \\
a_1 + a_2 + \cdots + a_n x &= 0
\end{aligned}
\tag{42}
$$

The coefficients are then obtained by inserting x, solving the simultaneous equations, and applying the normalization condition:

$$
\sum_i a_i^2 = 1
\tag{43}
$$

Equation 43 derives from the LCAO-MO approximation, use of normalized AOs (i.e., $\int \phi^2 \, d\tau = 1$), and the requirement that molecular orbitals also be normalized (i.e., $\int \psi^2 \, d\tau = 1$). Thus,

$$
\int \psi^2 \, d\tau = 1 \qquad \text{(normalization of MOs)}
\tag{44}
$$

$$
\int \left(\sum_i a_i \phi_i \right)^2 d\tau = 1
\tag{45}
$$

$$
\sum_i a_i^2 \int \phi_i^2 \, d\tau = 1
\tag{46}
$$

$$
\int \phi_i^2 \, d\tau = 1 \qquad \text{(normalization of AOs)}
\tag{47}
$$

$$
\sum_i a_i^2 = 1
\tag{43}
$$

To illustrate the solution for MO coefficients, we will consider the allyl case, **16**. The secular determinant for allyl must be

16

Allyl π system
(cation, radical, or anion)

$$
\begin{vmatrix}
x & 1 & 0 \\
1 & x & 1 \\
0 & 1 & x
\end{vmatrix} = 0
$$

and the variation equations must be

$$
\begin{aligned}
a_1 x + a_2 \qquad\quad &= 0 \\
a_1 + a_2 x + a_3 \; &= 0 \\
a_2 + a_3 x &= 0
\end{aligned}
$$

The roots of the determinant are

$$
x = -\sqrt{2}, \; x = 0, \; x = \sqrt{2}
$$

Applying $x = -\sqrt{2}$, the variation equations become

$$-\sqrt{2}\, a_1 + a_2 = 0$$

$$a_1 - \sqrt{2}\, a_2 + a_3 = 0$$

$$a_2 - \sqrt{2}\, a_3 = 0$$

From the first equation,

$$a_2 = \sqrt{2}\, a_1$$

Substitution into the third equation gives

$$a_1 = a_3$$

Normalization (eq. 43) gives

$$a_1^2 + a_2^2 + a_3^2 = 1$$

or

$$a_1^2 + (\sqrt{2}\, a_1)^2 + a_1^2 = 1$$

$$4 a_1^2 = 1$$

$$a_1 = \frac{1}{2}$$

Therefore,

$$a_2 = \frac{1}{\sqrt{2}}, \; a_3 = \frac{1}{2}$$

And

$$\psi_1 = \frac{1}{2}\, \phi_1 + \frac{1}{\sqrt{2}}\, \phi_2 + \frac{1}{2}\, \phi_3$$

Solution with $x = 0$ and $\sqrt{2}$ would give the other two wavefunctions for allyl.

This method for determining coefficients becomes unwieldy for larger systems, and an alternative method (the method of cofactors discussed later) is usually used. However, the alternative method does not work for systems with degenerate orbitals—orbitals of equal energy. In case of degenerate systems, we must use the present, longer method and consider the symmetry and arrangement of nodes in the MOs (see p. 8). To illustrate, the cyclopropenyl system has two degenerate antibonding MOs, **17** and **18**, for

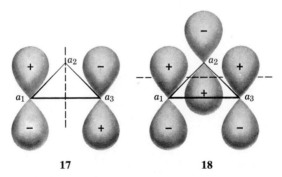

17   18

which $x = 1$. That **17** and **18** are degenerate is evident from the fact that both orbitals have two nodes. The secular determinant for the cyclopropenyl

π system is

$$\begin{vmatrix} x & 1 & 1 \\ 1 & x & 1 \\ 1 & 1 & x \end{vmatrix} = 0$$

and

$$x = -2, x = 1, x = 1$$

The variation equations are

$$a_1 x + a_2 + a_3 = 0$$
$$a_1 + a_2 x + a_3 = 0$$
$$a_1 + a_2 + a_3 x = 0$$

Inserting one solution for x, where $x = 1$, gives

$$a_1 + a_2 + a_3 = 0$$
$$a_1 + a_2 + a_3 = 0$$
$$a_1 + a_2 + a_3 = 0$$

Additional conditions must be provided if these equations are to be solved. Observation of MO **17** shows that a node passes through atom 2 and

$$a_2 = 0$$

Therefore, from the previous variation equation,

$$a_1 + a_3 = 0$$
$$a_1 = -a_3$$

Normalization gives

$$a_1^2 + a_2^2 + a_3^2 = 1$$
$$a_1^2 + 0 + a_1^2 = 1$$
$$a_1 = \frac{1}{\sqrt{2}} \phi_3$$

And

$$\psi_{17} = \frac{1}{\sqrt{2}} \phi_1 - \frac{1}{\sqrt{2}} \phi_3$$

For MO **18**, coefficients a_1 and a_3 must be equal, since, as stated earlier, each MO must be symmetric (or antisymmetric) with respect to existing molecular symmetry elements and cyclopropenyl has a plane of symmetry between atoms 1 and 3; they must be equivalent. Therefore,

$$a_1 = a_3$$

From the variation equation for $x = 1$

$$a_1 + a_2 + a_3 = 0$$

and we get

$$a_1 + a_2 + a_1 = 0$$
$$a_2 = -2a_1$$

Normalization gives

$$a_1^2 + a_2^2 + a_3^2 = 1$$

$$a_1^2 + (-2a_1)^2 + a_1^2 = 1$$

$$a_1 = \frac{1}{\sqrt{6}}$$

And

$$\psi_{18} = \frac{1}{\sqrt{6}} \phi_1 - \frac{2}{\sqrt{6}} \phi_2 + \frac{1}{\sqrt{6}} \phi_3$$

Symmetry conditions can be applied to simplify coefficient determination for any suitable molecule: for butadiene,

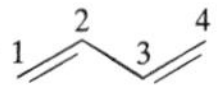

$$a_1 = \pm a_4 \qquad a_2 = \pm a_3$$

or for methylenecyclopropene,

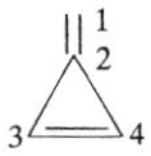

$$a_3 = \pm a_4$$

However, for all molecules except those with degenerate MOs, it is simpler to determine coefficients by the method of cofactors.[3] This method consists of applying the following three equations:

$$\frac{a_i}{a_1} = + \frac{\text{cofactor } i}{\text{cofactor } 1} \quad \text{if} \quad i = \text{odd} \tag{48}$$

$$\frac{a_i}{a_1} = - \frac{\text{cofactor } i}{\text{cofactor } 1} \quad \text{if} \quad i = \text{even} \tag{49}$$

$$a_1^2 = \frac{1}{\sum (a_i/a_1)^2} \tag{50}$$

The cofactor of the ith term is the secular determinant with the ith column and one row removed; the top row is usually chosen, but this is not necessary (refer to the following example or Appendix 1). Equation 50 can be derived from the normalization condition by dividing by a_1^2:

$$\frac{\sum a_i^2}{a_1^2} = \frac{1}{a_1^2} \tag{51}$$

$$a_1^2 = \frac{1}{\sum (a_i/a_1)^2} \tag{50}$$

As an example of the use of the method of cofactors, we will now determine the coefficients of the four AOs in each of the four MOs of 1,3-butadiene. The first step is to determine the general form of the ratios of the cofactors from the determinant for this molecule (p. 15):

$$\frac{a_1}{a_1} = 1$$

$$\frac{a_2}{a_1} = -\frac{\begin{vmatrix} 1 & 1 & 0 \\ 0 & x & 1 \\ 0 & 1 & x \end{vmatrix}}{\begin{vmatrix} x & 1 & 0 \\ 1 & x & 1 \\ 0 & 1 & x \end{vmatrix}} = \frac{-(x^2 - 1)}{x^3 - 2x}$$

$$\frac{a_3}{a_1} = +\frac{\begin{vmatrix} 1 & x & 0 \\ 0 & 1 & 1 \\ 0 & 0 & x \end{vmatrix}}{x^3 - 2x} = \frac{x}{x^3 - 2x} = \frac{1}{x^2 - 2}$$

$$\frac{a_4}{a_1} = -\frac{\begin{vmatrix} 1 & x & 1 \\ 0 & 1 & x \\ 0 & 0 & 1 \end{vmatrix}}{x^3 - 2x} = \frac{-1}{x^3 - 2x}$$

Now inserting one of the solutions for x (-1.62) gives the ratios a_i/a_1 (Table 1.1) and applying eq. 50 gives a_i. The remaining coefficients can then be determined from the a_i/a_1 ratios. Repeating for each x gives:

$$\psi_1 = 0.37\phi_1 + 0.60\phi_2 + 0.60\phi_3 + 0.37\phi_4$$
$$\psi_2 = 0.60\phi_1 + 0.37\phi_2 - 0.37\phi_3 - 0.60\phi_4$$
$$\psi_3 = 0.60\phi_1 - 0.37\phi_2 - 0.37\phi_3 + 0.60\phi_4$$
$$\psi_4 = 0.37\phi_1 - 0.60\phi_2 + 0.60\phi_3 - 0.37\phi_4$$

The magnitude of the coefficient gives the contribution of each AO to each MO and allows a sketch of the MOs to be made; we have done this qualitatively in Figure 1.8, although it could, of course, be done quantitatively. The dotted lines in Figure 1.8 follow the signs of the lobes of each AO and illustrate bonding and antibonding interactions. Points at which the dotted line crosses the solid line are nodes corresponding to antibonding interactions.

Table 1.1 *Determination of a_i for ψ_1 of 1,3-Butadiene, $x = -1.62$*

i	$\dfrac{a_i}{a_1}$	$\left(\dfrac{a_i}{a_1}\right)^2$	a_i
1	1.00	1.00	0.37
2	1.62	2.62	0.60
3	1.62	2.62	0.60
4	1.00	1.00	0.37
		$\Sigma = 7.24$	

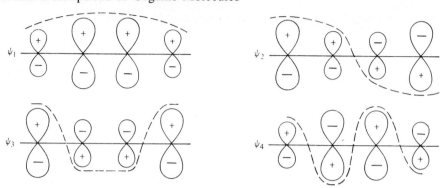

Fig. 1.8   *The molecular orbitals of butadiene.*

In summary, the HMO method results from application of the LCAO-MO approximation and the Variation Principle to the Schrödinger equation to obtain the variation equation. Certain assumptions are then made to simplify solution of the secular determinant in terms of the two parameters α and β. Application of the HMO method involves two steps:

1.   Solution of the secular determinant for E

$$|H_{ij} - ES_{ij}| = 0 \tag{33}$$

2.   Substitution of these values of E into the variation equation to yield a_j

$$\sum_j a_j (H_{ij} - ES_{ij}) = 0 \qquad i = 1, 2, \ldots, n \tag{30}$$

The Hückel molecular orbital energy values have been used in the treatment of many problems; the first such example was the calculation of the relative energies of cyclic conjugated polyenes to give the important $(4n+2)$ rule [monocyclic conjugated hydrocarbons containing $(4n+2)\pi$ electrons are exceptionally stable and aromatic—see Chapter 2].[6,7] It is important to realize that before Hückel's pioneering work, no suitable theory existed for such fundamental chemistry as the properties of double bonds and of benzene. The Kekule representation of benzene, which contains rapidly alternating single and double bonds, was, of course, untenable. Hückel's separation of σ and π electrons and his parametric treatment of the integrals from the secular equations must be regarded as among the truly significant advances in organic chemistry. Other applications include calculation of resonance energies (delocalization energy discussed previously), predictions of orientation in electrophilic aromatic substitution, estimates of the stabilities of charge transfer complexes, and calculations of ionization potentials and polarographic oxidation potentials.[3-5, 8-10]

The MO coefficients are also extremely useful and can be used to calculate such quantities as electron density and charge density at various atoms in a molecule, bond order between atoms, and free valence:[3-5]

Electron density

$$q_i = \sum_\mu n_\mu a_{\mu i}^2 \tag{52}$$

where n_μ is the number of electrons in ψ_μ

Charge density

$$\delta_i = 1 - q_i \tag{53}$$

π bond order
$$p_{ij} = \sum_{\mu} n_{\mu} a_{\mu i} a_{\mu j} \tag{54}$$

Free valence
$$F_i = 1.732 - \sum_i p_{ij} \tag{55}$$

Electron density and charge density have been used to predict the position of nucleophilic or electrophilic attack on a molecule. Bond orders have been used to estimate bond length; an increase in bond order should lead to a decrease in bond length. The 1.732 of eq. 55 is the maximum possible π bonding order to any atom; this number comes from the bond order to the central atom of trimethylenemethane. If an atom has less than this maximum amount of π bonding order between it and adjacent atoms, it is said to have unused bonding power or free valence and is considered to be reactive in proportion to the magnitude of this free valence.

Solutions of eq. 52 through 55 for electron density, charge density, π bond order, and free valence for butadiene follow. First, electron density:

$$q_1 = n_1 a_{11}^2 + n_2 a_{21}^2 + n_3 a_{31}^2 + n_4 a_{41}^2$$
$$q_1 = 2(0.37)^2 + 2(0.60)^2 + 0(0.60)^2 + 0(0.37)^2$$
$$q_1 = 1.00$$
$$q_2 = n_1 a_{12}^2 + n_2 a_{22}^2$$
$$q_2 = 2(0.60)^2 + 2(0.37)^2 = 1.00$$

The symmetry of butadiene demands that $q_1 = q_4$ and $q_2 = q_3$. Charge density is given by $1 - q_i$ and equals zero for each position of butadiene.

The π bond order between atoms 1 and 2 is given by

$$p_{12} = n_1 a_{11} a_{12} + n_2 a_{21} a_{22}$$
$$p_{12} = 2(0.37)(0.60) + 2(0.60)(0.37)$$
$$p_{12} = 0.89 = p_{34}$$

And between atoms 2 and 3,

$$p_{23} = n_1 a_{12} a_{13} + n_2 a_{22} a_{23}$$
$$p_{23} = 2(0.60)(0.60) + 2(0.37)(-0.37) = 0.45$$

As would be expected from the usual manner of drawing butadiene with double bonds between atoms 1 and 2 and between 3 and 4, there is more π bond order between these atoms than between atoms 2 and 3; however, this calculation does indicate significant π overlap or conjugation between atoms 2 and 3.

The free valence of atom 1 is given by

$$F_1 = 1.732 - p_{12} = 1.732 - 0.89 = 0.84$$

and for atom 2

$$F_2 = 1.732 - p_{12} - p_{23} = 0.39$$

Thus, atoms 1 and 4 of butadiene are predicted to be more reactive than atoms 2 and 3; this order is normally observed.

Some Simplifying Observations

Calculation of HMO energies and wavefunctions for large molecules can be a very tedious process, since large determinants must be solved. For this reason, it is usually necessary to use group theory to reduce the size of the determinants involved, or to use a computer to solve the large determinants. We will not go into group theory here; however, its application to MO problems is actually quite simple and several excellent expositions of the topic are available (see Appendix 3 for a simple application of symmetry).[4,9–12]

It is possible to derive a great deal of information concerning HMO energies and wavefunctions simply by determining the number and arrangement of atoms in a molecule. As mentioned earlier, the energy of a particular MO is directly related to the number of nodes in the MO. Thus, to derive the symmetry of an MO (arrangement of positive and negative lobes) and the relative energy of an MO, it is only necessary to add nodes to each orbital.[13] For example, the allyl system has three MOs; the first has no nodes, the second one, and the third two:

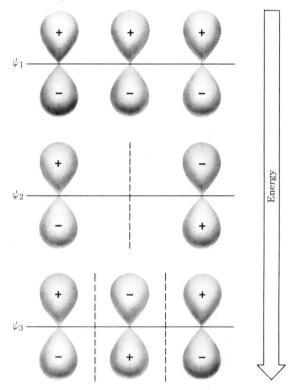

This type of treatment is most easily done if only the sign of the "top" lobe of each $2p$ orbital is represented; thus, for allyl:

$$E = 2B + 0AB = 2B$$

$$E = 0B + 0AB = 0$$

Energy

$$E = 0B + 2AB = 2AB$$

The relative energies of the MOs can easily be determined if the number of bonding (B) and antibonding (AB) interactions are counted: above, ψ_1 has 2B, ψ_2 has 0B and 0AB, and ψ_3 has 2AB.

That cyclic conjugated molecules will have degenerate MOs can be seen by the fact that nodes can be drawn across the rings in two different ways to give the same number of bonding and antibonding interactions (see the cyclopropenyl MOs **17** and **18**). For example, for benzene:

$$E = 6AB$$

$$E = 2AB$$

$$E = 2B$$

$$E = 6B$$

And we can see that benzene will have a degenerate pair of bonding MOs and also a degenerate pair of antibonding MOs.

A rule that aids in the quantitative calculation of energies (x values) from solution of the HMO determinants is that the sum of coefficients of β (x values) must equal zero:

$$E_\mu = \alpha - x_\mu \beta \tag{56}$$

$$\sum_\mu x_\mu = 0 \tag{57}$$

Thus, the final x value of a solution can be obtained by difference. Alternatively, eq. 57 can be used as a check on a complete calculation.

Consideration of the properties of a class of compounds known as *alternant hydrocarbons* permits a further reduction in tedious calculations. An alternant hydrocarbon (AH) is defined as any planar, conjugated hydrocarbon in which the carbons can be divided into starred sets such that no starred or unstarred carbon is adjacent to another of its kind.[8,13] For example, butadiene and benzene are alternant hydrocarbons but fulvene is nonalternant:

Alternant   Alternant   Nonalternant

The energy levels of an alternant hydrocarbon are arranged symmetrically about α. Thus, for an AH, the energy levels occur in pairs such that

$$E_\mu \text{ (of } \psi_\mu) = \alpha + x_\mu \beta \tag{58}$$

$$E_{-\mu} \text{ (of } \psi_{-\mu}) = \alpha - x_\mu \beta \tag{59}$$

Fig. 1.9 *Arrangement of energy levels for acyclic AHs.*

For butadiene then, if the energy of ψ_1 has been calculated, the energy of ψ_4 is obtained by changing the sign of x_μ (similarly for ψ_2 and ψ_3):

$$E_1 = \alpha + 1.62\beta \quad \text{and} \quad E_4 = \alpha - 1.62\beta$$

This pairing property of AHs leads to easily recognizable patterns for the energy levels. The pattern for acyclic polyenes is given in Figure 1.9. Notice that all of the odd AHs, *which must, of course, be ionic or radical species,* have a nonbonding MO. Cyclic AHs follow a different pattern (Figure 1.10) in that degenerate pairs can occur.

Fig. 1.10 *Arrangement of energy levels for cyclic AHs.*

Except for a few cases, the energy level arrangements for non-AHs follow no simple pattern. Monocyclic compounds with an odd number of carbons are an exception; the pattern for these nonalternant compounds, and for the alternant monocyclic compounds also, can be derived in the following manner (illustrated in Figure 1.11): (*a*) Draw the cyclic structure as a geometrical figure with point down, (*b*) represent α with a dotted line through the middle of the figure, and (*c*) determine the relative positions of the energy levels for each compound given by the positions of the corners of the figure. Using these energy level arrangements, we can see that all of the monocyclic hydrocarbons with $(4n+2)\pi$ electrons have all of the bonding orbitals filled, a so called closed-shell configuration. Hückel recognized that the closed-shell electronic configuration would impart exceptional stability to a molecule and was, thus, able to make his now famous prediction regarding the aromaticity of $(4n+2)$ systems. We will consider the topic of aromaticity in detail in Chapter 2.

This pairing property of AHs also reduces the number of calculations required to determine coefficients of the wavefunctions. For AHs, the

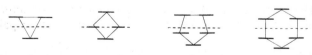

Fig. 1.11 *Energy level arrangements for alternant and nonalternant monocyclic hydrocarbons.*

antibonding orbital of any "pair" (i.e., $+x$ and $-x$) will differ from the bonding orbital only in that the signs of the unstarred coefficients will be reversed. For butadiene,

$$\psi_1 = 0.37\phi_1 + 0.60\phi_2 + 0.60\phi_3 + 0.37\phi_4$$

$$\psi_4 = 0.37\phi_1 - 0.60\phi_2 + 0.60\phi_3 - 0.37\phi_4$$

Since carbons 2 and 4 are unstarred, ψ_4 is obtained simply by changing the signs of ϕ_2 and ϕ_4 in ψ_1.

1.3 The Perturbational Molecular Orbital Method[8,14]

The perturbational molecular orbital (PMO) method is in practice a very simple and powerful tool for the calculation of stabilities and reactivities of organic molecules. The object of the approach is to calculate directly the differences between systems rather than to treat systems separately and then determine the differences. As an illustration, consider the effect on the aromaticity of benzene of replacing one carbon with nitrogen. One approach to this problem is to calculate the absolute energies of pyridine and benzene and then to compare them; since the difference between the two molecules is small in relation to their total energies, a very precise theoretical method is required. The PMO method, on the other hand, treats the problem by determining the perturbation of the benzene energy that results upon changing a carbon to a nitrogen. Because of this direct approach, the PMO method can be of a relatively approximate nature and still give accurate results.

In addition, the PMO method is developed in terms of the properties of alternant hydrocarbons. As we have just seen, the energy levels and wavefunctions of AHs adhere to simple patterns. This regularity of behavior extends to other properties and will allow us to do many important calculations in a matter of seconds. The PMO method as applied to organic systems through the properties of AHs has in large part been developed by M. J. S. Dewar.[8]

According to the PMO method, molecular perturbations can be expressed as intramolecular (in terms of changes in the Coulomb integral $\delta\alpha$ and the resonance integral $\delta\beta$) as in eq. 60 or intermolecular as in eq. 61.

$$\delta E = \sum_i q_i \delta\alpha_i + 2 \sum_i \sum_j p_{ij}\delta\beta_{ij} \tag{60}$$

$$\delta E_\mu = \frac{a_{\mu r}^2 b_{\rho s}^2 \beta_{rs}}{E_\mu - F_\rho} \tag{61}$$

In eq. 60 the subscripts i and j refer to the interacting atoms, q_i is the electron density (eq. 52), and p_{ij} is the bond order (eq. 54). Equation 61 is considered to apply to the union of two molecules R and S through atoms r and s, **19**. E_μ and F_ρ refer to the energies of the interacting MOs, ψ_μ and ψ_ρ, while $a_{\mu r}$ and $b_{\rho s}$ refer to the MO coefficients at positions r and s in ψ_μ and ψ_ρ, respectively. β_{rs} represents the amount of π overlap between atoms r and s.

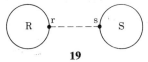

19

The relationship for the intermolecular interaction of two degenerate MOs is simpler and more frequently utilized:

$$\delta E = \varepsilon \pm a_{\mu r} b_{\rho s} \beta_{rs} \tag{62}$$

Equation 62 is simply a quantitative expression for the splitting that results when two degenerate MOs interact, **20**.

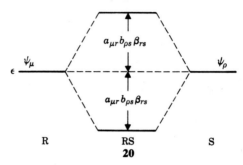

R   RS   S
20

There are many potential uses of these PMO equations. We will consider two types of application only; however, these two types permit examination of many different chemical problems. The two applications to be considered are: (*a*) intermolecular union of odd AHs and (*b*) intramolecular union of AHs.

Intermolecular Union of Odd Alternant Hydrocarbons

As pointed out previously, an alternant hydrocarbon is one in which the carbons of a π system can be divided into starred and unstarred sets such that no starred or unstarred carbon is adjacent to another of its kind. For odd AHs, there should be more starred than unstarred atoms.

All odd AHs have nonbonding molecular orbitals (NBMOs). The chemistry of the odd AHs is in large part determined by the nature of this NBMO. As we will show, the distribution of charge density throughout the molecule and combination with other molecules is controlled solely by the NBMO.

All NBMOs of odd AHs have certain common characteristics which permit ready determination of their form. *First, the coefficients of NBMOs at*

The NBMO of benzyl

21

each unstarred position are zero. For example, the NBMO of benzyl **21** must be

$$\psi_0 = a_2 \phi_2 + a_4 \phi_4 + a_6 \phi_6 + a_7 \phi_7$$

where

$$a_1 = a_3 = a_5 = 0$$

Second, the sum of the NBMO coefficients at each position about an unstarred position must equal zero:

$$\sum_i^{about\ 0} a_{0i} = 0 \tag{63}$$

For benzyl **21** then,

$$a_2 + a_4 = 0 \text{ (about } a_3)$$
$$a_4 + a_6 = 0 \text{ (about } a_5)$$

and

$$a_2 + a_6 + a_7 = 0 \text{ (about } a_1)$$

A third property that can be used is the normalization condition:

$$\sum a_i^2 = 1 \tag{43}$$

Use of the three properties just mentioned leads to a method for determination of NBMO coefficients in three simple steps:

1. Star alternant atoms.

2. Assign arbitrary values to the coefficients on the starred atoms such that the sum about each unstarred position equals zero.

3. Apply the normalization condition.

To illustrate, we will determine the NBMO coefficients of benzyl:

Step 1.

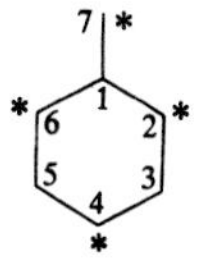

Step 2.

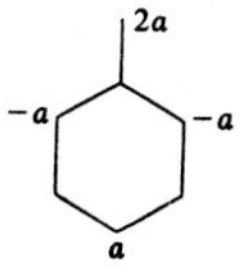

Step 3.

$$(-a)^2 + a^2 + (-a)^2 + (2a)^2 = 1$$
$$7a^2 = 1$$
$$a = \frac{1}{\sqrt{7}}$$

Therefore,

$$\psi_0 = -\frac{1}{\sqrt{7}}\phi_2 + \frac{1}{\sqrt{7}}\phi_4 - \frac{1}{\sqrt{7}}\phi_6 + \frac{2}{\sqrt{7}}\phi_7$$

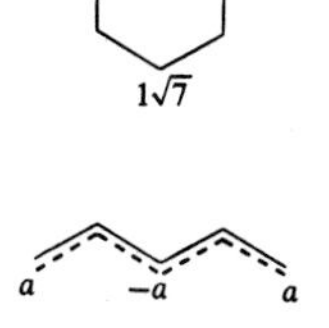

Similarly, pentadienyl is

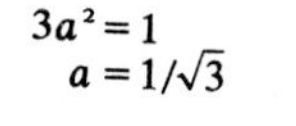

$$3a^2 = 1$$
$$a = 1/\sqrt{3}$$

These NBMO coefficients can be used to treat several different problems. In an even AH or an odd AH radical, the electron density is unity at all positions. The electron density at each starred atom i in an odd AH anion is given by $1+a_{0i}^2$ and in the cation by $1-a_{0i}^2$; electron density at the unstarred positions is unity.

To return to the original subject of this section, the union of two odd AHs, we need only consider the NBMOs of the two fragments to be joined. For example, if two allyl radicals are joined to give hexatriene, the energy

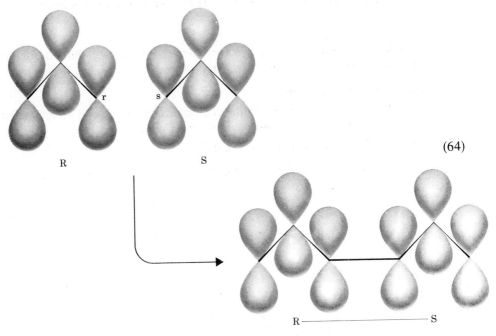

(64)

change of this union *will be controlled by the interaction of the two NBMOs.* Note that the union is a hypothetical one, since chemically the union of two allyl radicals will give 1,5-hexadiene not 1,3,5-hexatriene. The "split" of the allyl *bonding* MOs results in no net energy change; the energy gain is cancelled by an equal energy loss (Figure 1.12). The energy gain resulting from filling the new bonding MO formed from the NBMOs is given by the PMO equation for intermolecular union of degenerate orbitals

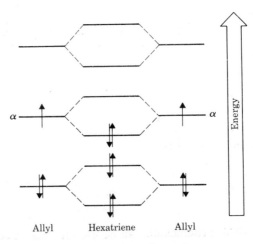

Fig. 1.12 *Energy diagram for the union of two allyl radicals.*

(eq. 62) as

$$\delta E = 2\Sigma a_{0r}b_{0s}\beta \tag{65}$$

where a_{0r} and b_{0s} are the NBMO coefficients on atoms r and s (the newly bonded atoms 3 and 4) of fragments R and S. Solution for these coefficients shows that they are equal to $1/\sqrt{2}$ and the energy change is β.

$$a \overset{\frown}{\quad\quad} a$$

$$2a^2 = 1$$

$$a = \frac{1}{\sqrt{2}}$$

$$\delta E = 2\left(\frac{1}{\sqrt{2}}\right)\left(\frac{1}{\sqrt{2}}\right)\beta = \beta$$

Both ends of the allyl radicals could be joined to give benzene:

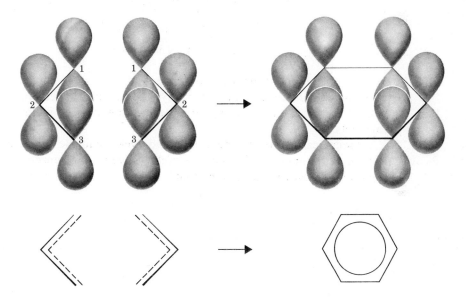

In this case the energy change equals 2β,

$$\delta E = 2(a_{01}b_{01} + a_{03}b_{03})\beta$$

$$\delta E = 2\left[\left(\frac{1}{\sqrt{2}}\right)\left(\frac{1}{\sqrt{2}}\right) + \left(\frac{-1}{\sqrt{2}}\right)\left(\frac{-1}{\sqrt{2}}\right)\right]\beta$$

$$\delta E = 2\beta$$

and benzene is seen to be more stable than hexatriene, because the new bonding MOs for benzene are more stable than those for hexatriene.

These calculations can be made even more simple by using a *hypothetical* simplest odd AH, a one carbon π unit or "methyl." Such a unit will have an NBMO coefficient of unity, since

$$a_{01}^2 = 1$$

$$a_{01} = 1$$

To illustrate the use of this π methyl group, which we will symbolize with a heavy dot •, consider its union with the pentadienyl system:

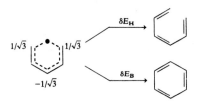

This union can give hexatriene if one bond is formed to methyl, δE_H, or it can give benzene if two bonds are formed to methyl, δE_B. Calculation of the energy changes now only requires the coefficients of one fragment, pentadienyl, since a_{01} for methyl is unity. Again, benzene is calculated to be more stable than hexatriene. Note that in this calculation

$$\delta E_H = 2\left(\frac{1}{\sqrt{3}}\right)\beta = \frac{2}{\sqrt{3}}\beta$$

$$\delta E_B = 2\left(\frac{1}{\sqrt{3}} + \frac{1}{\sqrt{3}}\right)\beta = \frac{4}{\sqrt{3}}\beta$$

the energy difference between benzene and hexatriene differs from the foregoing calculation in which two allyl radicals are joined; in using this method, it is necessary to compare molecules obtained by joining the same fragments.

A similar calculation for butadiene and cyclobutadiene shows the linear polyene to be more stable than the cyclic polyene; for benzene and hexatriene the opposite situation was found:

$$\delta E_B = 2(1/\sqrt{2})\beta = \sqrt{2}\beta$$

$$\delta E_C = 2\left(\frac{1}{\sqrt{2}} - \frac{1}{\sqrt{2}}\right)\beta = 0$$

To summarize, the intermolecular union of two odd AHs, one of which is methyl, can be treated by the use of

$$\delta E = 2\sum_r a_{0r}\beta \tag{66}$$

and knowledge of the NBMO coefficient (a_{0r}) for the atom or atoms joined to methyl. The coefficients should be chosen such that the most favorable energy change results. For example, the union of methyl and allyl to give butadiene (discussed previously) could give an energy change of $\sqrt{2}\,\beta$ or $-\sqrt{2}\,\beta$ depending on the end of allyl to which methyl was joined; the positive coefficient should be chosen.

Intramolecular Union of Alternant Hydrocarbons

Insertion of new bonds in alternant hydrocarbons produces an energy change that is readily calculable by use of eq. 67 (a simplified form of the PMO equation for intramolecular union, eq. 60) and knowledge of the special properties of AHs.

$$\delta E = 2p_{rs}\, \beta \qquad (67)$$

The π bond order (p_{rs}) between two starred atoms or between two unstarred atoms in an even AH or an odd-AH radical is zero, in an odd-AH anion is $a_{0r}a_{0s}$, and in an odd-AH cation is $-a_{0r}a_{0s}$.

Consider formation of a new bond in cyclooctatetraene to give pentalene:

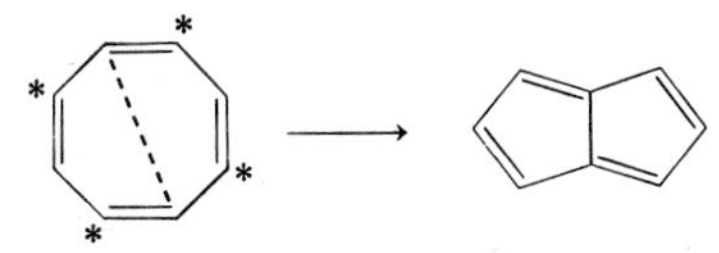

This new bond must be between like atoms (starred or unstarred) and the π bond order for this bond must equal zero. On this basis, pentalene is no more stable (or aromatic) than cyclooctatetraene.

Similar reasoning shows that fulvene **22** is no more stable than hexatriene.

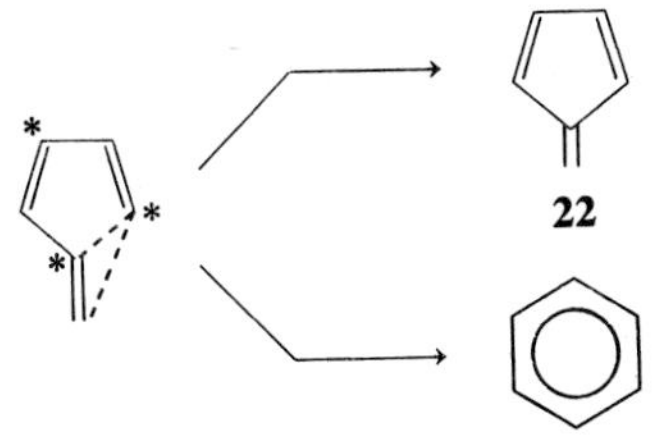

Note the bond order between starred and unstarred atoms (to give benzene) cannot be derived from the principle stated previously, which applies only to union of like atoms; bond orders for such cases can be obtained only by performing a complete HMO calculation. However, problems of this type can be simply treated by use of the intermolecular equations presented in the previous section. Another example of intramolecular union is the formation of naphthalene and azulene from cyclodecapentaene:

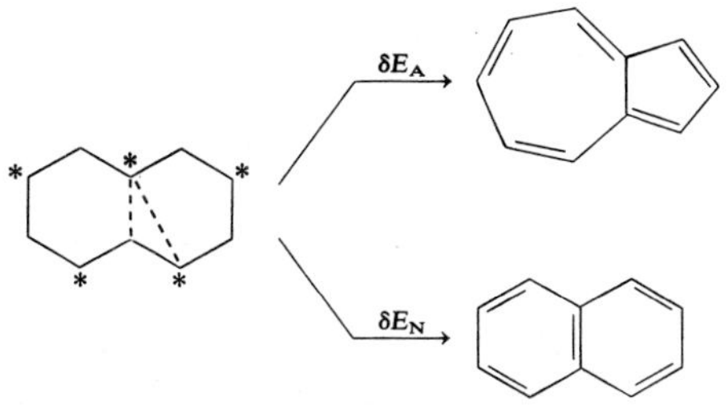

δE_A is zero since like atoms are joined, but no prediction can be made regarding δE_N. This latter comparison can be made by use of an intermolecular union, however:

$$\delta E = 4a\beta$$

$$\delta E = 6a\beta$$

Charged systems can also be treated by the intramolecular method. For example, the cyclopentadienyl anion is seen to be more stable than the linear anion, but the cyclic cation is less stable than the linear form:

$$\delta E^{+} = -2\left(\frac{1}{\sqrt{3}}\right)\left(\frac{1}{\sqrt{3}}\right) = -\frac{2}{3}\beta$$

$$\delta E^{-} = 2\left(\frac{1}{\sqrt{3}}\right)\left(\frac{1}{\sqrt{3}}\right) = \frac{2}{3}\beta$$

It should be noted that these simple calculations, and those in the previous section, give good qualitative predictions of the chemical properties of AHs, predictions that are usually superior to those derived from HMO theory. The PMO method has been used in the treatment of electrophilic aromatic substitution, nucleophilic aliphatic substitution, aromaticity of conjugated hydrocarbons, pericyclic reactions, heteroatom substituent effects in conjugated systems, elimination–addition reactions, and π-complex stability; we will consider the first four of these applications of the PMO method in Chapter 2.

1.4 Advanced Methods—*ab initio* and Semiempirical[8–10,15–19]

It is not the intent of this text to present the advanced theoretical methods in any detail. The following material is presented for the elucidation of the reader who wishes to have some grasp of the mathematical difficulties inherent in the quantum mechanical method and who also wishes to have an overview of the methodology of the advanced methods. This material is not necessary to the development of the subsequent sections.

The advanced methods are of two primary types, exemplified by the Roothaan and the Pople methods; all variations require the use of computers. The Roothaan method is used by theoreticians to treat small molecules and is one of the purest treatments available in that it utilizes few approximations (therefore *ab initio* or "from first principles") and does not distinguish between σ and π electrons. Recent developments and the use of large computers have permitted application to larger organic species. The Pople method involves the use of several approximations and is further distinguished from the Roothaan method by the use of empirical parameters (therefore semiempirical) to treat several difficult portions of the calculation. The Pople method and its variations include a large percentage of the calculations done today on conjugated organic molecules. In addition there have been several recent applications of semiempirical methods to treatment of all valence electrons.[19,20] The following two sections are brief accounts of the Roothaan and Pople methods.

The Roothaan Method

The primary difference between this derivation and the one used for the HMO equations is that we must now consider the form of the Hamiltonian operator. The result will be a variation equation and a secular determinant just as for the HMO method; however, the integrals in the Roothaan equation will not be treated as parameters of fixed value, but will be solved. H is usually written as

$$H = \sum_i \frac{\hbar^2}{2m} \nabla_i^2 - \sum_i \sum_m \frac{Z_m e^2}{r_{im}} + \sum_{i<j} \sum \frac{e^2}{r_{ij}} \tag{68}$$

where the symbol ∇_i^2 represents the Laplacian operator,

$$\frac{\partial^2}{\partial x_i^2} + \frac{\partial^2}{\partial y_i^2} + \frac{\partial^2}{\partial z_i^2} \tag{69}$$

m is the mass of the nucleus, Z_m is the atomic number of nucleus m, e is the charge of an electron, r_{im} is the distance from electron i to the nucleus m, and r_{ij} is the distance from electron i to electron j. The first term represents the kinetic energy of electrons, the second term gives the potential energy due to attraction between the nucleus m and the electron i, and the third term gives the mutual repulsion between electrons i and j.

According to the Born–Oppenheimer approximation, the nucleus is considered to be stationary with the electrons moving about it; thus, the first two terms in the Hamiltonian are functions only of the coordinates of one electron i. The third term, however, is a function of the coordinates of two electrons, i and j, and is difficult to treat. Most methods utilize an approximate one-electron operator known as the *Hartree–Fock operator*:

$$H \cong F = H^{\text{core}} + \sum_j (2J_j - K_j) \tag{70}$$

where H^{core} derives from the one-electron portion of eq. 68 and represents the interaction of a nucleus with an electron, J_j represents the Coulomb repulsion between electrons, and K_j results from the Pauli principle and corrects the total electronic repulsion for correlation in the movements of electrons with like spin. The segments J_j and K_j approximate the two-electron part of the Hamiltonian operator. To illustrate the use of the Hartree–Fock operator and the form of J_j and K_j, consider the total energy of a system that is obtained upon solution of the Schrödinger equation

$$F\psi = E\psi \tag{71}$$

Multiplying by ψ and integrating over all space yields an expression for E,

$$E = \frac{\int \psi F \psi \, d\tau}{\int \psi^2 \, d\tau} \tag{72}$$

which for normalized MOs reduces to

$$E = \int \psi F \psi \, d\tau \tag{73}$$

If we use F as given by eq. 70 and for ψ use a wavefunction that obeys the Pauli exclusion principle (usually the Slater determinants are used), the Hartree–Fock energy equation is obtained:

$$E = 2 \sum_m E_m + \sum_m \sum_n (2J_{mn} - K_{mn}) \tag{74}$$

where E_m results from H^{core} and is given by

$$E_m = \int \psi_m(i) H^{core} \psi_m(i) \, d\tau_i \tag{75}$$

and the Coulomb integral is

$$J_{mn} = \int\int \psi_m(i) \, \psi_m(i) \frac{e^2}{r_{ij}} \psi_n(j) \, \psi_n(j) \, d\tau_i \, d\tau_j \tag{76}$$

and the exchange integral is

$$K_{mn} = \int\int \psi_m(i) \, \psi_n(i) \frac{e^2}{r_{ij}} \psi_m(j) \, \psi_n(j) \, d\tau_i \, d\tau_j \tag{77}$$

The subscripts m and n refer to molecular orbitals while i and j refer to electrons.

The Coulomb integral can now be seen to represent the interaction between an electron in ψ_m and one in ψ_n. The exchange integral measures the result of electrons with identical spins exchanging orbitals.

Direct solution of Hartree–Fock equations is very difficult even for small systems. In an effort to improve this situation, Roothaan[21] approximated the Hartree–Fock orbitals that are given in tabular form with LCAO orbitals as in the HMO method:

$$\psi_\mu = \sum_j a_{\mu j} \phi_j \tag{78}$$

Application of the variation principle yields a set of secular equations that superficially resemble the HMO equations

$$\sum_j a_\mu (F_{ij} - E_\mu S_{ij}) = 0 \qquad i = 1, 2 \ldots, n \tag{79}$$

where

$$|F_{ij} - ES_{ij}| = 0 \tag{80}$$

$$F_{ij} = H_{ij}^{core} + \sum_k \sum_l P_{kl}[(ij, kl) - \tfrac{1}{2}(ik, jl)] \tag{81}$$

$$(ij, kl) = \int\int \phi_i(1) \, \phi_j(1) \frac{e^2}{r_{12}} \phi_k(2) \, \phi_l(2) \, d\tau_1 \, d\tau_2 \tag{82}$$

$$H_{ij}^{core} = \int \phi_i H^{core} \phi_j \, d\tau \tag{83}$$

$$P_{kl} = 2 \sum a_{\mu k} a_{\mu l} \tag{84}$$

Equations 79 through 84 are the Roothaan equations. A problem arises in their solution since ψ_μ must be known in order to solve eqs. 78 through 80 to obtain ψ_μ. Obviously, ψ_μ cannot be known before its solution is effected. The answer to this dilemma is to apply an iterative procedure referred to as the *self-consistent field* (SCF) *method*: (*a*) Assume a trial eigenfunction (a so-called basis set) $\psi_\mu^0 = \Sigma a_{\mu i}^0 \phi_i$; (*b*) solve, in order, eqs. 81, 80, and 79; and (*c*) repeat steps (*a*) and (*b*) with the new coefficients from the solution of eq. 79 until the coefficients and the energies converge, that is, until they become self-consistent.

The major problem in solution of the Roothaan equations lies with solution of the four center integrals (ij, kl). Until recently, this difficulty restricted calculations to diatomic and linear triatomic molecules; however, much larger molecules are now being studied. For example, *ab initio* calculation of the relative energies, conformations, and mechanisms of stabilization of simple alkyl cations has been done.[20]

The Pople Method

The usual method for circumventing the mathematical problems that restrict the Roothaan method is to go to a semi-empirical treatment. If some of the integrals are fixed by experiment, this will provide an accurate treatment and will even permit some integrals to be dropped. Various semi-empirical treatments have been applied with excellent success to conjugated molecules, and more recently with limited success to all-valence-electron studies.

Pople made the approximation that different AOs will not overlap.[22]

$$\phi_i \phi_j \, d\tau = 0 \tag{85}$$

This approximation is usually known as neglect of differential overlap and will cause all three and four center integrals (ii, kl) and (ij, kl) to go to zero. Removal of these terms from F_{ij} (eq. 81) gives:

$$F_{ii} = H_{ii}^{core} + \tfrac{1}{2}P_{ii}(ii, ii) + \sum_{j \neq i} P_{ij}(ii, jj) \tag{86}$$

$$F_{ij} = H_{ij}^{core} - \tfrac{1}{2}P_{ij}(ii, jj) \qquad i \neq j \tag{87}$$

It can be shown that H_{ii}^{core} can be replaced with W_i, the ionization potential of an electron in an isolated $2p$ orbital. Also,

$$P_{jj} = q_j \tag{88}$$

$$P_{ij} = p_{ij} \tag{89}$$

$$H_{ij}^{core} = \beta_{ij}^c \tag{90}$$

Therefore, the final form of the matrix terms F_{ij} for the Pople method are written as:

$$F_{ii} = W_i + \tfrac{1}{2}q_i(ii, ii) + \sum_{j \neq i}(q_j - c_j)(ii, jj) \tag{91}$$

$$F_{ij} = \beta_{ij}^c - \tfrac{1}{2}p_{ij}(ii, jj) \tag{92}$$

where c_j is the number of electrons in the AO ϕ_j.

The Pople equations (eqs. 79, 80, 91, and 92) are also solved by an SCF procedure. The integrals (ii, ii) and (ii, jj) are determined from experiment by relating them to physical parameters such as ionization potentials and electron affinities. We need not be concerned with W_i, since molecules are built from atoms, and W_i will cancel when the energy of the electrons on the component atoms is subtracted from that of the molecules. Finally, β_{ij} is treated as a simple parameter and adjusted to give agreement with experimental properties. Thus, a different set of βs will be obtained for each type of experiment, for example absorption and heats of formation.

In summary, organic chemists are primarily using two types of advanced SCF calculations today: *ab initio* and semi-empirical. The *ab initio* methods are based on the Roothaan equations and are derived from first principles. In contrast, the semi-empirical methods, as exemplified by the Pople method, ignore some of the more difficult parts of the Roothaan equations, give other parameters experimental values determined by their physical significance, and use an adjustable parameter to fit experiment. The semi-empirical methods are more common, generally agree better with experiment, and have been used to treat practically every problem imaginable concerning molecular stability and reactivity.

1.5 The Valence Bond and Resonance Methods

Valence bond (VB) theory was developed by Heitler and London in 1927 as the first quantum mechanical treatment of the hydrogen molecule.[1,15,23] The method resembles molecular orbital methods in that wavefunctions are constructed from combinations of atomic orbitals; however, rather than using a LCAO approach, the VB wavefunctions are derived by combining representations of contributing structures. For example, a wavefunction for the hydrogen molecule can be constructed by combining representations for **23**, **24**, and **25**.

$$H_A\text{--}H_B \qquad H_A^+ \; H_B^- \qquad H_A^- \; H_B^+$$

$$\textbf{23} \qquad\qquad \textbf{24} \qquad\qquad \textbf{25}$$

Thus, the wavefunction can be given as

$$\psi = \phi_A(1)\,\phi_A(2) + \phi_A(1)\,\phi_B(2) + \phi_A(2)\,\phi_B(1) + \phi_B(1)\,\phi_B(2) \qquad (93)$$

where the numbers represent electrons 1 and 2, and the first and last terms represent the ionic forms of the molecule.

The great majority of advances in theoretical organic chemistry have utilized molecular orbital methods; thus, we will not consider the valence bond method further. However, we will examine resonance theory, a nonrigorous offshoot of VB theory that is commonly used in introductory courses in organic chemistry.[24]

According to resonance theory,[24] if a molecule can be represented by more than one valence bond structure (e.g., **26** and **27**), this molecule will be

$$\textbf{26} \qquad\qquad\qquad \textbf{27}$$

resonance forms of benzene

especially stable because of the occurrence of electron delocalization or resonance. The restriction is imposed that contributing valence bond structures must be of approximately equal stabilities; for example, the ability to draw charged forms for neutral molecules does not indicate any enhanced stability. It is further concluded that molecules for which more than one valence bond structure or resonance form can be drawn are best represented as a resonance hybrid or blend of these structures.

Resonance theory can be of use in certain instances. For example, the following cations would be correctly predicted to be relatively stable since different resonance forms can be drawn for each (note the use of the double-headed arrow to indicate relation between resonance contributors):

$$CH_2 = CH - CH_2{}^+ \longleftrightarrow {}^+CH_2 - CH = CH_2$$

The latter example is a postulated intermediate from an electrophilic aromatic substitution. Resonance theory finds its primary use for treatment of these reactions.

Although resonance theory does provide a simple method for examining organic molecules, its basic premise is unfortunately not correct: the property of having more than one resonance form does not always mean that a molecule will be more stable. The principle works well for charged species, since the ability to draw various resonance forms indicates that charge delocalization by conjugation can occur. However, for neutral molecules the property of having more than one resonance form is of no consequence; such a molecule may actually be quite unstable. For example, cyclobutadiene has more than one resonance form, but it is less stable than butadiene which has only one resonance form. For monocyclic molecules, such as cyclobutadiene, the $(4n + 2)$ rule can be used to predict stability, but what can be concluded about molecules such as butalene **31** or pentalene **32** that have more than one resonance form and to which the $(4n + 2)$ rule does not apply? Other techniques must be applied.

In conclusion, resonance theory is useful for certain situations such as considering the effects of substituents on ions; however, the method does have severe limitations. The PMO method is more generally applicable and is also easy to apply if a few equations are remembered. In the following chapter, we will consider applications and comparisons of these theories.

References

1. L. Pauling, "The Chemical Bond," Cornell University Press, Ithaca, N.Y., 1967.
2. H. B. Gray, "Chemical Bonds," Benjamin, Menlo Park, Calif., 1973.
3. J. D. Roberts, "Molecular Orbital Calculations," Benjamin, New York, 1962.
4. A. Streitwieser, Jr., "Molecular Orbital Theory for Organic Chemists," Wiley, New York, 1961.
5. A. Liberles, "Introduction to Molecular-Orbital Theory," Holt, Rinehart, and Winston, New York, 1966.
6. E. Hückel, *Z. Phys.*, **70,** 204 (1931).
7. E. Hückel and H. G. Gilde, *J. Chem. Ed.*, **49,** 2 (1972).
8. M. J. S. Dewar, "The Molecular Orbital Theory of Organic Chemistry," McGraw-Hill, New York, 1969.
9. R. Daudel, R. LeFebvre, and C. Moser, "Quantum Chemistry," Interscience, New York, 1959.
10. L. Salem, "Molecular Orbital Theory of Conjugated Systems," Benjamin, New York, 1966.
11. F. A. Cotton, "Chemical Applications of Group Theory," 2nd ed., Wiley, New York, 1971.
12. M. Orchin and H. H. Jaffe, "Symmetry, Orbitals, and Spectra," Wiley, New York, 1971.
13. W. C. Herndon and E. Silber, *J. Chem. Ed.*, **48,** 502 (1971).
14. W. B. Smith, *J. Chem. Educ.*, **48,** 749 (1971).
15. C. Sandorfy, "Electronic Spectra and Quantum Chemistry," Prentice-Hall, Englewood Cliffs, N.J., 1964.
16. J. N. Murrell, "The Theory of the Electronic Spectra of Organic Molecules," Wiley, New York, 1968.
17. F. L. Pilar, "Elementary Quantum Chemistry," McGraw-Hill, New York, 1968.
18. R. Daudel and D. Sandorfy, "Semiempirical Wave-Mechanical Calculations on Polyatomic Molecules," Yale Univ. Press, New Haven, Conn., 1971.
19. J. A. Pople and D. L. Beveridge, "Approximate Molecular Orbital Theory," McGraw-Hill, New York, 1970.
20. P. C. Hariharan, L. Radom, J. A. Pople, and P. v. R. Schleyer, *J. Amer. Chem. Soc.*, **96,** 599 (1974).
21. C. C. J. Roothaan, *Rev. Mod. Phys.*, **23,** 69 (1951).
22. J. A. Pople, D. P. Santry, and G. A. Segal, *J. Chem. Phys.*, **43,** 129 (1965).
23. W. Heitler and F. London, *Z. Phys.*, **44,** 455 (1927).
24. G. W. Wheland, "Resonance in Organic Chemistry," Wiley, New York, 1955.

Bibliography

Chemical Bonding

L. Pauling, "The Chemical Bond," Cornell Univ. Press, Ithaca, N.Y., 1967.

H. B. Gray, "Chemical Bonds," Benjamin, Menlo Park, Calif., 1973.

Molecular Orbital Theory, Introductory

J. D. Roberts, "Molecular Orbital Calculations," Benjamin, New York, 1962.

A. Streitwieser, Jr., "Molecular Orbital Theory for Organic Chemists," Wiley, New York, 1961.

A. Liberles, "Introduction to Molecular-Orbital Theory," Holt, Rinehart and Winston, New York, 1966.

E. Hückel and H. G. Gilde, *J. Chem. Educ.*, **49**, 2 (1972).

Perturbational Molecular Orbital Method

M. J. S. Dewar and R. C. Dougherty, "The PMO Theory of Organic Chemistry," Plenum, New York, 1974.

Molecular Orbital Theory, Advanced

M. J. S. Dewar, "The Molecular Orbital Theory of Organic Chemistry," McGraw-Hill, New York, 1969.

R. Daudel, R. LeFebvre, and C. Moser, "Quantum Chemistry," Interscience, New York, 1959.

L. Salem, "Molecular Orbital Theory of Conjugated Systems," Benjamin, New York, 1966.

F. L. Pilar, "Elementary Quantum Chemistry," McGraw-Hill, New York, 1968.

R. Daudel and D. Sandorfy, "Semiempirical Wave-Mechanical Calculations on Polyatomic Molecules," Yale Univ. Press, New Haven, Conn., 1971.

J. A. Pople and D. L. Beveridge, "Approximate Molecular Orbital Theory," McGraw-Hill, New York, 1970.

P. C. Hariharan, L. Radom, J. A. Pople, and P. v. R. Schleyer, *J. Amer. Chem. Soc.*, **96**, 599 (1974).

Problems

---

1. Show the proper spatial arrangement of the $2p$ orbitals and hydrogens of the following molecules:

$$CH_2=CH_2, \quad \text{⌲}, \quad \text{⌲},$$

$$\text{⬠}, \quad CH_2=C=CH_2, \quad CH\equiv C-C\equiv CH$$

2. Show by use of a drawing why there is a strong restriction to rotation about $C=C$ but not $C-C$ of 1,3-butadiene.

3. How many π molecular orbitals and π electrons will the following have:

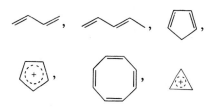

4. Show why the ground state of O_2 should be a triplet (i.e., have two unpaired electrons).

5. Derive the pattern of MO energy levels for hexatriene by use of the pairing principle.

6. Utilize the relationship between the number of nodes and energy to derive orbital symmetries for the following molecules:

$$\begin{array}{c} CH \\ \| \quad (CH)_n \qquad n = 1 \text{ to } 4 \\ CH \end{array}$$

7. Calculate the HMO wavefunctions and corresponding energy values for the following molecules:

(a)   anion, cation, and radical

(b)   anion, cation, and radical

(c)

(d)

(e)

8. Calculate the delocalization energies for the molecules in problem 7.

9. Calculate, by use of the method of cofactors, the MO coefficients for the bonding MO of the allyl cation; using this information and the properties of AHs, derive the MO coefficients of the other two MOs.

10. Calculate the electron densities (q_i) and bond orders (p_{ij}) for the molecules in problem 7.

11. Derive the $(4n + 2)$ rule by utilizing the principles presented in Figures 1.10 and 1.11.

12. Apply the PMO equations to compare the stabilities of (a) naphthalene, azulene, and cyclodecapentaene and (b) octatetraene, cyclooctatetraene, benzocyclobutadiene, and pentalene.

13. Determine the charge density for each atom in the benzyl, benzhydryl, and triphenylmethyl anions (only NBMO coefficients are required).

14. Calculate the change in butadiene MO energies resulting from a dependence of β on bondlength if $\beta_{12} = 1.1\beta$ and $\beta_{23} = 0.9\beta$ (use PMO equations for δE).

15. Calculate the eigenvalues and eigenfunctions for H_2 using Roothaan's method, given the following data:

$$(11, 11) = 0.625 \qquad (11, 22) = 0.468$$
$$(11, 12) = 0.371 \qquad (12, 12) = 0.255$$
$$S_{12} = 0.678 \qquad H_{11}^c = -1.043$$
$$H_{12}^c = -0.842$$

(all in electron volts).

APPENDIX 1
Solution of Determinants

Determinants can be broken down into smaller determinants by summing the multiples of each term in the top row times its cofactor; the cofactor of the jth term is the determinant with the jth column and the top row removed. To illustrate, consider the following determinant:

$$\begin{vmatrix} X & 1 & 0 \\ 1 & X & 1 \\ 0 & 1 & X \end{vmatrix} = X \begin{vmatrix} X & 1 \\ 1 & X \end{vmatrix} - 1 \begin{vmatrix} 1 & 1 \\ 0 & X \end{vmatrix} + 0 \begin{vmatrix} 1 & X \\ 0 & 1 \end{vmatrix}$$

Note that the even terms are always negative. The 2×2 determinant is solved simply by cross-multiplying and subtracting multiples:

$$\begin{vmatrix} X & 1 \\ 1 & X \end{vmatrix} = X^2 - 1$$

To complete the solution of the foregoing determinant, we have

$$X(X^2 - 1) - 1(X) = 0$$
$$X^3 - 2X = 0$$
$$(X^2 - 2)X = 0$$

$$X = \pm\sqrt{2}; 0$$

APPENDIX 2
Solution of Numerical Equations

It is sometimes difficult to obtain the roots of numerical equations such as

$$X^4 - 4X^2 + 2X + 1 = 0$$

Such equations can be readily solved by first plotting the graph of the function to obtain approximate roots and then applying Newton's method for improving the approximate values.

For example, if the above equation is set equal to Y

$$Y = X^4 - 4X^2 + 2X + 1$$

and Y plotted against X (choose whole numbers for X first), the resulting curve crosses the X-axis four times, at $X = 1$ and approximately 1.5, -0.3, and -2.1. The approximate values can be improved by use of Newton's method, which can be stated as

$$X_{\text{improved}} = X_{\text{approx}} + \frac{Y}{dY/dX}$$

choosing $X = -2.1$ from the above graphic approximation gives

$$X_{\text{imp}} = -2.1 - \frac{1.4}{9.5} = -2.2$$

This process can be repeated until a root of desired accuracy is obtained (usually two cycles are sufficient for our work). When solving the HMO equations, it should be recalled that $\Sigma X_j = 0$ and that there will be no imaginary roots.

Another method is to find one or more roots by inspection (try whole numbers) then factor them out. For the example above,

$$X^4 - 4X^2 + 2X + 1 = 0$$

and by inspection

$$X = 1$$

or

$$\frac{X^4 - 4X^2 + 2X + 1}{X - 1} = X^3 + X^2 - 3X - 1$$

Then the latter cubic equation can be solved with less difficulty.

The quadratic equation is also useful:

$$aX^2 + bX + c = 0$$

$$X = \frac{-b \pm (b^2 - 4ac)^{\frac{1}{2}}}{2a}$$

APPENDIX 3
A Simple Application of Symmetry

Application of even the simplest symmetry elements can greatly simplify the calculation of HMO energy values. For example, the determinant for cyclobutadiene is

$$\begin{vmatrix} X & 1 & 0 & 1 \\ 1 & X & 1 & 0 \\ 0 & 1 & X & 1 \\ 1 & 0 & 1 & X \end{vmatrix} = 0$$

and the variation equations are

$$a_1X + a_2 + a_4 = 0$$
$$a_1 + a_2X + a_3 = 0$$
$$a_2 + a_3X + a_4 = 0$$
$$a_1 + a_3 + a_4X = 0$$

Now if we consider the plane of symmetry as shown by the dotted line

since each wavefunction must be symmetric or antisymmetric with respect to this plane of symmetry, the following possibilities exist:

$$a_1 = a_4$$
$$a_2 = a_3$$

or

$$-a_1 = a_4$$
$$-a_2 = a_3$$

This gives two sets of equations from the Variation Equations.

First set:

$$a_1X + a_1 + a_2 = 0$$
$$a_2X + a_1 + a_2 = 0$$

Second set:

$$a_1X + a_2 - a_1 = 0$$
$$a_2X + a_1 - a_2 = 0$$

Considering the first set of equations, we see that

$$a_2 = -(a_1X + a_1)$$

Substitution into the second equation of the first set gives

$$a_1X^2 + 2a_1X = 0$$

or

$$X^2 + 2X = 0$$

which upon solution gives

$$X = 0, -2$$

Treatment of the second set of equations in this fashion gives

$$X^2 - 2X = 0$$
$$X = 0, 2$$

Application of this same method to benzene results in two cubic equations, obviously a significant improvement over solution of a 6×6 determinant.

2

Applications of Theoretical Methods

Applications of the methods considered in Chapter 1 are numerous and cannot be covered in detail in the limited space available here. We will, however, study four specific applications to illustrate the utility of the theoretical approach to organic chemical problems and to further familiarize the student with the available methods. Major emphasis will be placed on use of the PMO method because of its ease of applicability and general usefulness.

2.1 Aromaticity

There have been several definitions of aromaticity ranging from the original based on smell, to a propensity to undergo substitution rather than addition

upon reaction with electrophiles,

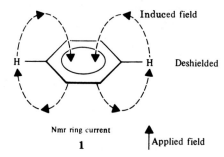

$$\text{(1)}$$

to a more recent definition based on the presence or absence of an nmr ring current (**1**).[1-5] We define an aromatic molecule as one that is more stable than its most stable valence bond resonance contributor.

Induced field

Deshielded

Nmr ring current

1

Applied field

The special properties of aromatic molecules are due to resonance: the extensive delocalization or interaction of π electrons throughout a π system that imparts stability to this system. For example, the reaction of eq. 1 results in substitution because substitution regenerates the delocalized and stable π system, whereas addition does not.

The importance of resonance in a molecule is usually discussed in terms of resonance energy: the actual energy of the molecule minus the bond energy of the most stable valence bond contributor. For example, the resonance energy of benzene is equal to its actual energy minus the bond energy of cyclohexatriene in which the single and double bonds are localized (i.e., the most stable valence bond contributor). Unfortunately, there has also been a great deal of confusion concerning the calculation of resonance energies since the localized model is just a model, and many methods have evolved for calculating the bond energies of these hypothetical localized bonds. Most methods have been designed to determine the energy of a "pure" localized bond and the extra stabilization of nonaromatic conjugated molecules such as butadiene relative to such a model has been considered to be resonance stabilization.[6,7] These methods result in very large resonance energies for truly aromatic molecules and in many instances for nonaromatic molecules. The delocalization energy (DE) obtained from the HMO method by use of eq. 2 is such a method.

$$DE = E_\pi(\text{calculated or experimental}) - E_{loc} \qquad (2)$$

In contrast to these methods, aromaticity has also been treated by contrasting the molecule in question with a localized structure that *includes*

the type of interaction occurring between π segments of a linear polyene.[8-10]
Linear polyenes are quite different from aromatic systems in that there is
bond alternation in linear polyenes and none in aromatic molecules. Furthermore, the heats of formation of linear polyenes are found to be a simple
additive function of the number of bonds

$$\Delta H_f = n(E_{C-C}) + (n+1)(E_{C=C}) + (2n+4)E_{C-H} \tag{3}$$

where n represents the number of carbon–carbon single bonds in the
polyene. Molecules with alternating bond lengths and additive bond energies
are defined as *nonaromatic*, and the bonds in such molecules are defined as
localized bonds. It should be emphasized that there can be significant π
character in the single bonds of such molecules; the critical point is that the
amount of π character is a constant for bonds in a nonaromatic molecule
but not for the bonds in an aromatic molecule. Heats of formation of
aromatic molecules bear no simple relation to numbers of bonds. An
aromatic compound is then defined as one in which the bonds are not
localized and the heat of formation is greater than that for the most stable
valence bond contributor. If the energy of the localized model is calculated
including some π-electron delocalization, calculated resonance energies will
be much lower than those calculated on the basis of the completely localized
model discussed earlier. While it might seem arbitrary to favor one of these
approaches over the other, such is not the case since inclusion of some
electron delocalization in the model structure gives resonance energies that
correlate more closely with chemical properties; we will refer to resonance
energies calculated in this fashion as Dewar (after M. J. S. Dewar of the
University of Texas in Austin) resonance energies (DRE).[8-10]

For determination of DRE, the energy of the localized model can be
obtained by determining the energies of the various localized bond types
(e.g., CH_2=CH and CH—CH) in linear polyenes by use of eq. 3. The value
of localized bonds in terms of HMO β values are given in Table 2.1.
Localized bond energies in terms of kilocalories per mole have also been
derived:

$$DRE = \Delta H_f^{loc} - \Delta H_f \tag{4}$$

$$\Delta H_f^{loc}(kcal/mole) = 7.435 \text{ (no. of C)} - 0.605 \text{ (no. of H)} \tag{5}[9]$$

For example, an HMO calculation gives E_π of 8.00β for benzene. The
localized model for benzene has three double bonds and three single bonds,
and thus from Table 2.1, E_{loc} is 7.62β. The DRE is then calculated to be

Table 2.1 *HMO Bond Energies for Localized Carbon–
Carbon Bonds*[10]

Type of Bond	Calculated Energy (β)
H_2C=CH	2.00
HC=CH	2.07
H_2C=C	2.00
HC=C	2.11
C=C	2.17
HC—CH	0.47
HC—C	0.44
C—C	0.46

0.38β. The HMO method gives excellent agreement with experiment when applied in this manner. Or in terms of heats of formation, ΔH_f^{loc} is calculated to be 41.0 kcal/mole, and ΔH_f is observed to be 19.8 kcal/mole; the result from eq. 4 is a DRE of 21.2 kcal/mole.

Our treatment of aromaticity will be based on the DRE:

> *Aromatic molecules have positive DREs, antiaromatic molecules have negative DREs, and nonaromatic molecules have DREs of approximately zero.*

Dewar gives another definition that aids in treating aromaticity. An *essential single or double bond* is defined as one that remains single or double in every possible uncharged valence bond structure.[8] As we will see, an essential bond must be localized and cannot contribute to aromaticity. Thus, a molecule that is composed only of essential bonds fits the definition of nonaromatic and is further classified as a *classical hydrocarbon*. In contrast, a molecule that is not entirely composed of essential bonds cannot be represented by one valence bond structure and is classified as a *nonclassical hydrocarbon*. Compound **2** is a classical hydrocarbon but **3** and **4** are

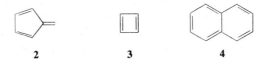

2   3   4

nonclassical. Although a classical hydrocarbon can only be nonaromatic, a nonclassical hydrocarbon can be aromatic, nonaromatic, or antiaromatic.

We will now consider several different classes of conjugated hydrocarbons, in each case applying the definitions discussed above and comparing the predictions of the various theoretical methods.

Benzenoid Hydrocarbons

The benzenoid hydrocarbons are all nonclassical and, thus, potentially aromatic. Table 2.2 presents experimental and calculated (HMO and semi-empirical) DREs for a series of benzenoid hydrocarbons. The DEs are also given. Although many different values for DEs may be found depending upon the method used for calculating the energy of the localized model, the values in Table 2.2 are of typical magnitude. It is important to remember that both the experimental DRE and DE are determined by the difference between an experimental heat of formation and a *calculated* heat of formation for a localized model; thus, experimental DREs and DEs in this table are not true experimental numbers. There is good agreement between the three methods for calculating DREs, and since ΔH_f^{loc} is the same for all three, it must be assumed that the semi-empirical and HMO methods work well for the prediction of heats of formation of benzenoid molecules.

The calculations of Table 2.2 confirm the earlier statement that the DE measure of aromaticity gives higher estimates of resonance energy than the DRE measure. The benzenoid hydrocarbons fit all criteria of aromaticity. For example, upon reaction with electrophiles, substitution, not addition, is observed. Since both the DEs and DREs of Table 2.2 are consistent with all

Table 2.2 *DREs and DEs for Benzenoid Hydrocarbons*

Compound	DRE (kcal/mole)			DE (kcal/mole)[6]
	Experimental[9]	Semi-empirical[9]	HMO[10]	
	21	21	21	36
	33	33	30	61
	43	42	36	84
	49	49	42	91
	63	54	44	109
	64	64	52	117[7]
	68	68	53	126[7]

these molecules being aromatic, a test of the relative utilities of the two methods is not provided in this instance.

Application of the PMO method also leads to the prediction that benzenoid hydrocarbons should be especially stable. For example, consider anthracene and the related compound **5**. Anthracene is predicted to be more stable than **5**.

$$(6)$$

5

$$\delta E = 2 \sum_r a_{0r} \beta_{rs} \tag{1.65}$$

$$\delta E_{AN} = 2(a + a + 2a)\beta = 8a\beta$$

$$\delta E_s = 2(a + 2a)\beta = 6a\beta$$

Several resonance structures can be drawn for all of the benzenoid hydrocarbons; however, as is shown in the next section, more than one resonance structure can be drawn for many nonaromatic and antiaromatic compounds as well.

Monocyclic Conjugated Hydrocarbons

The monocyclic conjugated hydrocarbons are those species to which Hückel's famous $(4n+2)$ rule applies; this rule states that planar monocyclic conjugated hydrocarbons with $(4n+2)$ π electrons, where n is an integer, will be aromatic.[2] The rule derives from the $(4n+2)$ systems having closed-shell electronic arrangements; that is, the bonding MOs are filled, and the antibonding and nonbonding MOs are empty. The prediction of aromaticity and stability for such unlikely looking molecules as **6** and **7** and the

$$\boxed{}\; X^- \qquad\qquad \boxed{}\; X^-$$

6   **7**

subsequent verification of these predictions has been one of the major successes of theoretical chemistry. Similarly the Hückel rule and the justification for it provide a rationale for the nonaromatic character of molecules, such as cyclobutadiene and cyclooctatetraene, which do not have $(4n+2)$ π electrons but which *a priori* appear so similar to benzene.

Table 2.3 presents DREs for the annulenes (monocyclic hydrocarbons of the general formula $[-CH{=}CH-]_m$), all of which are nonclassical. From these data we can see that those species with $(4n+2)$ π electrons are predicted to be stable or aromatic while those with $4n$ π electrons are

Table 2.3 DREs for Annulenes

[n]-Annulene	DREs (kcal/mole)	
	Semi-empirical[9]	HMO[10]
4	−17	−59
6	21	21
8	−10	−27
10	10	14
12	4	−16
14	3.5	13
16	2.8	−10

predicted to be relatively unstable or antiaromatic. Cyclooctatetraene has been shown to exist in rapidly interconvening "tub" forms with alternating long single bonds and short double bonds. Presumably this nonplanar form

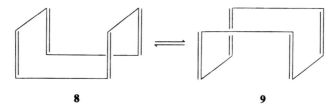

8   **9**

avoids the destabilizing antiaromatic interaction. Similarly, cyclobutadiene appears to exist in a rectangular, singlet form (see Figure 1.10).[12,13]

[10]-Annulene has reportedly been prepared at low temperatures in three different forms, **10** through **12**.[14,15] Significantly, these molecules were found to be quite unstable. Strong destabilization results in **10** and **11** because of angle strain; angles are 144° in a regular decagon. For structure **12,** there is severe crowding of the two hydrogens inside the ring. Apparently, these destabilizing factors are greater than the aromatization effect of a cyclic, 10-π-electron system. The bridged [10]-annulene **13** has been synthesized

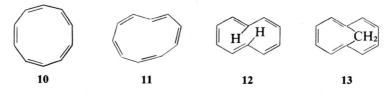

10   **11**   **12**   **13**

and does show aromatic character.[16] In this case, the bridge removes the interaction between the inside hydrogens.

The HMO and semi-empirical methods give different predictions for the larger annulenes. The semi-empirical method predicts approximately equal stabilities for all annulenes above 10, while the HMO method retains the alternation of aromatic–antiaromatic. Experiment supports the latter prediction.[2] For example, [18]-annulene undergoes electrophilic substitution, and [14]- and [18]-annulene exhibit nmr ring currents consistent with aromaticity, whereas the [16]- and [24]-annulenes do not. Also, [24]-annulene has alternating bond lengths as expected for a nonaromatic molecule.

Resonance theory is of no use for predictions of annulene aromaticity, since more than one resonance form can be drawn for all the annulenes. The PMO method does offer a rapid and accurate method for this study. All of the annulenes can be constructed by use of the hypothetical "methyl radical" and treated by use of eq. 1.65:

$$\delta E = 2 \sum a_{0r}\beta_{rs}$$

Thus, benzene is predicted to be more stable than hexatriene **14** while cyclooctatetraene is predicted to be less stable than its corresponding linear polyene **15**. The $(4n+2)$ rule can be demonstrated easily in this manner.

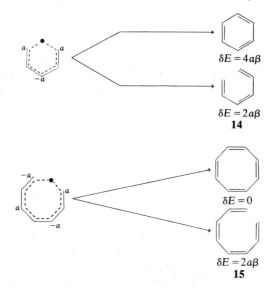

To this point, we have considered the annulenes that are even AHs; the cyclic, non-AHs constitute another class of conjugated hydrocarbons that may be aromatic. Such species as the cyclopentadienide anion and the tropylium cation fit the $(4n+2)$ rule, of course, and are known to be exceptionally stable. There is a problem in treating these charged systems as we did the annulenes, since there are no classical odd AHs to serve as localized models. Since the acyclic ions are nonclassical, **16**, there will be no

16

constant energy value for each double and single bond as there was for the linear polyenes. Dewar has treated this problem by defining the cyclic ion as aromatic if it has a greater π-bond energy than the acyclic counterpart after correcting for the extra σ bond in the cyclic molecule:[8]

$$E_{arom} = E_{cyclic} - E_{acyclic} - \sigma\text{-bond correction}$$

Table 2.4 presents these aromatic energies for the three smallest cyclic odd AHs. These data agree with the $(4n+2)$ rule and with experiment; the

Table 2.4 *Calculated (Semi-empirical) Aromatic Energies for Cyclic Ions*[8]

Cyclic Ion	E_{arom}(kcal/mole)	
	Cation	Anion
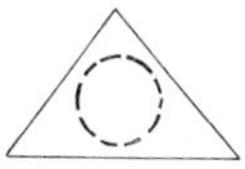 (triangle)	42.4	−35.9
 (pentagon)	−16.8	30.2
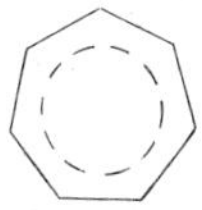 (heptagon)	21.4	−5.8

cyclopropenyl cation, the cyclopentadienyl anion, and the cycloheptatrienyl cation (the tropylium ion) are known to be unusually stable.

The PMO method can be applied to the study of odd AHs by use of the equation for an intramolecular perturbation:

$$\delta E = 2p_{rs}\beta \tag{1.67}$$

Recall that

$$p_{rs} = 0 \qquad \text{for a radical} \tag{7}$$

$$p_{rs} = a_{0r}a_{0s} \qquad \text{for an anion} \tag{8}$$

$$p_{rs} = -a_{0r}a_{0s} \quad \text{for a cation} \tag{9}$$

Applying eq. 1.67 to the cyclopentadienyl system shows that the anion is more stable than the acyclic ion, while the cation is less stable:

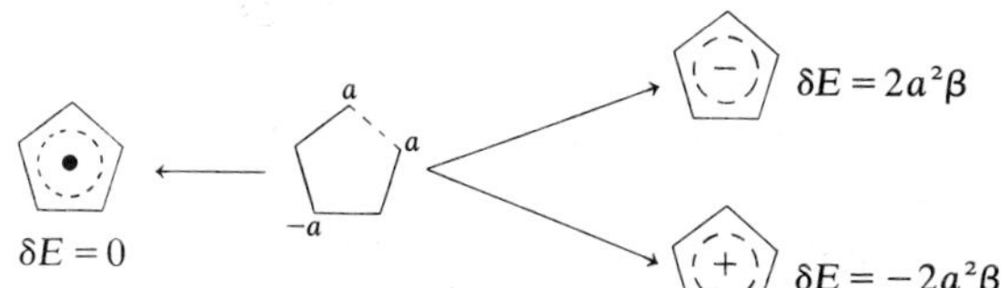

If the other ions are treated in the same manner, the $(4n+2)$ rule again is observed. It is also of interest that cyclic radicals are predicted to be no more stable than the acyclic radicals; this prediction agrees with the experimental observation that the cyclic radicals possess no exceptional stability.

Even Nonalternant Hydrocarbons

The even non-AHs include a large number of conjugated molecules that have received much theoretical and synthetic attention. These molecules are of interest because most are nonclassical and thus predicted to be aromatic

Table 2.5 *DREs and DEs for Even Nonalternant Hydrocarbons*

Compound	DE[7] (kcal/mole)		DRE (kcal/mole)	
	Exptl	HMO	Semi-empirical[8]	HMO[10]
Pentalene **17**	—	25.8	−6.5	−7.5
Calicene **18**	—	30.8	—	18.8
Fulvene **19**	11	15.4	1.0	−0.5
Heptafulvene **20**	14	20.9	0.5	−1.1
Heptalene **21**	—	38.0	−2.1	−2.7
Azulene **22**	30–32	35.3	4.2	12.4
Sesquifulvalene **23**	—	41.3	4.1	14.5
Fulvalene **24**	20	29.2	2.5	−17.7
Heptafulvalene **25**	29	42.0	2.3	−10.8

by resonance theory, **17**. Furthermore, both experimental and HMO DEs

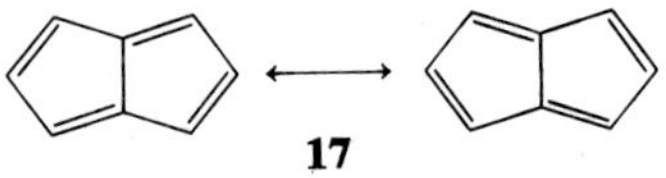

17

indicate that many of the even non-AHs will be aromatic (Table 2.5). Some members of this class also have resonance forms which contain two $(4n+2)$ segments within the same molecule, **18**.

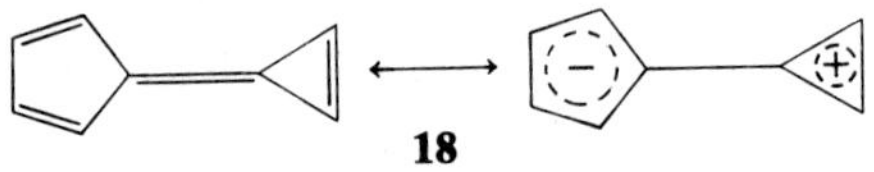

18

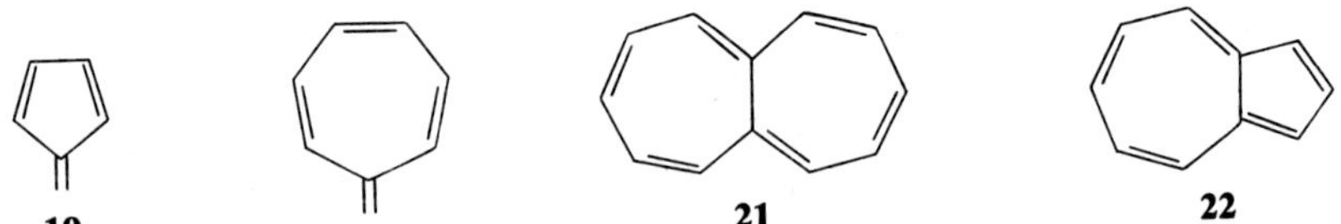

19   **20**   **21**   **22**

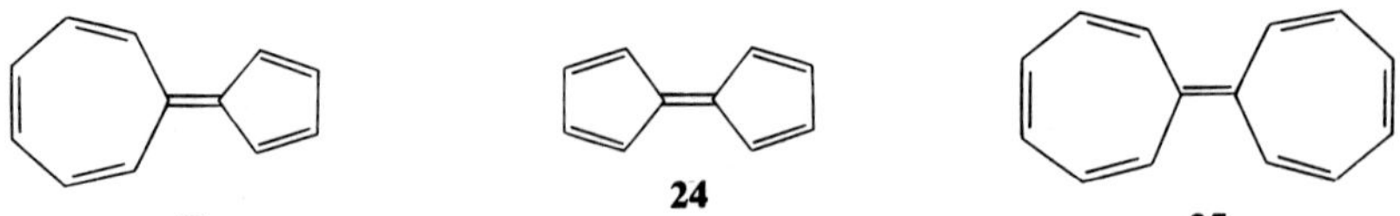

23   **24**   **25**

Of the molecules in Table 2.5, all have been synthesized except pentalene **17**, calicene **18**, and sesquifulvalene **23**; some simple derivatives of calicene and sesquifulvalene have been prepared.[2–4] Examination of the properties of these molecules shows that only azulene **22** is clearly aromatic, while the derivatives of calicene and sesquifulvalene are probably aromatic. The other molecules are all unstable and exhibit no aromatic properties. In this instance, the DE method for estimating resonance energies is misleading, predicting large resonance energies for molecules that have been shown to be unstable. The DRE method agrees well with experiment.

The PMO method provides much insight into the properties of the non-AHs. Since the bond order between two starred atoms is zero in an even AH (Chapter 1), fulvene **19** and heptafulvene **20** can be seen to be no more stable than the corresponding linear polyenes; *e.g.* **26**. Actually, since the

$$\delta E = 2p_{rs}\beta = 0$$

26

fulvenes are classical molecules, the π bonds must be localized, and these molecules are seen to be nonaromatic without the necessity of a calculation.

Similarly, pentalene **17**, heptalene **21**, and azulene **22** can be constructed by joining two starred atoms in the corresponding annulene; *e.g.*, **27** to **17**

27   **28**

and **28** to **22**. Thus, these molecules will also be no more stable than the corresponding annulenes, and since pentalene and heptalene are constructed from antiaromatic 4*n* annulenes ([8]- and [12]-annulenes, respectively), pentalene and heptalene would be expected to be antiaromatic. On the other hand, azulene is formed from the aromatic [10]-annulene and would thus be predicted to be aromatic.

It is also important to note that the transannular bonds in pentalene, heptalene, and azulene are essential bonds and thus are localized and do not contribute to aromaticity. Again a calculation is seen to be unnecessary.

The bonds in the fulvalenes (calicene **18**, sesquifulvalene **23**, fulvalene **24**, and heptafulvalene **25**) are also all essential bonds, and these molecules would be predicted to be nonaromatic. In view of the DREs of calicene and sesquifulvene, however, it seems that the (4*n* +2) nature of the segments of these molecules in the charged form overcomes this restriction.

Biaryls and Styrenes

According to our use of the term essential bond, the extent of conjugation between fragments separated by an essential bond will be identical to that in an acyclic polyene such as butadiene. Since the phenyl rings in biphenyl **29** and stilbene **30** are separated by essential bonds, these molecules would be

29

30

expected to be no more aromatic than two completely separate benzenes. The data in Table 2.6, a collection of DREs for benzenoid compounds containing essential bonds, shows that the above prediction is correct: π fragments separated by essential bonds are independent as regards the extensive conjugation of aromaticity. The DREs for these molecules depend

Table 2.6 *Experimental DREs for Benzene Derivatives*[9]

Compound	Total DRE	DRE per ϕ Unit
ϕ—H	21.2	21.2
ϕ—CH=CH$_2$	21.3	21.3
ϕ—CH=CH—ϕ	42.1	21.0
ϕ—ϕ	43.6	21.8

solely on the number of phenyl rings; no enhanced aromaticity results upon making larger π systems that are separated by essential bonds.

There is conjugation and delocalization between these so-called separated units, however. An estimate of this nonaromatic conjugation can be derived from the observation that there is a 5-kcal barrier to rotation about the single bond of *trans*-butadiene.[9] It is important to note that the shorter single bond in butadiene, as compared to saturated hydrocarbons, is probably not due entirely to conjugation, but must be in part the result of bonding between two sp^2-hybridized carbons rather than between two sp^3-hybridized carbons.[8]

This section on aromaticity has given a brief treatment of the subject. Many systems of interest, primarily those containing heteroatoms, were not examined. A detailed treatment of aromaticity, both from a theoretical and synthetic viewpoint, is given in refs. 2 through 10. The basic ideas and tools for the study of aromaticity have been presented, however; with this in mind, several problems dealing with classes of compounds not covered directly are given at the end of the chapter.

2.2 Electrophilic Aromatic Substitution

As noted in the introduction to this chapter, reaction of an electrophile with an aromatic molecule results in substitution of the electrophile for a hydrogen of the aromatic molecule (eq. 1). The various experimental studies of the mechanism of this reaction are discussed in Chapter 4. In this section, we will examine two types of theoretical treatments of the mechanism: (*a*) the isolated molecule approximation and (*b*) the localization energy procedure.

Isolated Molecule Approximation[7]

According to the isolated molecule approximation, an approaching electrophile will attack the position of an aromatic molecule which has the "most available" π electrons. Application of the procedure involves calculation of various electron availability parameters for the isolated aromatic molecules. Several measures of electron availability have been proposed; of these the free valence index F_i is typical:

$$F_i = 1.732 - \sum_i p_{ij} \tag{1.55}$$

The free valence parameter is intended to measure residual π-bonding power, since F_i is determined by the difference between the maximum

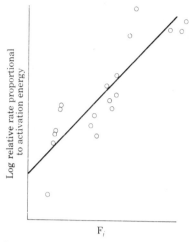

Fig. 2.1 *A plot of log relative rate against F_i for the electrophilic substitution of benzenoid hydrocarbons.*[8]

possible π bond order for any carbon atom (1.732 for the central carbon of trimethylenemethane) and the π bond order of the carbon in question. Figure 2.1 illustrates the fair agreement between F_i and the log of rates of reaction for a series of benzenoid hydrocarbons.

The determination of free valence is a relatively time-consuming process, since a complete MO calculation must be performed to obtain the coefficients of the wavefunctions and the bond orders. The localization procedures provide more accurate and rapid methods for treating this problem.

Localization Energy Procedures

Reactivity is controlled by the energy difference between the ground state and the transition state; that is, by the activation energy. The transition state is, of course, transitory, and we can only make educated guesses as to its structure. It is frequently assumed (see Chapter 3 for a complete discussion) that similar reactions may be compared by means of the energy differences between ground state and intermediate (in this case, the σ complex). In other words, the intermediate, about which much is known, is used as a model of the transition state to derive an energy difference proportional to the energy of activation. On this basis, a calculation of the energy difference between ground states and the σ complex should give a good approximation of the activation energy for electrophilic aromatic substitution. This approach is the basis of the localization energy procedure which derives its

$$\text{(benzene)} + E^+ \xrightarrow{\text{slow}} \text{(} \sigma \text{ complex)} \longrightarrow \text{(product)} + H^+$$

σ complex

name from the fact that a π bond is "localized" upon going from the reactant to the σ complex.

$$E_a \; \alpha \; \delta E_{loc} = E_{\sigma \, complex} - E_{arom} \tag{10}$$

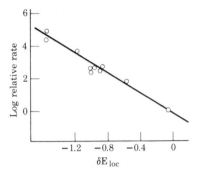

Fig. 2.2 *Plot of log relative rate for nitration of benzenoid hydrocarbons against δE_{loc} from a semi-empirical method.* (Reprinted with permission of McGraw-Hill, Ref. 8, p. 298.)

The localization energy has been calculated both by HMO[7] and semi-empirical[8] methods with the latter method giving somewhat better results; data for the semi-empirical treatment of nitration of benzenoid hydrocarbons are plotted in Figure 2.2.

A calculation of δE_{loc} by either HMO or semi-empirical methods requires the expenditure of much time. On the other hand, the PMO method can be applied to this problem in a matter of minutes and with comparable accuracy. Any benzenoid hydrocarbon can be constructed by union of the "methyl radical" with a π system:

The energy change for this aromatization or *delocalization* process can be calculated by use of eq. 1.6 or in this instance by use of eq. 11. The

$$\delta E = 2\sum_r a_{0r}\beta \tag{1.6}$$

$$\delta E_{deloc} = 2(a_{0i} + a_{0j})\beta \tag{11}$$

substitution reaction proceeds most rapidly when E_a or δE_{loc} is small; a low δE_{deloc} (the reverse of δE_{loc} but we can ignore signs) will then correspond to a fast reaction. Thus, of the following reactions, electrophilic attack at position 1 of naphthalene is preferred:

$$a = \frac{1}{\sqrt{3}} \qquad a = \frac{1}{\sqrt{3}} \qquad \qquad \delta E_{deloc} = 2\left(\frac{2}{\sqrt{3}}\right)\beta = 2.31\beta$$

$$a = \frac{1}{\sqrt{11}} \qquad a = \frac{2}{\sqrt{11}} \qquad \qquad \delta E_{deloc} = 2\left(\frac{3}{\sqrt{11}}\right)\beta = 1.81\beta$$

$$a = 2/\sqrt{8} \qquad a = 1/\sqrt{8} \qquad \qquad \delta E_{deloc} = 2\left(\frac{3}{\sqrt{8}}\right)\beta = 2.12\beta$$

The resonance method for treating electrophilic aromatic substitution is actually a localization energy procedure, since it involves comparison of stabilities of various σ complexes. Thus, naphthalene is predicted to be substituted most easily at position 1 since this ion, **31**, has two resonance forms in which the aromatic sextet of the adjacent ring remains intact while the ion from substitution at position 2, **32**, has only one of these favored

31   **32**

resonance forms. This method is quite accurate at predicting the position of substitution and is especially good for substituted benzenes (e.g., nitrobenzene), but it cannot be used to compare different hydrocarbons as can the PMO method.

2.3 Nucleophilic Aliphatic Substitution

Certain alkyl derivatives, primarily halides and esters, undergo ready substitution of a nucleophile for the attached group (Chapter 4).

$$R-X + \text{nucleophile} \rightarrow R-\text{nucleophile} + X^- \qquad (12)$$

Nucleophile $= C_2H_5OH, H_2O, OH^-, N_3^-, Br^-,$ and so on

$$X = -Cl, -Br, -OSO_2Ar, \text{ and so on}$$

These reactions appear to proceed through a transition state in which the carbon undergoing substitution is positively charged; that is, the transition state resembles a carbocation **33**. Molecular orbital theory can be applied to

33

these reactions by calculating the delocalization energies for arylmethyl derivatives going to arylmethyl carbocations:

$$Ar-CH_2-Cl \xrightarrow{\delta E_{deloc}} Ar-CH_2^+ + Cl^-$$

$$\downarrow \xrightarrow{Nuc^-} Ar-CH_2-Nuc \qquad (13)$$

Since there is no conjugation between the CH_2Cl group and the aryl group, the delocalization energy is essentially the difference between the π energies of the cation and the unsubstituted arene (eq. 14).

$$\delta E_{deloc} = E_{ArCH_2^+} - E_{ArH} \qquad (14)$$

If, as before, we assume that E_a for a reaction such as eq. 13 is approximated by the energy change on going to the first intermediate, then a relationship between δE_{deloc} and E_a would be predicted. In fact, plots of log

k for reaction 13 against δE_{deloc} calculated either by HMO[7] or semi-empirical methods[8] are observed to be linear.

The PMO method can also be applied to calculate δE_{deloc} for arylmethyl derivatives, but first we must derive a new equation.[8] The combination of an even hydrocarbon such as benzene plus a methyl to give the benzyl carbonium ion is difficult to treat using the PMO equations; however, the scheme presented in eq. 15 permits this calculation. The term needed is δE_{TR}, the energy resulting from combination of an even hydrocarbon T with methyl:

$$T(\text{even}) \xrightarrow[\delta E_{\text{TR}}]{+\text{methyl}} R(\text{odd}) \xrightarrow[\delta E_{\text{RS}}]{+\text{methyl}} S(T\text{—}CH\text{=}CH_2)$$

$$+\text{—}CH\text{=}CH_2$$

$$\delta E_{\text{TS}}$$

$$(15)$$

The energy required to take T to S is simply 2β since an essential double bond has been added,

$$\delta E_{\text{TS}} = 2\beta \tag{16}$$

The energy required to take R to S is given by

$$\delta E = 2 \sum_{\text{r}} a_{0\text{r}} \beta_{\text{rs}} \tag{17}$$

$$\delta E_{\text{RS}} = 2 a_{0\text{r}} \beta \tag{18}$$

for the union of two odd AHs.

The relationship between the three combinations is given by

$$\delta E_{\text{TS}} = \delta E_{\text{TR}} + \delta E_{\text{RS}} \tag{19}$$

Assigning TS and RS negative signs for exothermic processes and summing according to eq. 19 gives the final expression for δE_{TR}.

$$\delta E_{\text{TR}} = 2\beta(a_{0\text{r}} - 1) \tag{20}$$

The energy change δE_{TR} can be considered to be a general expression for δE_{deloc} of the process given by eq. 13. Since δE_{deloc} is assumed to be proportional to E_a,

$$\delta E_a \propto \delta E_{\text{deloc}} = \delta E_{\text{TR}} \tag{21}$$

smaller NBMO coefficients should give a smaller E_a and a faster reaction. This is observed; for example, 1-naphthylmethyl chloride **35** reacts more rapidly than benzyl chloride **34**.

2.4 Pericyclic Reactions

A pericyclic reaction is defined as a reaction in which bonds are made and broken in a single, concerted, and cyclic transition state.[17] Such reactions are largely unaffected by solvent changes (unless reactants are charged) or by catalysis and involve no nucleophilic or electrophilic components. Pericyclic reactions can be divided into four subclasses: electrocyclic reactions, cycloadditions, sigmatropic rearrangements, and cheletropic reactions:

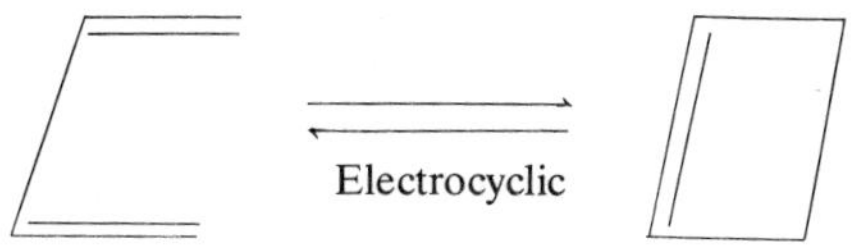

Electrocyclic (22)

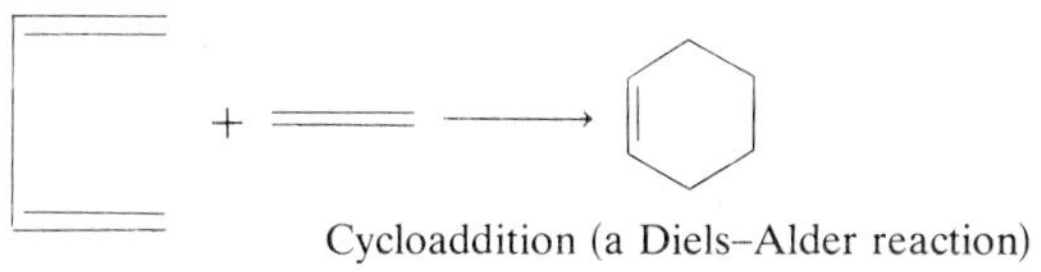

Cycloaddition (a Diels–Alder reaction) (23)

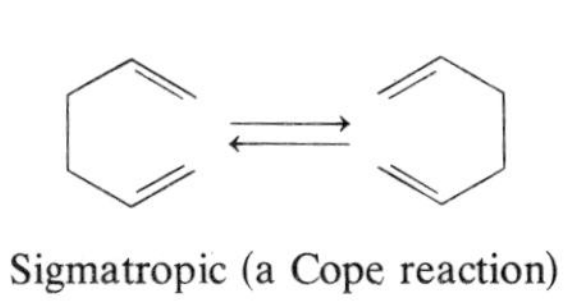

Sigmatropic (a Cope reaction) (24)

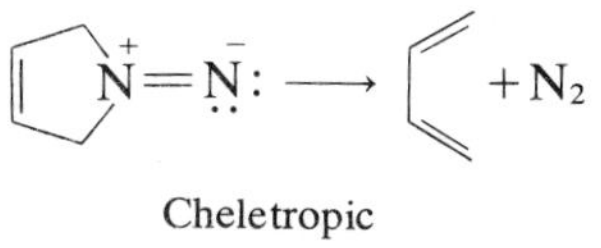

Cheletropic (25)

These reactions include some of the best known and most important reactions of organic chemistry and until recently were poorly understood. For example, consider the following reactions:

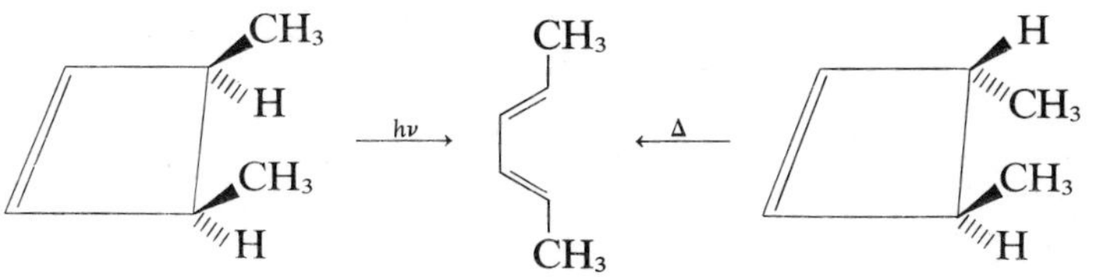

(26)

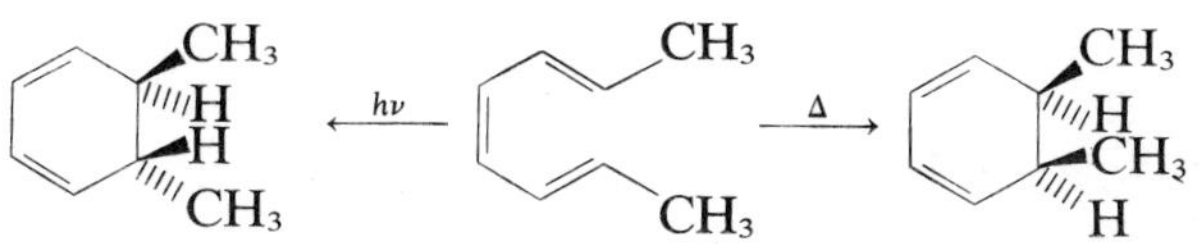

(27)

(28)

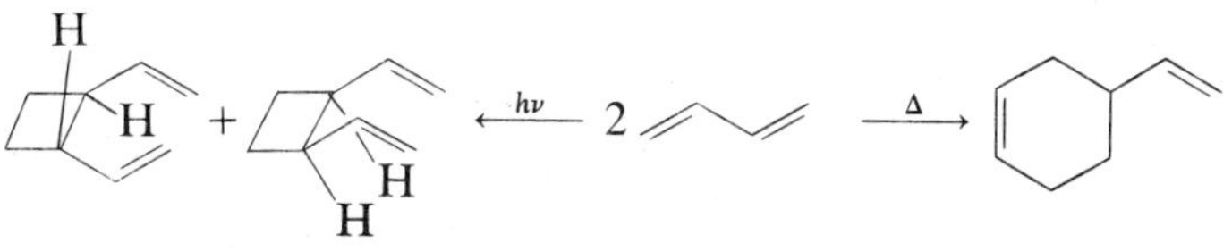

(29)

An explanation for these confusing changes in stereochemistry and products (or lack of products) was first advanced by Woodward and Hoffman[18] in 1965 and has since been developed into a complete theory by these and other workers.[8,17-26] There are now three theories in use by chemists for treating pericyclic reactions, and all three have been asserted to be the "fundamental" principle governing such reactions; we will present briefly each of these three principles. In practice, the behavior of the pericyclic reactions can be predicted (empirically, but not theoretically) solely on the basis of the number of bonds formed and broken in the reaction. These simple relationships will be given after the theoretical discussion.

Conservation of Orbital Symmetry

Woodward and Hoffmann based their revolutionary work on one basic and all-encompassing principle: *concerted reactions occur with conservation of orbital symmetry*; that is, concerted reactions occur readily when the orbital symmetry characteristics of reactants and products are the same, and only with difficulty when the symmetry characteristics differ.[17] Thus, to determine whether a reaction is allowed, it is necessary to determine the symmetry of the pertinent orbitals of reactants and products, and then connect orbitals of like symmetry; if the occupied orbitals of the reactants can be converted to the occupied orbitals of the products, the reaction is symmetry-allowed; if not, it is symmetry-forbidden. It is not necessary to know the quantitative form of the MOs since symmetry is the controlling characteristic; thus, the simple relationship between number of nodes and energy can be applied to derive MO symmetries (Chapter 1).

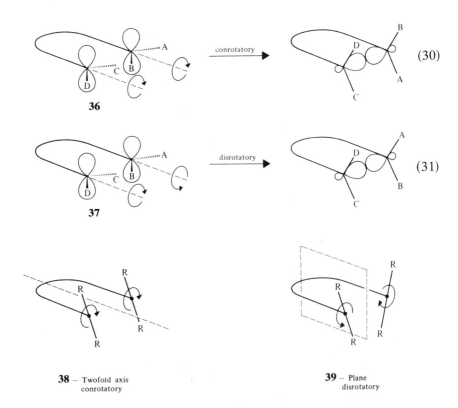

38 – Twofold axis conrotatory	**39** – Plane disrotatory

To illustrate this principle, we will construct a "correlation" diagram for the conversion of butadiene into cyclobutene. First, two possible types of ring closure, conrotatory and disrotatory, must be defined. The conrotatory motion involves two clockwise rotations (when viewed end on) of the terminal groupings (and orbitals), while the disrotatory involves one clockwise and one counterclockwise rotation, **36** and **37**. Examination of structures **38** and **39**, which are partially completed rotations, reveals that the conrotatory motion takes place with a twofold axis of symmetry and the disrotatory with a plane of symmetry.

To test for a possible disrotatory ring closure, we must classify reactant (butadiene) and product (cyclobutene) MOs as symmetric (S) or antisymmetric (A) with respect to a plane of symmetry (Figure 2.3).

Arrangement of the MOs in order of energy (the absolute energy relationship is not necessary) and connection of orbitals of like symmetry yields the correlation diagram for the disrotatory ring closure (Figure 2.4). We can see that the bonding MO ψ_2 of butadiene is forced to correlate with an antibonding level π^* of cyclobutene; therefore, this reaction is symmetry-forbidden and will not occur—more favorable alternatives are available.

To test for the possibility of a conrotatory ring closure, we must classify the same MOs with respect to a twofold axis of symmetry. The results of

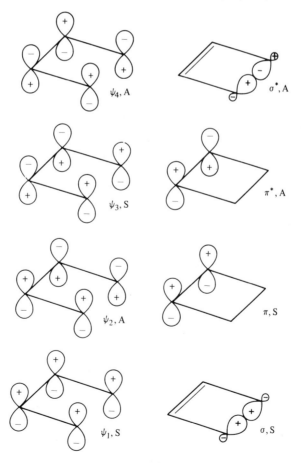

Fig. 2.3 *Classification of MOs of butadiene and cyclobutene with respect to a plane of symmetry: S, symmetric; A, antisymmetric.*

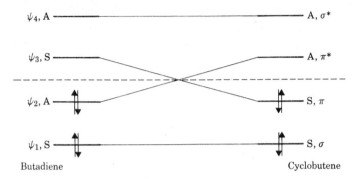

ψ_4, A ——————————————— A, σ^*

ψ_3, S ——————————————— A, π^*

ψ_2, A —————————————— S, π

ψ_1, S ——————————————— S, σ

Butadiene   Cyclobutene

Fig. 2.4 *Correlation diagram for the disrotatory ring closure of butadiene.*

these operations are shown in Figure 2.5. In this case, the bonding MOs of butadiene are converted into the bonding MOs of cyclobutene, and the reaction is symmetry allowed.

By constructing these correlation diagrams and applying the principle of conservation of orbital symmetry, we have shown that all thermal ring closures of butadienes or ring openings of cyclobutenes will occur in a conrotatory fashion; this is observed experimentally (see eq. 26). Substituents on the molecules will not change the symmetry of the π system and will have no effect on the stereochemistry of the reaction. If the butadiene or cyclobutene is constricted such that the conrotatory motion cannot occur, reaction could take place by a nonconcerted radical cleavage-closure or, if sufficient energy is available, by the symmetry-forbidden pathway.

If the same treatment is applied to 1,3,5-hexatriene, the disrotatory ring closure is seen to be symmetry-allowed (see eq. 27). The following rule can be derived in this manner for *thermal* electrocyclic reactions of k π-electron polyenes:

$$k = 4n + 2 \quad \text{disrotatory} \tag{32}$$
$$k = 4n \quad \text{conrotatory} \tag{33}$$

As can be seen from reactions 26 and 27, the photochemical electrocyclic reactions proceed with a reversal in stereochemistry relative to the thermal reactions; thus, the following rules hold for the *photochemical* reactions:

$$k = 4n + 2 \quad \text{conrotatory} \tag{34}$$

$$k = 4n \quad \text{disrotatory} \tag{35}$$

This stereochemical reversal can be explained on the basis of the correlation diagrams in Figures 2.4 and 2.5. For the photochemical case, light

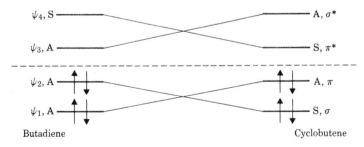

ψ_4, S ——————— A, σ^*

ψ_3, A ——————— S, π^*

ψ_2, A ——————— A, π

ψ_1, A ——————— S, σ

Butadiene   Cyclobutene

Fig. 2.5 *Correlation diagram for the conrotatary ring closure of butadiene.*

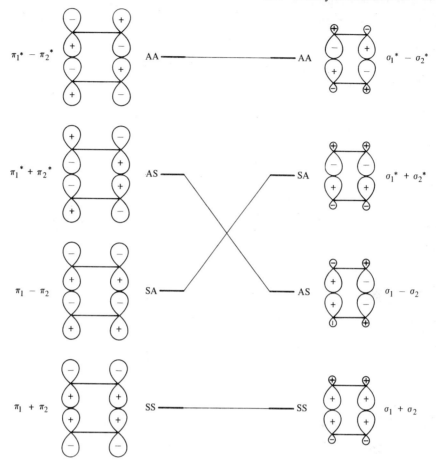

Fig. 2.6   *Correlation diagram for the symmetry-forbidden combination of two ethylenes. Symmetry elements used are the two perpendicular planes* $\begin{array}{c} \quad \\ \rule{1em}{0.5pt}\!\!\Big|\!\!\rule{1em}{0.5pt} \\ 1 \end{array} 2.$

absorption results in the promotion of an electron from the highest occupied MO (HOMO) to the lowest unoccupied MO (LUMO). For the conrotatory diagram (Figure 2.5), ψ_3, the newly occupied orbital, correlates with an even more unstable σ^* level. For the disrotatory diagram (Figure 2.4), ψ_3 correlates with a π-bonding level, and the photochemical ring closure of butadiene will therefore proceed in a disrotatory fashion.

We have discussed the electrocyclic ring closures of butadiene in order to illustrate the principle of conservation of orbital symmetry. Correlation diagrams can also be constructed for most of the other pericyclic reactions; however, this becomes a very tedious process in most cases. For example, Figure 2.6 gives the correlation diagram for the relatively simple cycloaddition of two ethylenes to give cyclobutane.

The Frontier Orbital Method[21,22]

During a chemical reaction, electrons vacate MOs corresponding to broken bonds to fill MOs of new bonds. According to frontier orbital theory (developed primarily by K. Fukui), this flow of electrons is most important between the HOMO and LUMO of interacting fragments, and *reactions occur such that there is maximum overlap between the HOMO and LUMO of the reacting species.* This overlapping, of course, is controlled by the symmetry of the orbitals.

For example, consider the cycloaddition of butadiene and ethylene to give cyclohexene, **40**.

$$\text{\Large{$\langle$}} + \| \xrightarrow{\Delta} \text{\Large{$\bigcirc$}} \tag{36}$$

40

This reaction is known as a [4+2] cycloaddition, since a molecule with four π electrons is added to a molecule with two π electrons. If the HOMO of butadiene and the LUMO of ethylene are chosen (the opposite choice will work also), these MOs can be seen to give positive overlap, **41**; the reaction is then said to be allowed.

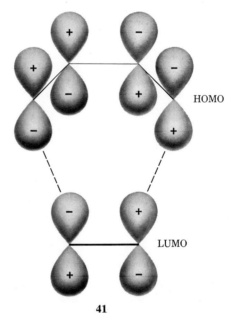

41

On the other hand, the thermal cycloaddition reaction of two ethylenes to give cyclobutane (a [2+2] cycloaddition) does not proceed as drawn because bringing the HOMO and LUMO together does not result in positive overlap, **42**.

Positive overlap could result in the above case if one lobe of the HOMO could somehow get to the top of the LUMO, **43**. While this may seem to be a highly unlikely possibility it has been observed to occur in certain instances and is known as an *antarafacial addition:* bond formations on opposite faces of a π system. Additions in which both new bonds are formed on the same face of the π system are known as *suprafacial additions* (as in **41**). The suprafacial and antarafacial labels are usually added as subscripts to derive a mechanistic label; thus, the addition depicted in **43** is designated a $[2_s + 2_a]$ cycloaddition.

From what we have seen so far, $[4_s + 2_s]$ cycloadditions and $[2_s + 2_a]$

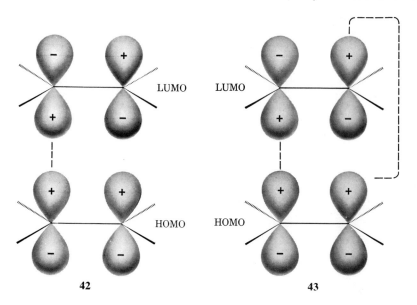

cycloadditions are allowed thermally while $[2_s + 2_s]$ are not (these observations will be generalized shortly). The frontier orbital method can be applied to many other reactions. For example the electrocyclic cleavage of a cyclobutene to a butadiene can be treated by considering the σ and π bonds being broken as separate, interacting fragments, **44**. In order for positive overlap to result, the σ bond must be cleaved in a conrotatory fashion. This result is in agreement with experiment and the prediction from the previous section.

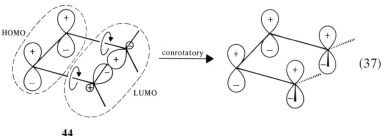

$$\tag{37}$$

44

Photochemical reactions can also be treated, since we need only consider a different HOMO or LUMO. For example, a $[2_s + 2_s]$ cycloaddition is allowed photochemically, since a new HOMO results from electron promotion, **45**.

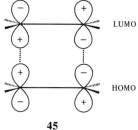

45

It normally is not necessary to restrict the HOMO to one reactant and the LUMO to the other, since either choice gives the correct answer. However, for those instances in which the direction of electron flow is obvious, the

HOMO should be assigned to the electron donor and the LUMO to the electron acceptor. Consider the reaction of N_2 with O_2, for example

$$N_2 + O_2 \rightarrow 2NO \qquad (38)$$

The electron flow should be to the more electronegative O_2; this process is forbidden by symmetry, **46**, since the sum of overlaps is zero (see Figure 1.2). On the other hand, electron flow from O_2 to N_2 is permitted by symmetry but difficult chemically, **47**.

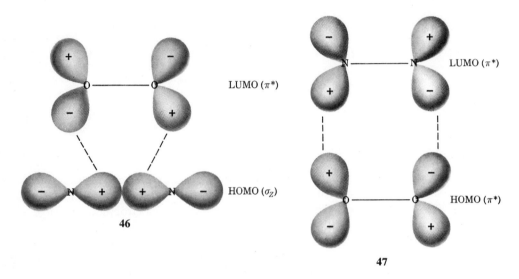

Similarly, the 1,4 addition of Br_2 to butadiene is nicely rationalized by frontier orbital theory as a flow of electrons from the HOMO of butadiene to the LUMO of bromine:

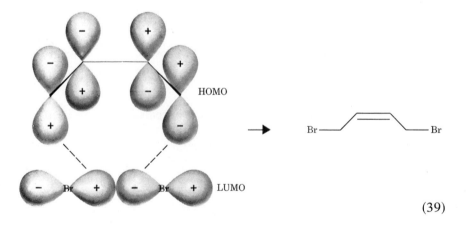

$$(39)$$

The Möbius–Hückel Concept

Dewar[8,24] and Zimmerman[23] have developed a method for treating pericyclic reactions based on the requirement that to occur, a thermal pericyclic reaction must pass through an aromatic transition state. To introduce this method, we must first examine a new concept, that of the Möbius system. The cyclic, conjugated π systems considered so far have all had an even number of sign inversions or antibonding interactions (e.g., **48** and **49**). Such systems, called Hückel systems, are aromatic when they

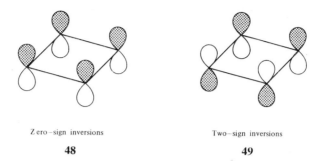

Zero–sign inversions   Two–sign inversions

48   **49**

contain $(4n+2)$ π electrons and are antiaromatic when they contain $4n$ π electrons (Section 2.1).

In contrast, the *Möbius system has an odd number of sign inversions.* For example, if we take a large, linear polyene, twist it 180°, and join the ends, the result is a cyclic polyene resembling a Möbius strip, with one sign inversion (**50**). The Möbius systems have closed shells when they contain $4n$

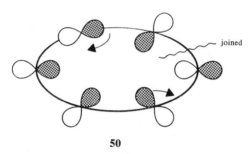

50

π electrons.[23] Thus, Hückel systems with $(4n+2)$ π electrons will be aromatic while Möbius systems with $4n$ π electrons will be aromatic.

Examination of the transition states for some of the pericyclic systems shows that some of them are Möbius systems—that is, they have an odd number of sign inversions (see Figure 2.7). If we return to the description of thermal pericyclic reactions as proceeding through aromatic transition states, the following rule results:

> *Thermal pericyclic reactions prefer Hückel geometries when $(4n+2)$ electrons are involved and Möbius geometries when $4n$ electrons are involved.*

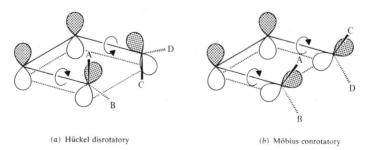

(a) Hückel disrotatory   (b) Möbius conrotatory

Fig. 2.7 (a) *Disrotatory Hückel-like closure of butadiene and* (b) *conrotatory Möbius-like closure of butadiene.*

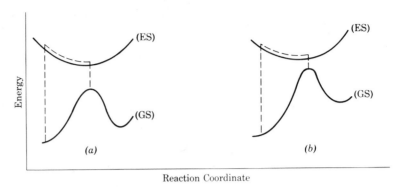

Fig. 2.8 *Potential energy curves for photochemical reactions via (a) aromatic and (b) anti-aromatic transition states.*

Thus, the conrotatory ring closure of butadiene is preferred (Figure 2.7*b*), since this geometry results in an aromatic transition state. Similarly, the $[2_s + 2_s]$ cycloaddition (**42**) is a $4n$ Hückel system giving an antiaromatic transition state and is forbidden, while the $[2_s + 2_a]$ (**43**) is a $4n$ Möbius system and is allowed.

Thermal reactions prefer aromatic transition states because such a transition state will be less energetic and easier to attain. In contrast, photochemical reactions prefer antiaromatic or less stable transition states. This is the case because a controlling factor in photochemical processes is *conversion of excited-state reactants into ground-state products*; that is, the dotted line in Figure 2.8 must be followed. For an aromatic transition state, there is a large energy gap between excited state and ground state while for an antiaromatic transition state, there is a small gap between excited and ground states.[24] The conversion of excited state to ground state is, therefore, easier for an antiaromatic transition state. In other words, by pushing up the transition state in curve GS (Figure 2.8*b*), the potential energy surfaces of excited state and ground state come closer together and the photochemical process is facilitated.

Since the photochemical pericyclic reactions prefer antiaromatic transition states, we can formulate the following rule, which is simply the reverse of that for thermal reactions:

> *Photochemical pericyclic reactions prefer Möbius geometries when $(4n + 2)$ electrons are involved and Hückel geometries when $4n$ electrons are involved.*

The PMO method is extremely useful for applying the Hückel–Möbius concept, since aromaticity or antiaromaticity can be predicted with ease for systems other than the monocyclic ones considered thus far.

Generalized Rules[17,25,26]

As mentioned at the beginning of this section, the allowedness and stereochemistry of a pericyclic reaction can be determined, without resort to theory, by application of some very simple rules. Before presenting these rules, we need to return briefly to the concepts of antarafacial and suprafacial. An antarafacial addition was defined as one in which new bonds are

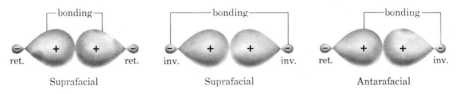

| ret. | ret. | inv. | inv. | ret. | inv. |
| Suprafacial | | Suprafacial | | Antarafacial | |

Fig. 2.9 *Suprafacial and antarafacial σ bond cleavage.*

formed on opposite faces of the π system, while the bond formations for a suprafacial addition are on the same face of a π system. Similarly, suprafacial and antarafacial cleavages of a σ bond can occur: A *suprafacial cleavage of a σ bond* is defined as one in which there is retention or inversion at both atoms; *an antarafacial cleavage of a σ bond* is one in which there is inversion at one atom and retention at the other atom. These cleavages are illustrated in Figure 2.9.

On the basis of the above definitions, the sigmatropic rearrangement depicted in eq. 40 can potentially proceed by an antarafacial (**51**) or suprafacial (**52**) cleavage of the σ bond. Similarly, the conrotatory ring

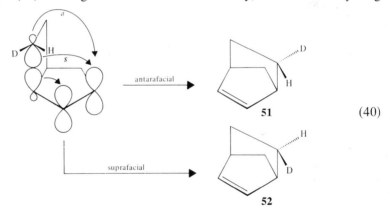

(40)

opening of cyclobutene can be seen to involve an antarafacial cleavage of the σ bond (**53**), while the disrotatory opening of the σ bond is suprafacial (**54**). The assignment of retention or inversion in this instance must be

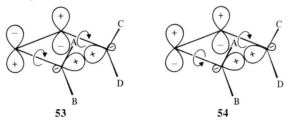

considered within the present context, since a planar sp^2-hybridized carbon is formed from the tetrahedral carbon.

The following, empirically derived rules are sufficient for the treatment of all pericyclic reactions:

> *A thermal, pericyclic reaction is symmetry allowed if the total number of suprafacial regroupings of the participating bonds is an odd number. If the number of suprafacial regroupings is even then the reaction is symmetry allowed under photochemical conditions.*

Remember that for electrocyclic reactions the following terms are synonymous: antarafacial–conrotatory and suprafacial–disrotatory.

In the following sections, we will consider some examples of pericyclic reactions and illustrate the use of the generalized rules. The number of examples given will be limited but should be sufficient to provide a thorough knowledge of the fundamentals involved. References 17, 25, and 26 contain many illustrative examples and also references to the examples given in the following sections.

Electrocyclic Reactions

According to the generalized rule, a conrotatory opening will be observed for the thermal reaction of cyclobutene because two bonds are broken and one must be an antarafacial (conrotatory) cleavage (**55**). Similarly the photochemical opening of this same molecule must involve a suprafacial cleavage of the bonds (**56** and eq. 41).

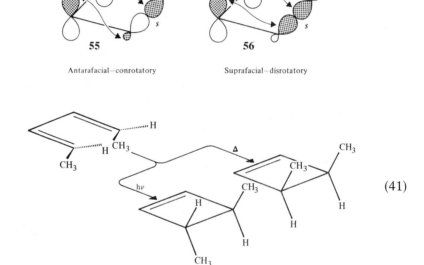

Among the more interesting electrocyclic reactions is the conversion of Dewar benzene into benzene:

$$\qquad (42)$$

The molecule of benzene is about 60 kcal/mole more stable than Dewar benzene. Thus, it would seem that the interconversion of the two molecules would be extremely facile; however, an activation energy of 23 kcal/mole is required to carry out the reaction![27] The reaction involves cleavage of two bonds, and according to the generalized rule, the opening must be conrotatory. Examination of eqs. 43 and 44 shows that the allowed conrotatory opening produces not benzene but an unlikely looking cyclohexatriene molecule with a *trans* double bond.

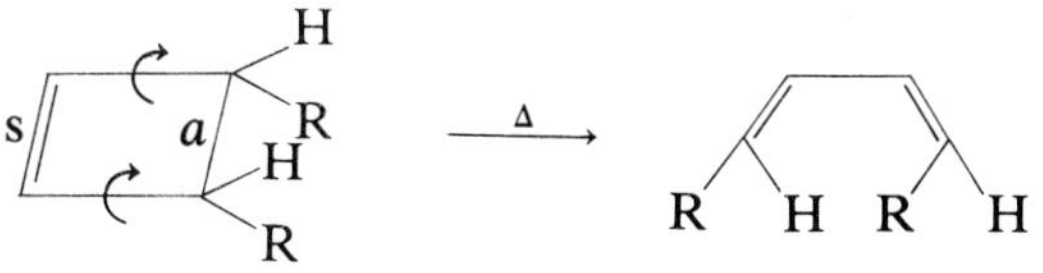

(43)

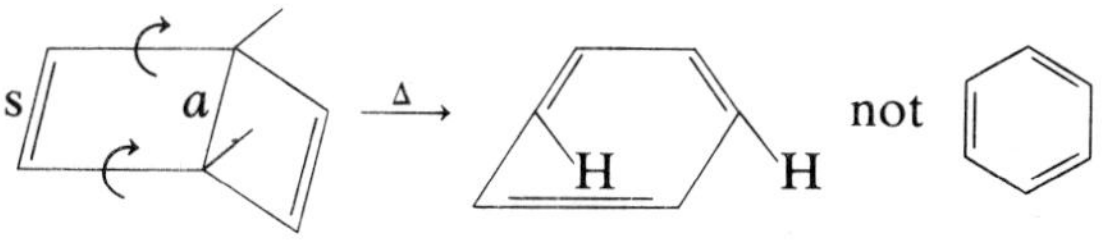

(44)

This reaction brings to mind the question of what happens when a reaction is forbidden on the basis of orbital symmetry. There appear to be three possibilities: (*a*) a nonconcerted reaction occurs to form a discrete intermediate, (*b*) enough energy is added to force the reaction to proceed via the symmetry-forbidden pathway, or (*c*) the reaction follows a symmetry-allowed pathway to give an excited state of the product.[28] There is evidence that the latter possibility occurs for the ring opening of Dewar benzene.[29]

The ring opening of cyclopropyl halides to allyl carbocations provides another interesting example of an electrocyclic reaction as shown in eq. 45. This reaction involves formation of the cyclopropyl cation and electrocyclic ring opening to give an allyl cation that is then attacked by solvent. Only one bond is broken in the opening, and this must be suprafacial or disrotatory (**57**). However, since only one of the two possible disrotatory openings is observed it seems that the ionization of X is accompanied by

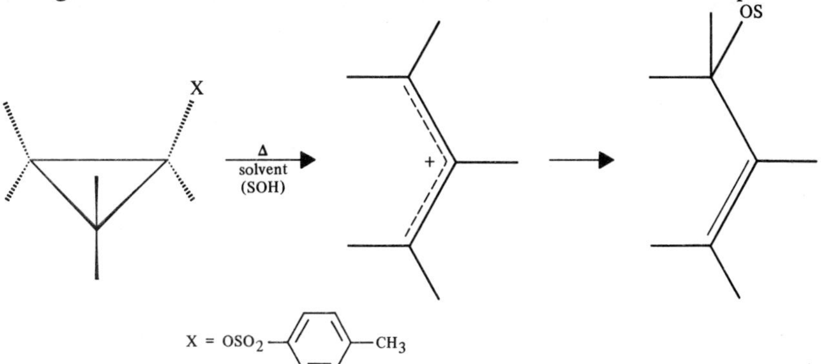

$$X = OSO_2 - \langle\ \rangle - CH_3$$

(45)

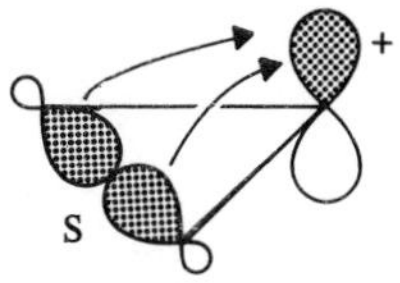

57

disrotatory

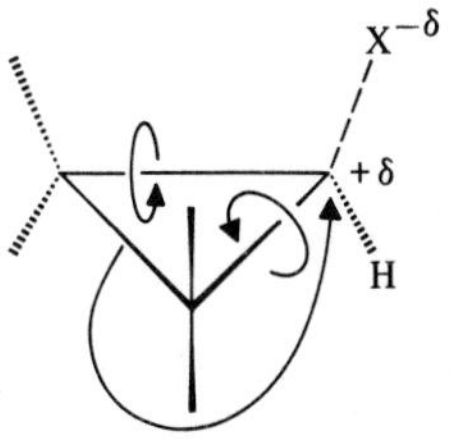

58

concurrent electrocyclic opening. The electrons in the C–C bond appear to flow to the backside of the C–X bond (**58**) and displace X^-.

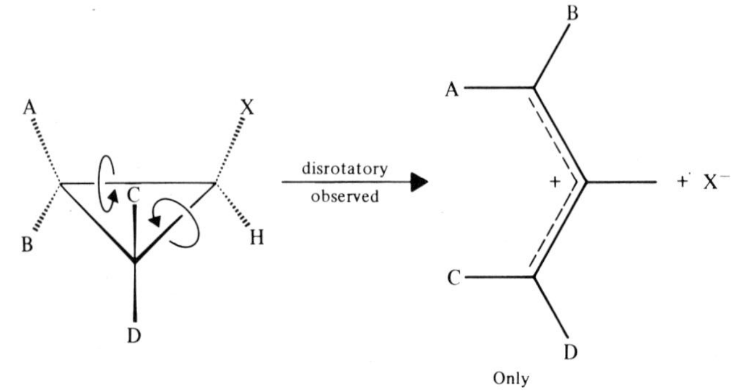

$$\xrightarrow[\text{observed}]{\text{disrotatory}} \qquad + \ X^- \qquad (46)$$

Only

In this case, the substituents A through D in eq. 46 will have a major effect, since A and C are forced together in the transition state. Thus, it is easy to see why **60** reacts 4500 times faster than **59**. Even more dramati-

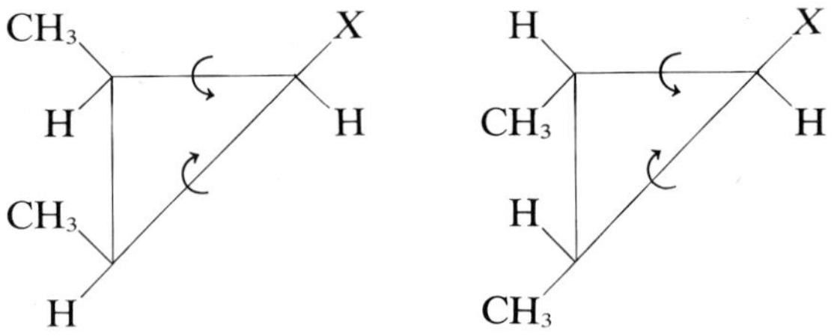

59   **60**

cally, **61** reacts 11,000 times as fast as **62**. This great rate difference is due to the required formation of a *trans*-allylic system in the simultaneous ionization-electrocyclic opening of **62**.

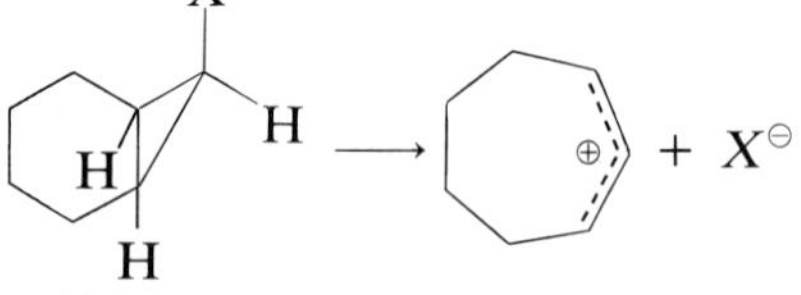

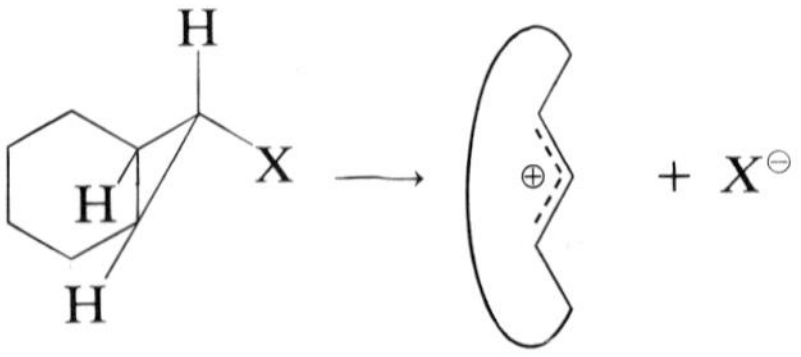

$$\longrightarrow \qquad + \ X^\ominus \qquad (47)$$

61

$$\longrightarrow \qquad + \ X^\ominus \qquad (48)$$

62

Cycloaddition Reactions

The generalized rules can also be used to treat cycloadditions handily. For example, the $[2_s + 2_s]$ cycloadditions, having an even number of suprafacial regroupings, are only allowed for a photochemical process:

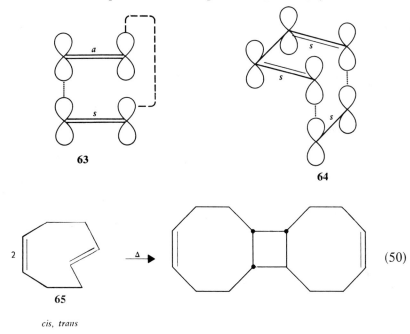

Similarly, the $[2_s + 2_a]$ and $[4_s + 2_s]$ processes are thermally allowed (**63** and **64**). The reaction of **65** provides an example of the $[2_s + 2_a]$ cycloaddition:

63

64

65

cis, trans

(50)

The addition of an alkene to bicycloheptadiene (norbornadiene),

(51)

is an example of a thermally allowed $[2_s + 2_s + 2_s]$ cycloaddition. Note the odd number of suprafacial regroupings. The cycloadditions discussed so far have involved π components only; however, σ components may participate also. The following example,

(52)

is actually a [4+2] cycloaddition in which two of the electrons come from a σ bond. This particular type of reaction is called an ene reaction and is referred to as a $[_\sigma 2_s +_\pi 2_s +_\pi 2_s]$ cycloaddition. Thus, a complete designation should also give the σ or π subscript.

To take matters one step further, the addition of acetylene derivatives to 1,4-cyclohexadiene is a photochemically allowed $[_\pi 2_s +_\pi 2_s +_\pi 2_s +_\pi 2_s]$ cycloaddition (the π bonds are independent):

$$X—C\equiv C—X \; + \; \text{(cyclohexadiene)} \; \xrightarrow{h\nu} \; \text{(product)} \tag{53}$$

This reaction provides a good example of an antiaromatic Hückel system ($4n, n = 2$), which is allowed only for photochemical processes, and illustrates the power of the Hückel–Möbius concept for treating rather complicated reactions (**66**).

66

The addition of tetracyanoethylene to heptafulvene, a $[_\pi 14_a +_\pi 2_s]$ process, involves the cleavage of eight bonds and thus must have an antarafacial regrouping to produce an odd number of suprafacial cleavages:

$$\text{(heptafulvene)} \; + \; \underset{CN}{\overset{CN}{C}}=\underset{CN}{\overset{CN}{C}} \; \xrightarrow{\Delta} \; \text{(product)} \tag{54}$$

This prediction is consistent with the product stereochemistry. The difficult antarafacial union is made possible by the fact that heptafulvene is not planar, but is twisted.

The electrocyclic reactions of the cyclopropyl cation were considered previously, and the positive center was essentially ignored in predicting the reaction stereochemistry. Cycloaddition reactions can also be treated by ignoring the positive center when the number of suprafacial regroupings are counted. For example, reaction 55 is thermally allowed since three suprafacial cleavages occur. This behavior is reasonable since the reaction is controlled by the number of electron pairs involved, not by the number of atomic centers involved.

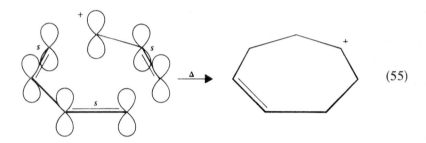

(55)

Similarly, a pair of electrons in an anion is counted as a suprafacial cleavage. Reaction 56 is then seen to be thermally allowed.

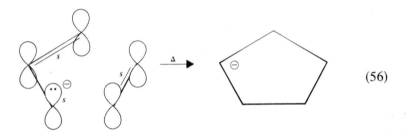

(56)

Sigmatropic Reactions

The general sigmatropic reaction is given as

$$
\begin{array}{ccc}
Z & & Z \\
| & & | \\
C-(C=C)_n & \rightarrow (C=C)_n - C
\end{array}
$$

(57)

and involves the migration of a σ bond to a new position within the system. These reactions are classified as $[i,j]$ order where i and j refer to the number of the atom to which each end of the "migrating" bond joins. The initial σ bond is always assumed to join the number 1 atoms. For example, reaction (58) is a $[1,3]$ sigmatropic reaction,

$$
\underset{1 \quad\ 2 \quad\ 3}{\overset{\displaystyle D}{\overset{|}{CH_2}}-CH=CH_2} \xrightarrow[\substack{i=1 \\ j=3}]{[1,3]} \underset{1 \quad\ 2 \quad\ 3}{CH_2=CH-\overset{\displaystyle D}{\overset{|}{CH_2}}}
$$

(58)

while reaction (59) is a [3,3] migration.

$$
\begin{array}{ccc}
1 & 2 & 3 \\
CH_2-CH=CH_2 & & 1 \quad 2 \quad 3 \\
| & \xrightarrow[\substack{i=3 \\ j=3}]{[3,3]} & CH_2=CH-CH_2 \\
CH_2-CH=CH_2 & & CH_2=CH-CH_2 \\
1 & 2 & 3 \qquad 1 \quad 2 \quad 3
\end{array}
$$
(59)

The rule for treating these reactions is the same as for the previous two reaction types:

> there must be an odd number of suprafacial regroupings for thermal reactions and an even number for photochemical reactions.

Thus, the thermal [1,3] shift shown in reaction (60)

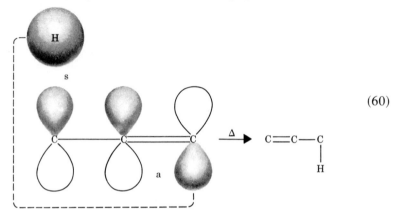

(60)

is allowed if there is one antarafacial cleavage; the steric requirements of this reaction are such that it is not a favorable process.

For example, compound **67** and some of its derivatives have been synthesized and shown to be stable at low temperatures. This stability was originally surprising because it seemed that tautomerism to give the more stable aromatic isomer would be rapid. The slowness of the isomerization is, of course, due to the [1,3]-hydrogen shift being symmetry-forbidden unless the hydrogen migrates to the opposite face of the π bond.

$$
\text{67} \quad \xrightarrow{25\,°C} \quad \text{CH}_3
$$
(61)

67

A migrating alkyl group that can invert eases this process:

$$
\xrightarrow{} \quad C=C-C \quad \text{Z (inverted)}
$$
(62)

and several such migrations have been observed.

A [1,4] sigmatropic shift in a cationic system such as **68** must also proceed

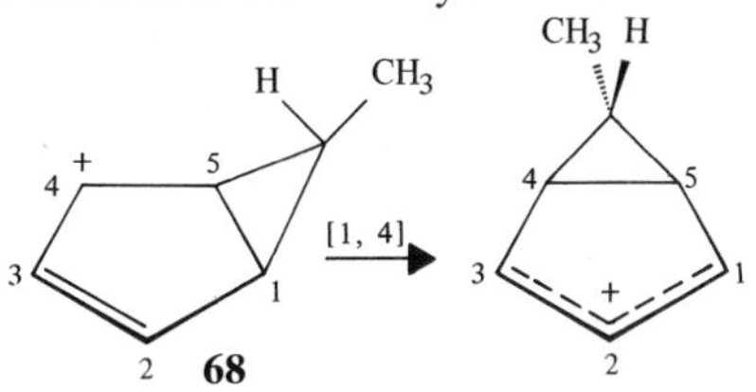

with inversion at the migrating center since only two *bonds* are broken—one suprafacial and one antarafacial; if there had been no inversion, the methyl would have been *endo* rather than *exo*.

A [1,5] migration can proceed with retention of configuration at the migrating center since three bonds are cleaved and all can be suprafacial:

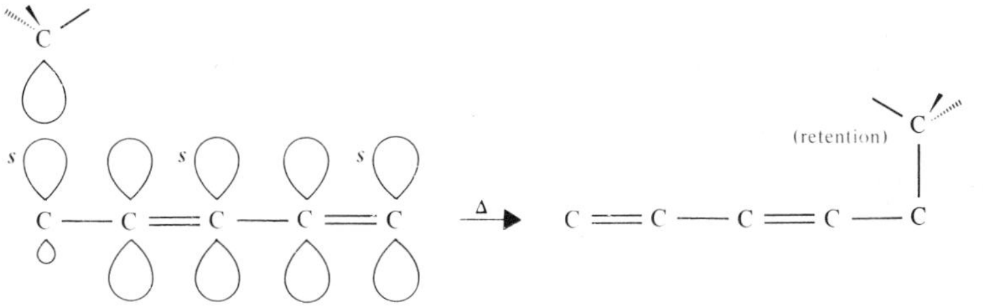

$$(63)$$

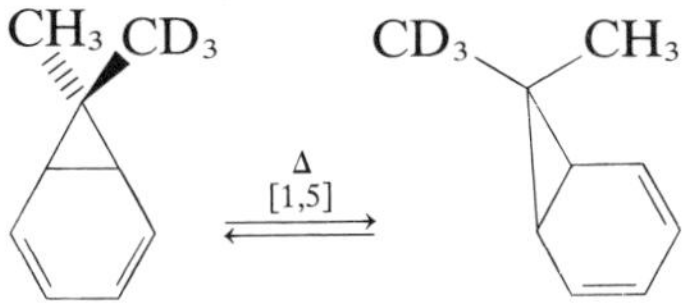

$$(64)$$

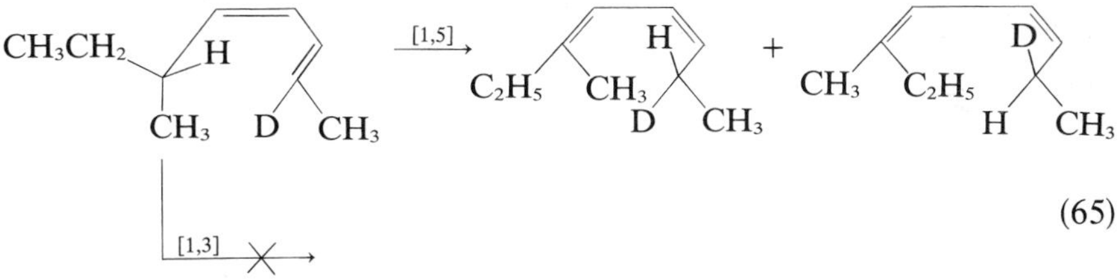

$$(65)$$

The Cope and Claisen rearrangements:

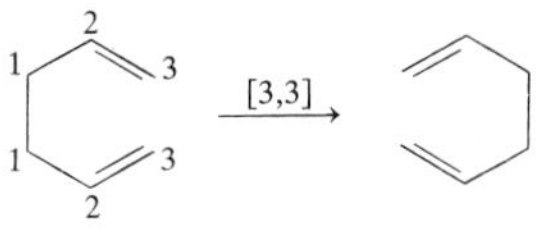

(Cope)   (66)

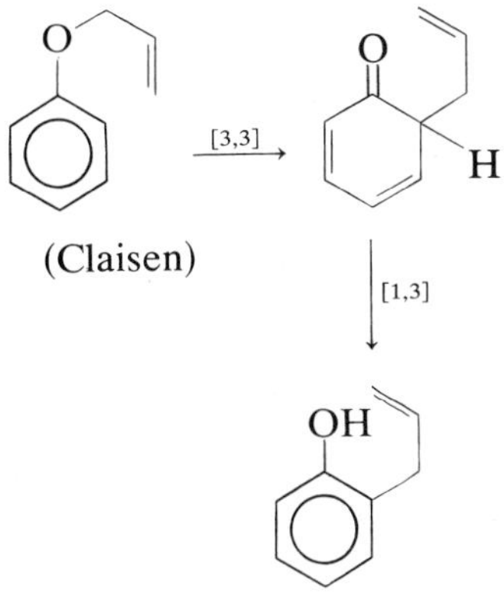

(Claisen)

$$(67)$$

are probably the best known sigmatropic rearrangements; both are examples of [3,3] sigmatropic rearrangements, and since three bonds are cleaved, all can proceed suprafacially.

The Cope rearrangement has been shown to proceed through a "chairlike" transition state (**69**) rather than the alternative "boatlike" transition state (**70**). This preference is reasonable, since **69** is an aromatic, Hückel

69	**70**	**71**

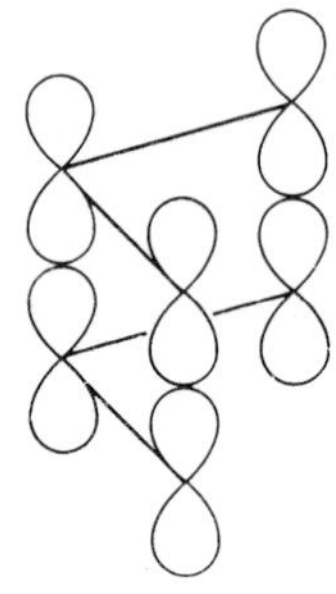

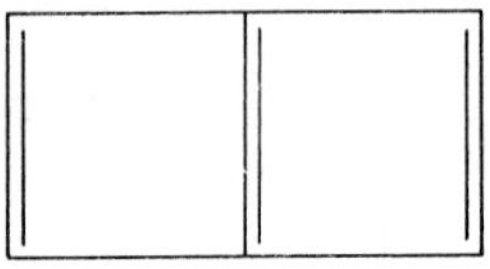

transition state, while **70** resembles butalene **71**, which has been shown by the PMO method to be antiaromatic.

The "chairlike" transition state is consistent with the double bond sterochemistry observed for this reaction.

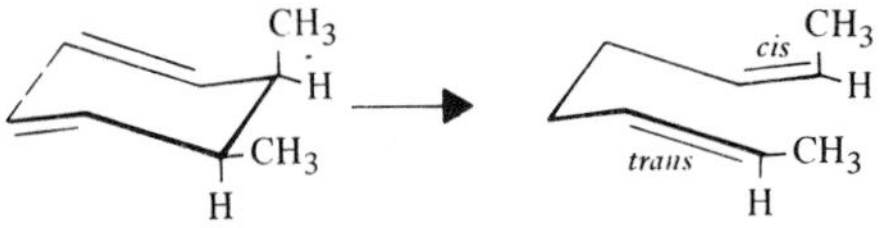

(68)

The Claisen rearrangement is known to give migration of the alkyl group to the p-position by a two-step process:

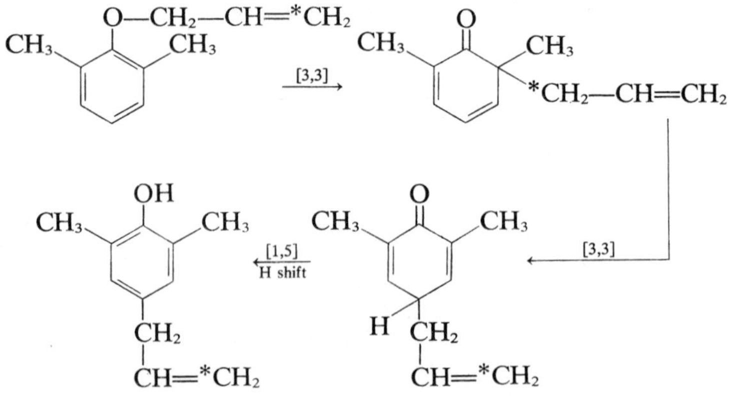

(69)

This sequence is consistent with a one-step [3,5] rearrangement being symmetry forbidden; sequential [3,3] rearrangements are allowed.

The corresponding, allowed [5,5] rearrangement has also been observed:

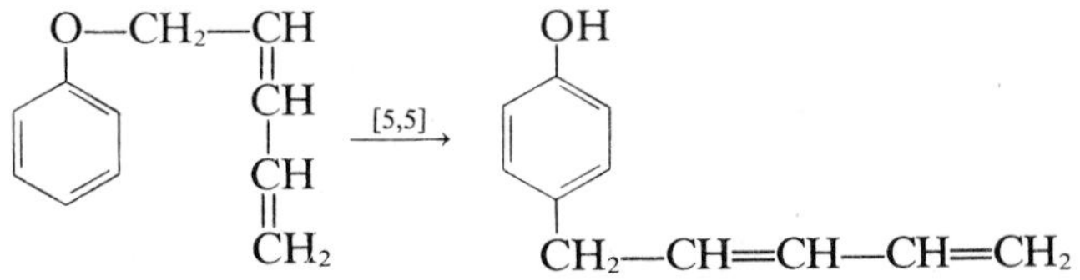

$$\text{(70)}$$

The Cope rearrangement can give products that are identical to the reactants (eq. 71); such a reaction is described as degenerate. The probability of the degenerate Cope rearrangement occurring can be enhanced by restricting the conformational possibilities. Reactions 71 through 73 and the reactions of molecule **74** represent a progression in this conformational restriction.

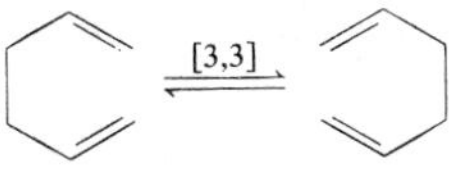

$$\text{[3,3]} \qquad \text{(71)}$$

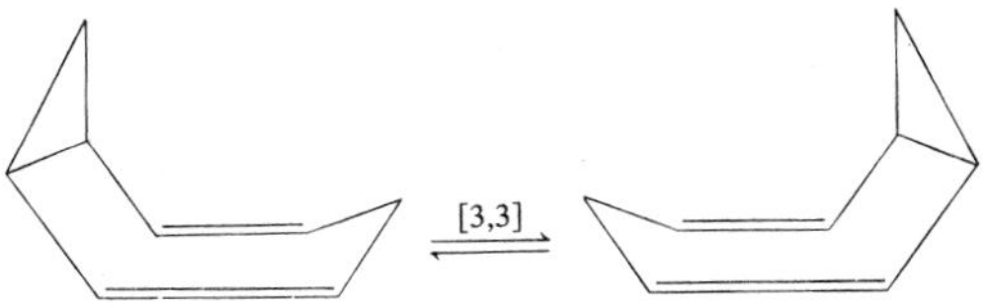

72   **73**

$$\text{[3,3]} \qquad \text{(72)}$$

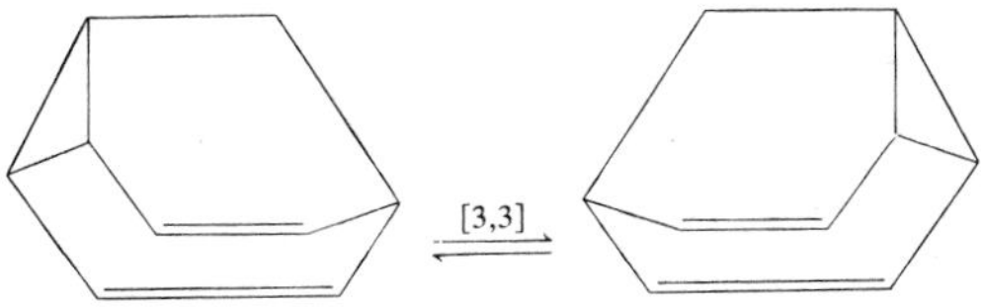

$$\text{[3,3]} \qquad \text{(73)}$$

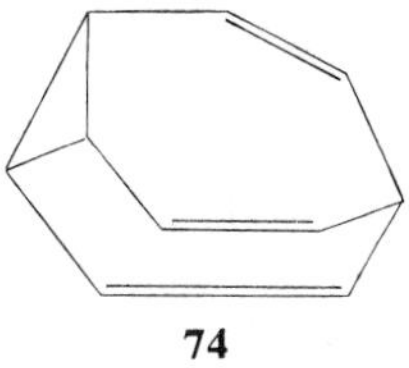

74

The rearrangement of **72** to **73** occurs at a rate of approximately 10^3/sec at 180°. The "ultimate" member of this group is bullvalene, **74**, since any of the ten carbons can be converted into any other carbon by means of [3,3] sigmatropic shifts. The nmr spectrum of this molecule shows only one peak and confirms the prediction of total equivalency.

References

1. L. M. Jackman and S. Sternhell, "Applications of Nuclear Magnetic Resonance Spectroscopy in Organic Chemistry," 2nd ed., Pergamon, Oxford, 1969.
2. P. J. Garratt, "Aromaticity," McGraw-Hill, New York, 1971.

3. D. M. G. Lloyd, "Carbocyclic Non-benzenoid Aromatic Compounds," Elsevier, New York, 1966.

4. J. P. Snyder, Ed., "Nonbenzenoid Aromatics," Vols. I and II, Academic Press, New York, 1969.

5. D. Ginsberg, Ed., "Nonbenzenoid Aromatic Compounds," Interscience, New York, 1959.

6. G. W. Wheland, "Resonance in Organic Chemistry," Wiley, New York, 1955.

7. A. Streitwieser, Jr., "Molecular Orbital Theory for Organic Chemists," Wiley, New York, 1962.

8. M. J. S. Dewar, "The Molecular Orbital Theory of Organic Chemistry," McGraw-Hill, New York, 1969.

9. N. C. Baird, *J. Chem. Educ.*, **48,** 509 (1971).

10. L. J. Schaad and B. A. Hess, Jr., *J. Chem. Educ.*, **51,** 640 (1974); B. A. Hess, Jr., and L. J. Schaad, *J. Amer. Chem. Soc.*, **93,** 305 (1971).

11. F. A. L. Anet, A. J. R. Bourn, and Y. S. Lin, *J. Amer. Chem. Soc.*, **86,** 3576 (1964).

12. P. Reeves, J. Henry, and R. Pettit, *J. Amer. Chem. Soc.*, **91,** 5888 (1969).

13. R. H. Grubbs and R. A. Grey, *J. Amer. Chem. Soc.*, **95,** 5765 (1973).

14. S. Masamune and N. Darby, *Acc. Chem. Res.*, **5,** 272 (1972).

15. E. E. van Tamelen, T. L. Burkoth, and R. H. Greeley, *J. Amer. Chem. Soc.*, **93,** 6120 (1971).

16. E. Vogel, in "Aromaticity, an International Symposium," Chemical Soc. Spec. Publ., No. 21, 113 (1967).

17. R. B. Woodward and R. Hoffman, "The Conservation of Orbital Symmetry," Academic Press, New York, 1970.

18. R. B. Woodward and R. Hoffman, *J. Amer. Chem. Soc.*, **87,** 395 (1965).

19. H. C. Longuet-Higgens and E. W. Abrahamson, *J. Amer. Chem. Soc.*, **87,** 2045 (1965).

20. H. E. Simmons and J. F. Bunnett, Eds., "Orbital Symmetry Papers," ACS Reprint Collection, Washington, D.C., 1974.

21. K. Fukui, *Acc. Chem. Res.*, **4,** 57 (1971).

22. R. G. Pearson, *Acc. Chem. Res.*, **4,** 152 (1971).

23. H. E. Zimmerman, *Acc. Chem. Res.*, **4,** 272 (1971).

24. M. J. S. Dewar, *Angew. Chem. Int. Ed. Engl.*, **10,** 761 (1971).

25. R. E. Lehr and A. P. Marchand, "Orbital Symmetry," Academic Press, New York, 1972.

26. T. L. Gilchrist and R. C. Storr, "Organic Reactions and Orbital Symmetry," Cambridge Univ. Press, London, 1972.

27. R. Breslow, J. Napierski, and A. H. Schmidt, *J. Amer. Chem. Soc.*, **94,** 5906 (1972).

28. M. J. S. Dewar, S. Kirschner, and H. W. Kollmar, *J. Amer. Chem. Soc.*, **96,** 7579 (1974).

29. P. Lechtken, R. Breslow, A. H. Schmidt, and N. J. Turro, *J. Amer. Chem. Soc.*, **95,** 3025 (1973).

Bibliography

Aromaticity

M. J. S. Dewar, "The Molecular Orbital Theory of Organic Chemistry," McGraw-Hill, New York, 1969.

A. Streitwieser, Jr., "Molecular Orbital Theory for Organic Chemists," Wiley, New York, 1962.

P. J. Garratt, "Aromaticity," McGraw-Hill, New York, 1971.

D. M. G. Lloyd, "Carbocyclic Non-benzenoid Aromatic Compounds," Elsevier, New York, 1966.

J. P. Snyder, Ed., "Nonbenzenoid Aromatics," Vols. I and II, Academic Press, New York, 1969.

D. Ginsberg, Ed., "Nonbenzenoid Aromatic Compounds," Interscience, New York, 1959.

N. C. Baird, *J. Chem. Educ.*, **48**, 509 (1971).

B. A. Hess, Jr., and L. J. Schaad, *J. Amer. Chem. Soc.*, **93**, 305 (1971).

Nucleophilic and Electrophilic Substitution

M. D. Bentley and M. J. S. Dewar, *J. Amer. Chem. Soc.*, **92**, 3991 (1970). See also the books by Dewar and Streitwieser referenced above.

Pericyclic Reactions

R. B. Woodward and R. Hoffmann, *J. Amer. Chem. Soc.*, **87**, 395 (1965).

H. C. Longuet-Higgens and E. W. Abrahamson, *J. Amer. Chem. Soc.*, **87**, 2045 (1965).

R. B. Woodward and R. Hoffmann, "The Conservation of Orbital Symmetry," Academic Press, New York, 1970.

K. Fukui, *Acc. Chem. Res.*, **4**, 57 (1971).

R. G. Pearson, *Acc. Chem. Res.*, **4**, 152 (1971).

H. E. Zimmerman, *Acc. Chem. Res.*, **4**, 272 (1971).

M. J. S. Dewar, *Angew. Chem. Int. Ed. Engl.*, **10**, 761 (1971).

R. E. Lehr and A. P. Marchand, "Orbital Symmetry," Academic Press, New York, 1972.

T. L. Gilchrist and R. C. Storr, "Organic Reactions and Orbital Symmetry," Cambridge Univ. Press, London, 1972. This work and the previous one give many examples and are written for the student.

Problems

1. By application of PMO theory determine whether:
 (*a*) anthracene or phenanthrene is more stable;
 (*b*) biphenyl or biphenylene is more stable;
 (*c*) the cyclopentadienyl anion or cation, the cycloheptatrienyl anion or cation, heptafulvene, cyclodecapentaene, or butalene is potentially aromatic.

2. Without using any calculations, other than application of the $(4n+2)$ rule, predict which of the following will be aromatic, nonaromatic, or antiaromatic.

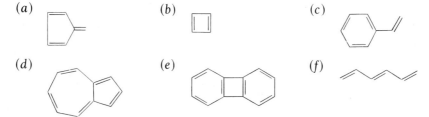

(*a*) (*b*) (*c*)

(*d*) (*e*) (*f*)

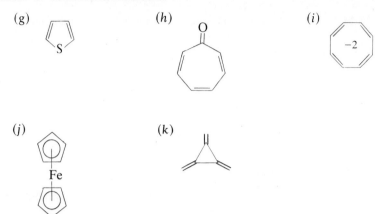

3. Repeat parts *c* through *e* of problem 8 of Chapter 1 for DREs instead of DEs.
4. Apply the intramolecular PMO equations to predict the stabilities of pentalene and heptalene relative to their respective annulenes.
5. The nmr spectrum of [18] annulene has two peaks, one at 1.1τ, which has twice the area of the other at 11.8τ. Explain this observation.
6. Arrange the following compounds in order of reactivity toward aqueous ethanol:

7. Predict the relative reactivities and orientation of nitration of the following molecules:

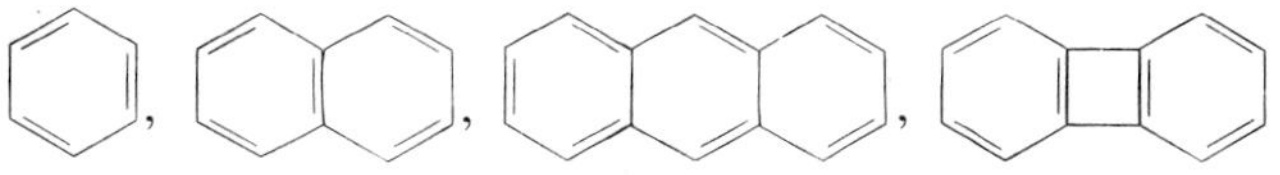

8. Is the following reaction allowed?

$$CH_2N_2 + CH_2=CH-CO_2Et \xrightarrow{\Delta}$$

9. Predict product stereochemistry for the following reactions:

(a)

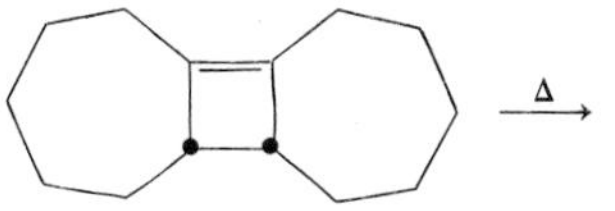

(b)

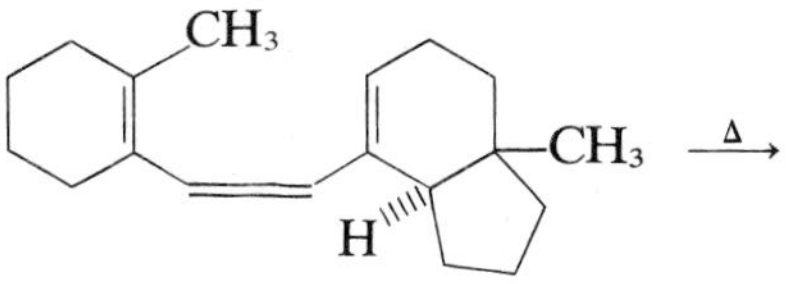

(c)

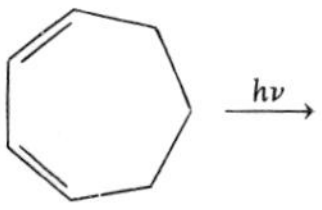

(d)

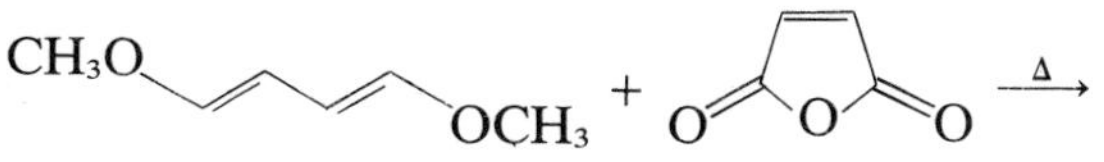

10. Show how the compound below can be used to test the allowedness of [1,3] and [1,5] sigmatropic shifts.

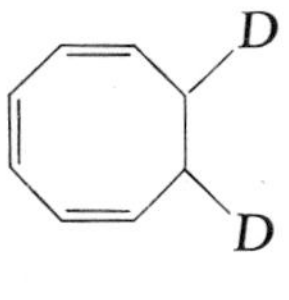

11. Rationalize the following:

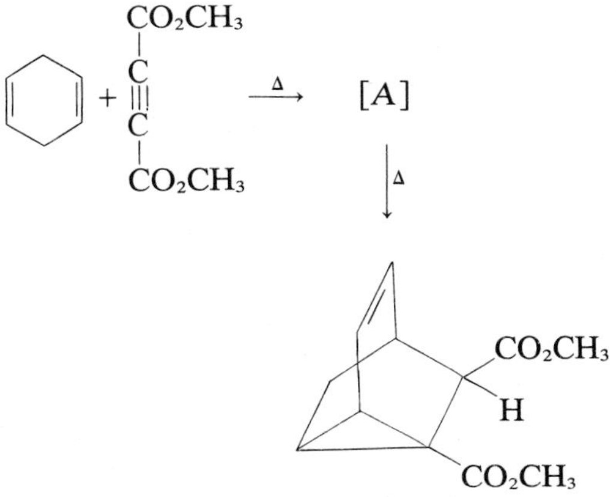

12. Predict products and stereochemistry for *thermal* electrocyclic ring opening of the following:

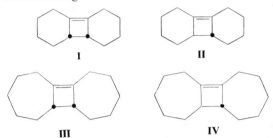

I   II

III   IV

Why does the following order of rates obtain: **IV, II ≫ I, III**?

13. Complete the following:

(a)

$$\xrightarrow{h\nu}$$

(b)

$$\xrightarrow{h\nu}$$

(c)

$$\xrightarrow{h\nu}$$

(d)

$$+ \xrightarrow{\Delta}$$

CH₃ CH₃

(e)

O—CH— CH₃

$$\xrightarrow{\Delta}$$

(f) 2 $$\xrightarrow{\Delta}$$

$$\xrightarrow{h\nu}$$

14. At 100°C, bullvalene, **74**, has only one peak in its nmr spectrum. Show why this is the case. At −25°C two peaks are observed, one at 4.3τ (6H) and the other at 7.9τ (4H). Explain.

15. Are there essential bonds in **V**? Is this an aromatic molecule?

V

16. The observed heats of formation of benzene, styrene, and naphthalene in kcal/mole are 19.8, 33.3, and 36.5. Use eq. 4 to calculate resonance energies for these molecules. Calculate the resonance energy of naphthalene in terms of β, given E_π calculated by the HMO method is 13.68β.

17. Apply the frontier orbital method and the Möbius–Hückel concept to predict the allowedness and stereochemistry of (a) a $[1,3]$ sigmatropic shift and (b) the electrocyclic ring opening of the cyclopropyl cation.

3

Determination of Organic Reaction Mechanisms

To many chemists, the real heart of chemistry is the study of mechanisms—the determination of the specific changes and interactions that occur as molecules undergo chemical reactions. To completely describe an organic reaction mechanism, we would need to know as a function of time the exact positions of all the atoms as the starting molecules are converted into product molecules. This is an idealized goal that can never be fully realized, since many of the changes take place too rapidly for any direct detection. The time scale for a molecular vibration or collision, for example, is in the range of 10^{-12} to 10^{-14} sec, much too rapid for standard spectroscopic detection. Virtually all information regarding reaction mechanisms comes by inference from indirect evidence. It is the job of the chemist to devise the proper experiments to generate the most conclusive possible evidence—to put good questions to the molecules and to interpret the answers.

Given the reliance upon indirect evidence, we see that a mechanism can never be truly proved. Rather, it should be said that evidence supports or is consistent with a particular mechanism. Of course, a particular piece of evidence can specifically disprove a given mechanism. The history of chemistry abounds with examples of "well-established" mechanisms that required major revision or dissolved completely in the light of a single piece of new evidence. A good reaction mechanism, therefore, must correlate all known facts about the reaction. Moreover, a good reaction mechanism will have predictive power; it should suggest new experiments that could test its validity and the validity of alternative mechanisms.

There are several reasons why the study of mechanisms holds such central importance in chemistry. For synthetic purposes, a knowledge of a reaction mechanism will often allow the reaction conditions to be selected for maximum product yield. For educational purposes, the student and researcher alike will find that a knowledge of reaction mechanisms can be used to correlate a large number of apparently different reactions; it is much easier and much more valuable to learn reaction mechanisms than to memorize the staggering number and variety of organic reactions. Finally, for intellectual purposes, the study of mechanisms represents the chemist's deepest probing into what molecules are really doing and why. Besides the utility of understanding reaction mechanisms, their determination is one of the most challenging areas of chemistry. In this chapter, we will review and discuss some of the important concepts and methods utilized in the determination of organic reaction mechanisms.

3.1 Proposing a Reasonable Mechanism

For most organic reactions, there undoubtedly will be many conceivable mechanisms. Deciding whether or not a potential mechanism is reasonable and worthy of further consideration and testing is one of the first important steps. Consider the following substitution reaction, for example:

$$CH_3I + OH^- \rightarrow CH_3OH + I^-$$

A conceivable mechanism for this (or for that matter, for any) reaction might be the disintegration of the starting molecules into atoms, followed by their reassembly into product molecules. It requires little chemical intuition to see that this is an absurd possibility:

$$CH_3I + OH^- \rightarrow C + 4H + O + I + e^- \rightarrow CH_3OH + I^-$$

A reasonable mechanism that should come to mind quickly would be the simple replacement of iodide by hydroxide:

$$CH_3 — I + OH^- \rightarrow CH_3 — OH + I^-$$

This should appear much more reasonable, since the features that are present in both the starting molecule and product (the three C—H bonds) are maintained intact. Even this simple mechanism is full of ambiguities. To fully understand this mechanism, the student should be asking questions like *when*, *where*, and *why*. Exactly *when* does the iodide ion leave—before, after, or while the hydroxide ion is attaching itself? *Where* does the hydroxide attach to the carbon atom, relative to the initial position of the iodine? And finally, *why* does this reaction take place at all, and especially *why* does it take place in the specific manner outlined in the proposed mechanism? These questions will give an idea of the kind and depth of information desired in a mechanistic study. There are, in fact, several possible reasonable mechanisms for this simple substitution reaction, and the major task will be to distinguish the correct mechanism from among the reasonable possibilities.

Although it might seem that deciding whether a potential mechanism is reasonable or not is a matter of experience or intuition, we can lay down some ground rules regarding the reasonableness of potential mechanisms.

1. *Occam's razor.* "A plurality is not to be posited without necessity" (by William of Occam, a fourteenth century British Franciscan and schoolman). This principle applies to any scientific explanation. The explanation (mechanism) should be as simple as possible, while still explaining all the available facts.

   The simple mechanism will normally suggest experimental tests that could demonstrate its validity. We will deal with the collection of such evidence shortly. As evidence becomes available, the facts may then demand refinements in the proposed mechanism or perhaps an entirely fresh start.

2. *In any multistep mechanism, each of the individual steps should be either unimolecular or bimolecular.*

   Our proposed mechanism for the substitution reaction above involved only a single mechanistic step and, therefore, would be considered a concerted reaction in the sense outlined in Chapter 2. All reactions may be arbitrarily divided into two categories: concerted reactions that take place in a single mechanistic step, which we have covered in detail in Chapter 2, and nonconcerted reactions that take place with more than one mechanistic step. Virtually all the remainder of this book will deal with nonconcerted reactions and their necessary consequence—reactive intermediates.

   When dealing with a multistep mechanism it is to be preferred that each of the steps be as simple as possible. Each step should be either unimolecular, involving a reaction of a single molecule, or bimolecular, involving the interaction of two molecules. Invoking any more than two molecules in a basic mechanistic step is disfavored because it is extremely unlikely that more than two molecules would be involved in a simultaneous collision. For example, in the gas phase at 1 atm, a two-body collision is about 1000 times more likely than a three-body collision.

3. *Each step should be energetically feasible.* Knowingly or unknowingly, it was mainly on this basis that we eliminated as absurd the mechanism that called for the breaking of methyl iodide and hydroxide into atoms. Using bond energy or other thermodynamic data, it can frequently be determined whether one mechanistic pathway is to be preferred over another. Structures that appear in both the substrate and the product will probably (but not necessarily) remain intact throughout the mechanism. Bonds that are broken in a particular step should usually be at least partially compensated by concurrent formation of new bonds.

4. *Each step should be chemically reasonable.* This final criterion probably requires the most experience and intuition. Ultimately, the student should develop a "feel" for what appears to be chemically reasonable and what is unreasonable. Basically, this criterion will draw upon a knowledge of similar chemical reactions. In general, the proposed mechanism should be in line with what is known for similar reactions.

Of course, reaction mechanisms that do not fit into the framework of previous knowledge occur occasionally, even frequently, and the observation of a departure from "normal" behavior represents an exciting opportunity. Such cases will naturally require a more substantial weight of evidence in order to establish a completely novel mechanistic scheme.

Thus far, we have only dealt with the pencil-and-paper aspects of determining organic reaction mechanisms—how to devise a reasonable hypothesis for experimental testing and what kind of questions need to be answered by the experiments. The specific experimental tests will vary substantially depending upon the type of reaction under study. All mechanisms must be at least supported by a careful product study, including stereochemical results where appropriate, and by consistent kinetic schemes. Within the areas of product studies and kinetic studies, there exists a vast variety of techniques and procedures for obtaining mechanistic evidence. Many of the techniques for probing mechanisms are rather standard, but in many cases the strongest evidence comes from unique experimental tests that have been specifically designed for the particular reaction under consideration. In the remainder of this chapter, we will deal with some of the standard methods of product studies, kinetic studies, and other techniques to show how mechanistic evidence is derived and utilized. It should be kept in mind that this will necessarily be an incomplete listing of mechanistic techniques, since there are really limitless ways in which a chemist may obtain information about molecules, and new techniques are always being developed.

3.2 Product Studies

The understanding of a reaction mechanism necessarily begins with a knowledge of exactly what the reaction is. That is, the structure of the starting materials must be well established and the exact structure of the products must be determined. Although this may appear trivial, it is essential to be certain of the initial and final points of the reaction before any of the action in between (i.e., the mechanism) can be considered.

Additionally, the purity of the starting materials and the products can be particularly significant. For example, the reaction of propylene with hydrogen bromide takes two completely different courses, depending upon whether or not traces of peroxide impurities are present (see Chapter 6). The arguments over the "correct" product lasted for several years before the importance of peroxides (a common type of impurity in many organic compounds) was demonstrated.

$$CH_3—CH=CH_2 + HBr$$

no peroxides → $CH_3—CH—CH_3$ 2-bromopropane
                     |
                     Br

with peroxides → $CH_3—CH_2—CH_2—Br$ 1-bromopropane

In general, the yield of product should be determined, if only to ascertain that the reaction under consideration is indeed the major reaction pathway.

Later experimental tests should show how the yield varies with reaction parameters such as temperature, pH, solvent, and so on. The presence or absence of side products should be carefully ascertained. Frequently the nature of the side products can be useful in deciding upon mechanistic possibilities, although it is also frequent that side products occur via a completely separate mechanistic pathway and, therefore, may be more distracting than helpful.

Stereochemistry

In determining the exact structures of the products, chemists often rely heavily upon the special stereochemical information generated by asymmetric carbon atoms. For example, the displacement of iodide ion from a primary or secondary alkyl iodide by hydroxide ion takes place with complete inversion of stereochemistry. This information could not have been obtained from a simple methyl iodide reaction, but was obtained by study of a dissymmetric molecule, such as (R)-2-iodobutane. The appearance of only

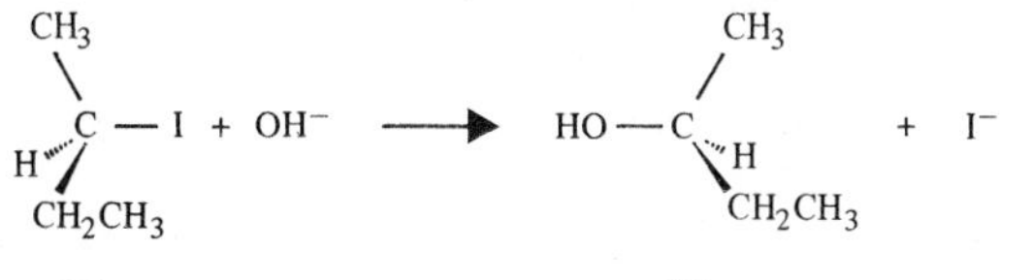

(S)-2-butanol as product demonstrates definitively the exact entry point of the hydroxide ion (opposite the leaving iodide ion).

Even when the starting materials are not optically active, wherever the products may possibly show stereochemistry, it should be sought. The addition of bromine to *cis*-2-butene yields racemic 2,3-dibromobutane, without any of the *meso* form. This conclusively demonstrates that the two new bromine atoms have added from opposite sides of the π bond, and any proposed mechanism must account for this:

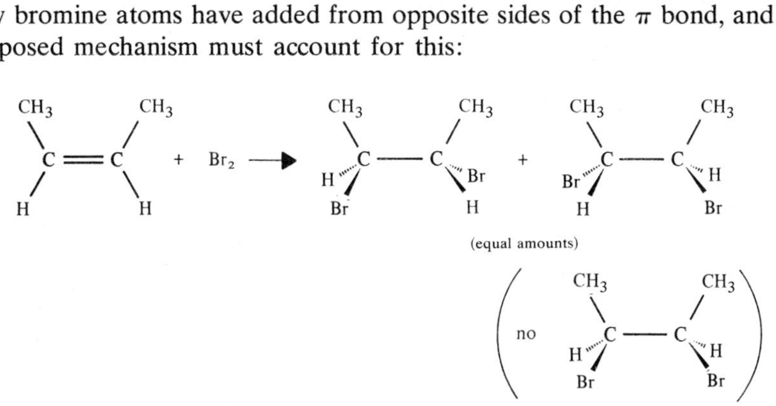

Isotopic Labeling

The use of isotopes represents a subtle but very effective method of determining specific information about the bonds that are involved in a reaction. Kinetic isotope effects are particularly valuable in this regard, and we shall deal with the technique later in this chapter. By simply utilizing isotopes as "labels," previously indistinguishable atoms may be readily distinguished. For example, benzene in deuterosulfuric acid rapidly incorporates deuterium:

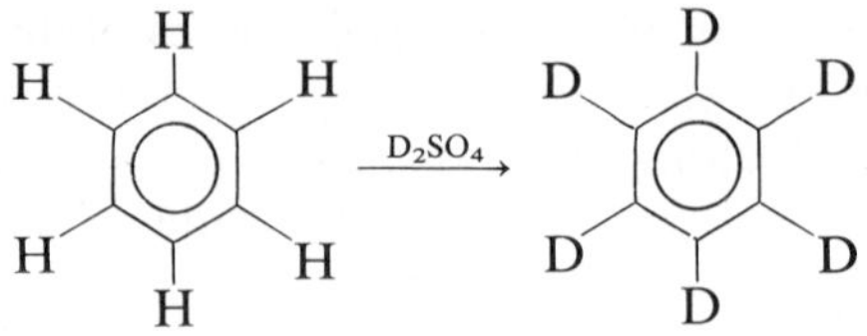

In normal sulfuric acid, this exchange would have been unobservable, but the isotope label, as distinguished by nmr, clearly shows that the acid hydrogens and benzene hydrogens are readily interconvertible (see Chapter 4 for this mechanism). The same conclusion, of course, could have been reached by considering the reaction of deuterobenzene with normal sulfuric acid.

After deuterium, the most frequently used isotopes for labeling purposes are carbon-13 (natural abundance 1.1% of all carbon) and radioactive carbon-14. Since ^{13}C has a nuclear spin, it can provide the basis for nmr spectroscopy (sometimes called cmr). This technique has become quite useful for structure determinations, since instrumentation has advanced to the stage where a natural abundance of ^{13}C is sufficient for routine spectra, and the ^{13}C chemical shifts are extremely sensitive to structural changes. For example, ^{13}C absorptions can be assigned for virtually all the carbons in steroids,[1] while this would scarcely be possible with proton nmr. For purposes of mechanistic studies, this gives a ^{13}C label the advantage of being quickly and unambiguously located. Most other isotopic labels are not so readily located spectroscopically, and usually require more cumbersome detection techniques.

The establishment of benzyne as an intermediate was most forcefully demonstrated by carbon-14 labeling.[2] By labeling one carbon atom, the displacement of iodide ion from iodobenzene by amide ion was shown to be a much more complex reaction than it might appear:

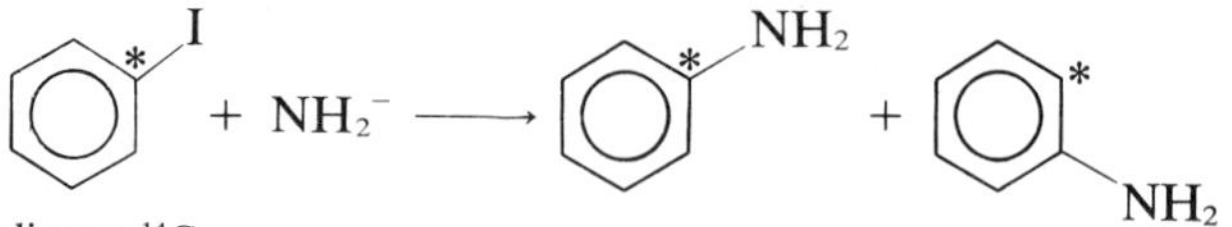

*indicates ^{14}C

(equal amounts)

The location of ^{14}C in the products showed that the original position of iodine attachment *and the adjacent positions* became attached equally to the incoming amide group. This required the postulation of the symmetrical intermediate called benzyne (see Chapter 5 for details):

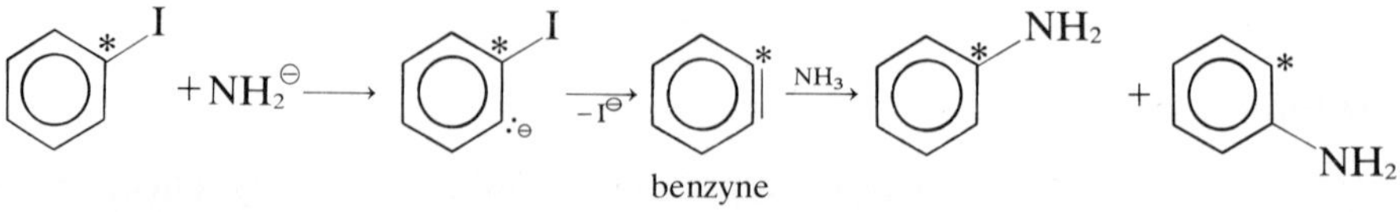

benzyne

It should be noted at this point that the beautiful simplicity of this kind of labeling technique belies the extreme experimental difficulties. Synthesis of the specifically labeled iodobenzene was a delicate procedure, and the careful degradation of the product to ascertain the location of the radioactivity was very difficult. In general, isotopic labeling is among the most laborious (and expensive) of procedures, but often it gives information obtainable in no other way.

3.3 Kinetics

By far the largest amount of mechanistic evidence is obtained by considering the effects of varying reaction parameters upon the course of a reaction. In some cases, these variations will lead to a change in the product yield or in the composition of the product mixture, but most often the effects of these variables are made quantitative by measuring the rate of the reaction (i.e., the kinetics).

Experimental Techniques

The rate of reaction is simply a rate of change of concentration, defined with respect to any of the products or reactants, as shown for the hypothetical reaction below:

$$A + B \rightarrow C + 2D$$

$$\text{Rate} = -\frac{d[A]}{dt} = -\frac{d[B]}{dt} = +\frac{d[C]}{dt} = +\frac{1}{2}\frac{d[D]}{dt}$$

A measurement of the rate of a reaction, therefore, requires a knowledge of the concentration of at least one of the species involved as a function of time. There are two fundamental methods of obtaining this information: by removing aliquots of the reaction mixture for analysis at selected intervals or by constantly or intermittently analyzing some physical feature of the reaction mixture that can be related to concentration. The first category includes such methods as gas chromatographic analysis, titration, treatment with specific analytical reagents, and other methods that are essentially destructive of the reaction mixture. The second category, in which the reaction mixture is monitored by some spectroscopic or other physical feature (e.g., volume or pressure for reactions involving gases or conductivity for reactions in solution) has the advantage of not disturbing the reaction so that as many measurements as desired, even continuous monitoring, can be performed.

The range of reaction rates that can be studied covers an immense time scale. The long end of the time scale is governed only by convenience—reactions running for weeks or months are not uncommon. Using standard spectroscopic techniques, reactions as fast as a minute can be routinely measured, requiring only quick hands for rapid mixing and an output onto a fast-moving strip chart recorder. Using special equipment and an oscilloscope display, the fast reaction techniques of stopped flow, temperature jump, and flash photolysis can be used to follow reactions occurring in the millisecond (10^{-3} sec) or even microsecond (10^{-6} sec) range. Current work with switched lasers has made available very brief light pulses that can interrogate molecules on time scales of nanoseconds (10^{-9} sec) and even into the picosecond (10^{-12} sec) range.*

---

* These laser pulses are not only short in duration, but are also physically short. Since light travels 3×10^8 m/sec, a 10 psec pulse would be only 3 mm long!

Kinetic Order of a Reaction

Once a reaction rate is determined under a given set of conditions, reaction parameters are then varied to ascertain the effect upon the reaction rate. The first parameter to be varied is normally the concentration of the individual reactants. In general, one would expect reaction rate to depend upon reactant concentration, since the fewer the reactant molecules, the fewer the necessary collisions for product formation. In a reaction involving A, B, and C, the general rate expression is:

$$\text{Rate} = k[A]^a[B]^b[C]^c$$

where k is called the rate constant (it is not truly a constant, since it is a function of temperature, solvent, and other specific conditions). A reaction that obeys such a kinetic equation would be said to be a order in A, b order in B, c order in C, and $(a+b+c)$ order overall. The exponents a, b, and c are usually 0, 1, or 2, but they do not necessarily correlate with the coefficients in the balanced equation of the reaction, nor are they necessarily integers.

Integrated Rate Laws

Since rate information is normally obtained in terms of concentration and time, integrated forms of rate equations are particularly useful. For the most common cases of zero-, first-, and second-order kinetics, the rate laws and integrated rate laws are as follows:

Zero-order

$$-\frac{d[A]}{dt} = k_0$$

$$\int_{[A]_0}^{[A]} d[A] = -\int_0^t k_0 \, dt$$

$$[A] = [A]_0 - k_0 t$$

First-order

$$-\frac{d[A]}{dt} = k_1[A]$$

$$\ln\left(\frac{[A]_0}{[A]}\right) = k_1 t$$

$$\text{or} \quad [A] = [A]_0 e^{-k_1 t}$$

Second-order

$$-\frac{d[A]}{dt} = k_2[A]^2$$

$$\frac{1}{[A]} = \frac{1}{[A]_0} + k_2 t$$

These equations describe different functions of concentration that might be linear with time, and this is exactly how kinetic order is determined. If the concentration of a reactant drops in direct proportion to the time of the reaction, the reaction must be zero-order in that reactant. If the natural logarithm (or reciprocal) of the concentration plots as a straight line versus time, the reaction is first(or second)-order with respect to that reactant. The slope of the line in each of these cases then provides the rate constant.

Besides the three rate laws above, another frequently encountered rate equation involves a first-order dependence on two different reactants. This rate law cannot be directly integrated unless there is some additional information relating the two different concentrations:

$$-\frac{d[A]}{dt} = k_2[A][B]$$

There are two standard ways of dealing with this situation. The first method involves the adjustment of the concentrations such that [A] is proportional to [B] at all times during the reaction. For example, if the reaction is

$$A + B \rightarrow C + D$$

then with initially equal concentrations of A and B, the concentrations should be maintained equal throughout the reaction. The rate law then simplifies to a second-order rate equation.

If [A] = [B] at all times,

$$-\frac{d[A]}{dt} = k_2[A]^2 \qquad \text{(second-order)}$$

The second technique involves the use of a large excess of one of the reactants such that this reactant concentration is essentially constant during the reaction. The rate law then appears first-order, with a pseudo-first-order rate constant that incorporates the second-order rate constant and the constant concentration of the excess reactant.

For $[B]_0$ = constant during the reaction (excess),

$$-\frac{d[A]}{dt} = k_1[A] \qquad \text{(pseudo-first-order)}$$

where $k_1 = k_2[B]_0$.

This second technique is the more generally applicable, since it may be used when even more than two reactants are involved in the rate equation, simply by using all but one reactant in excess. By switching the reactants that were kept in excess, the kinetic order with respect to all the reactants may be determined.

Mechanistic Implications

Having learned to handle the mathematics of the common rate laws and to determine kinetic order, the crucial aspect still remains—how does this information pertain to the mechanism? The answer to this question may be summarized in two major principles.

1. *The kinetics of a reaction will be the same as that of the slowest elementary step in the mechanism (the rate-determining step)*. This should seem reasonable enough, since a reaction consisting of several

mechanistic steps cannot proceed any faster than its slowest step. If the first step is slow, the later rapid steps must simply wait for the first step to turn out its needed product. If any of the later steps are rate-determining, then the initial fast reactions will simply build up the intermediate products (perhaps reversibly) until the slow step can gradually utilize them. Occasionally, more than one of the steps may be comparably slow, in which case the kinetics will depend on those steps in a more complex manner. In most cases, however, the overall reaction will proceed at the rate of the slowest individual step. This simplifies matters, since even a complex, multistep reaction mechanism will normally show the kinetics of only one of the elementary mechanistic steps. This also limits the obtainable information, however, since the rate law only provides information regarding the rate-determining step. Other methods are normally required to gather information about steps other than the rate-determining step. Additionally, this principle should point out the importance of breaking down a mechanism into elementary steps.

2.   *The kinetic order of any elementary step is the same as the molecularity of that step.* As an example, the direct reaction of A+B to give an intermediate complex X must be second-order, first-order in both A and B. Since this is an elementary step, the reaction proceeds by simple collision of A and B together, and the frequency of collision (and hence rate of reaction) should be directly proportional to the concentration of both A and B:

$$A + B \rightarrow X$$

$$Rate = k_2[A][B]$$

This principle further simplifies matters, since we may now write a rate law for any elementary reaction by inspection. With these two principles, we can tackle any proposed mechanism by assigning or assuming a rate-determining step and deriving the expected rate law for comparison with the observed experimental rate law.

Let us consider a specific example, the Diels–Alder reaction, and derive the predicted rate laws for some possible mechanisms:

We have already considered the Diels–Alder reaction as a concerted [4+2] cycloaddition (see Chapter 2), but in the case of an acyclic diene, we must add an additional step to the mechanism. 1,3-Butadiene can exist as either a *cis* or *trans* rotational isomer, with only the *cis* form sterically capable of undergoing the cycloaddition. Thus, the simplest mechanism would consist of two steps: a unimolecular isomerization of *trans* to *cis* diene (undoubtedly reversible), followed by a bimolecular cycloaddition of *cis* diene to dienophile:

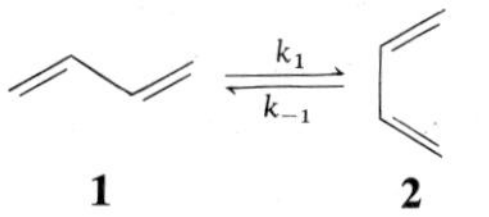

$$(1)$$

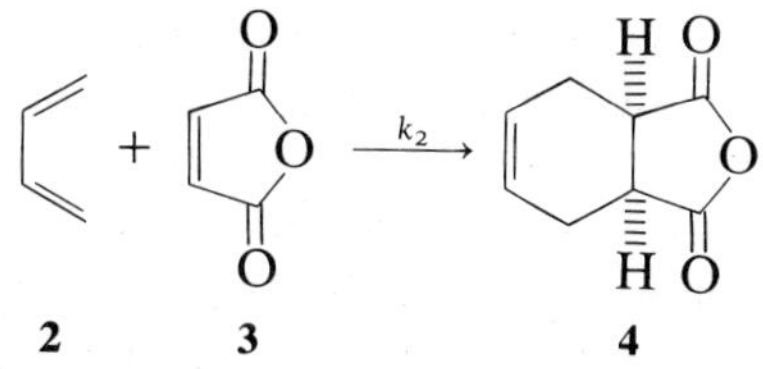

1   **2**

$$(2)$$

2   **3**   **4**

In progressing from starting materials toward products, either k_1 or k_2 may be rate-determining. We shall see that these two possibilities each lead to a different rate law.

Assuming the first step rate-determining:
If the initial isomerization were the slow step, then the overall rate would simply be the rate of the first step. This may be viewed as a situation in which every *cis* diene reacts with the dienophile as quickly as it is formed, and therefore the formation of **2** determines the overall rate:

$$\text{Rate} = k_1[\mathbf{1}] \qquad \text{(first-order)}$$

Assuming the second step rate-determining:
If the cycloaddition itself is the slow step, the overall rate will be the rate of step 2. This corresponds to the establishment of an equilibrium mixture of the isomers **1** and **2**, with the gradual removal of **2** by reaction with **3**. The rate law would then be second-order:

$$\text{Rate} = k_2[\mathbf{2}][\mathbf{3}] \qquad \text{(second-order)}$$

This rate law involves the concentration of the intermediate **2**, rather than that of the starting material **1**. The conversion can readily be made, however, by recognizing that the two exist in equilibrium as related by an equilibrium constant K. At equilibrium,

$$k_1[\mathbf{1}] = k_{-1}[\mathbf{2}]$$

$$K = \frac{k_1}{k_{-1}} = \frac{[\mathbf{2}]}{[\mathbf{1}]}$$

$$[\mathbf{2}] = \frac{k_1[\mathbf{1}]}{k_{-1}}$$

Substituting for [**2**],

$$\text{Rate} = \left(\frac{k_1 k_2}{k_{-1}}\right)[\mathbf{1}][\mathbf{3}]$$

$$\text{Rate} = k_2'[\mathbf{1}][\mathbf{3}]$$

$$k_2' = \left(\frac{k_1 k_2}{k_{-1}}\right)$$

Thus, the reaction would still be second-order, first-order in both **1** and **3**, but the observed second-order rate constant k_2' would actually be a combination of all three of the rate constants involved in the mechanism. If the equilibrium constant K could be measured independently, in the absence of **3**, then the absolute rate constant k_2 could be determined.

Thus, the Diels–Alder reaction might display either first- or second-order kinetics, depending upon which of the two steps is rate-determining. Experimentally, it is found that the rate law for the Diels–Alder reaction of 1,3-butadiene with maleic anhydride is second-order, first-order in each of the two reactants. In our proposed mechanism, this is consistent only with a rate-determining second step. It should be noted, however, that there exist other possible mechanisms that would be expected to show second-order kinetics, such as those shown below:

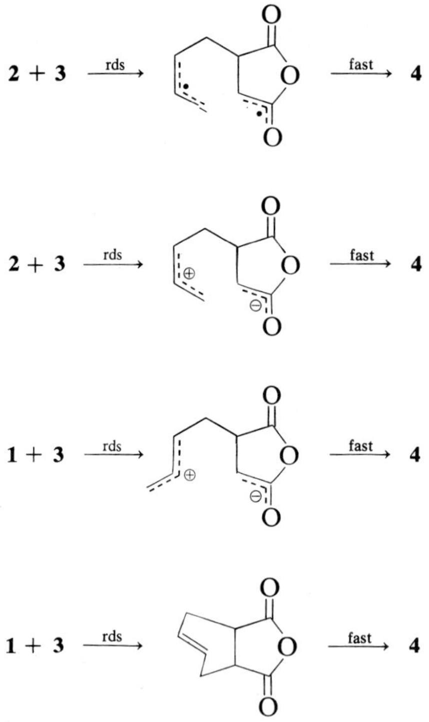

The observation of second-order kinetics, thus, is *consistent* with a concerted, rate-determining second step and definitely rules out a rate-determining first step, but it does not provide any information regarding the critical question of intermediates. Obtaining evidence for such intermediates, by product and kinetic studies, will be discussed later in this chapter and will be extensively covered for specific intermediates in Chapters 4 through 8.

The Steady-State Assumption

Although it is a simple matter to write the rate law for any elementary step in a reaction by referring to the molecularity of that step, it frequently occurs that the desired rate expression involves the concentration of some intermediate whose concentration is unknown. The concentration of that intermediate should preferably be expressed in terms of concentrations of starting materials, which are known initially and are likely to be the concentrations monitored to follow the reaction. This problem can often be overcome by application of the steady-state assumption: *any reactive intermediates involved in the mechanism quickly establish themselves at some low, constant, steady-state concentration.* We will now take another look at the rate expression for the Diels–Alder reaction, this time using the steady-state assumption for *cis*-butadiene (**2**) as a reactive intermediate.

$$1 \underset{k_{-1}}{\overset{k_1}{\rightleftharpoons}} 2$$

$$2 + 3 \xrightarrow{k_2} 4$$

Steady-state assumption:

$$\frac{d[2]}{dt} = 0$$

Therefore,

$$k_1[1] - k_{-1}[2] - k_2[2][3] = 0$$

$$[2] = \frac{k_1[1]}{k_{-1} + k_2[3]}$$

$$\text{Rate} = k_2[2][3]$$

$$\text{Rate} = \frac{k_1 k_2[1][3]}{k_{-1} + k_2[3]}$$

This rate law is completely general and has not involved assignment of any step as rate-determining. Thus, if both steps 1 and 2 were of comparable rate, this would be the appropriate rate law. We can readily derive the same two rate laws as before, depending upon the assumptions about relative reaction rates. If the first step is rate-determining (that is, intermediate **2** reacts as quickly as it is formed), then step 2 is much faster than the reverse of step 1:

$$k_2[2][3] \gg k_{-1}[2]$$

then

$$(k_{-1} + k_2[3]) \approx k_2[3]$$

$$\text{Rate} = k_1[1]$$

On the other hand, if the second step is rate-determining, then the equilibrium between **1** and **2** is established, and the second step will be slower than the reverse of step 1:

$$k_2[2][3] \ll k_{-1}[2]$$

then

$$(k_{-1} + k_2[3]) \approx k_{-1}$$

$$\text{Rate} = \left(\frac{k_1 k_2}{k_{-1}}\right)[1][3]$$

These two rate laws are the same two that we derived previously by consideration of the equilibrium between **1** and **2**. In general, the steady-state approach will be found the more useful because it is more generally applicable.

More Complex Kinetics

Although we have stated that zero-, first-, and second-order kinetics encompass the majority of organic reaction kinetics, more complex rate laws are occasionally encountered. Free radical chain reactions typically display a one-half-order dependence upon the concentration of initiator (see Chapter 6). Reactions catalyzed by acid or base often show kinetics of order higher than two—for example, first-order in each of two reactants and in catalyst. We will consider the base-catalyzed aldol condensation of acetone for which the following mechanism has been postulated with the second step rate-determining (see Chapter 5).

$$CH_3-\overset{\overset{O}{\|}}{C}-CH_3+OH^- \underset{k_{-1}}{\overset{k_1}{\rightleftharpoons}} CH_3-\overset{\overset{O}{\|}}{C}-CH_2^-+H_2O$$
$$\text{(AH)} \qquad\qquad\qquad \text{(A}^-\text{)}$$

$$AH+A^- \underset{k_{-2}}{\overset{k_2}{\rightleftharpoons}} CH_3-\overset{\overset{O}{\|}}{C}-CH_2-\overset{\overset{O^-}{|}}{C}(CH_3)_2$$
$$\text{(B)}$$

$$B+H_2O \underset{k_{-3}}{\overset{k_3}{\rightleftharpoons}} CH_3-\overset{\overset{O}{\|}}{C}-CH_2-\overset{\overset{OH}{|}}{C}(CH_3)_2+OH^-$$

$$\text{Rate} = k_2[AH][A^-] \qquad \text{(step 2 rds)}$$

Steady-state assumption for A^-:

$$\frac{d[A^-]}{dt} = k_1[AH][OH^-] - k_{-1}*[A^-] - k_2[AH][A^-] = 0$$

$$[A^-] = \frac{k_1[AH][OH^-]}{k_{-1} + k_2[AH]}$$

Assuming step 2 rate-determining:

$$k_2[AH] \ll k_{-1}$$

$$[A^-] = \left(\frac{k_1}{k_{-1}}\right)[AH][OH^-]$$

$$\text{Rate} = \left(\frac{k_1 k_2}{k_{-1}}\right)[AH]^2[OH^-] \qquad \text{(overall third-order)}$$

Examples of still more complex kinetics certainly exist, such as mixed kinetic orders in which the product may arise from more than one mechanistic pathway and rate laws involving concentrations in both numerator and denominator, as we would have had above if k_{-1} and $k_2[AH]$ were of comparable magnitude. At this point, however, the student should be able to write a rate law for a given mechanism, or alternatively, given a rate law, should be able to decide which potential mechanisms are consistent or inconsistent with it.

---

* In reactions of this type, [H_2O] is essentially constant and is usually included in the rate constant, k_{-1} in this case.

3.4 Reaction Energetics

In discussing mechanistic schemes, it is important to know something of the relative magnitudes of the various rate constants involved, not only to ascertain the rate-determining step, but also to allow simplification of the rate law in steady-state analyses. We will now consider those factors that make one rate constant larger than another—energetics.

Simple Collision Rate Theory

It has long been known that reaction rate constants vary as an exponential function of temperature, usually expressed in the form of the Arrhenius equation:

$$k = Ae^{-E_a/RT}$$

In this equation, A is known as the Arrhenius pre-exponential factor or frequency factor, and E_a is called the activation energy. Experimentally, a plot of $\ln k$ versus T^{-1} would be a straight line with slope $-E_a/R$ and intercept $\ln A$:

$$\ln k = \ln A - \frac{E_a}{RT}$$

The Arrhenius equation can be derived as follows from simple collision rate theory. The rate at which two molecules react may be expected to be (*a*) the *collision frequency* of the two molecules, which will be proportional to their concentrations, times (*b*) the probability that the bimolecular collision will occur with a suitable *orientation* for the desired reaction, times (*c*) the probability that the collision will impart sufficient *energy* to the molecules to undergo the particular reaction; that is,

Rate = (collision frequency)

× (orientation probability) × (energy probability)

The collision frequency is the factor that contains the concentrations of the reactants, as well as depends upon other factors such as molecular size and solvent effects. The orientation probability represents the fraction of all collisions that occur with the proper orientation for reaction. For example, the reaction of methyl iodide with hydroxide ion requires that the hydroxide ion collide with the carbon on the side opposite the iodine atom. Thus, most collisions will be ineffective in bringing about this substitution reaction, and the reaction will have a very low orientation probability. For a given reaction, we may assume the orientation probability to be constant (independent of temperature), but of course every different reaction will show its own characteristic orientation factor. The collision frequency and the orientation probability together comprise the frequency factor A and the concentration dependence.

The energy factor may be derived from the Maxwell–Boltzmann distribution law; the fraction of molecules (or collisions) that contain an energy of E_a or greater is given by

$$(\text{energy probability}) = e^{-E_a/RT}$$

In this regard, E_a is considered to be a critical *minimum* amount of energy that must be available in order for the reactants to go on to products. Thus, this simple collision rate theory arrives at the experimentally observed Arrhenius equation.

The exponential dependence of rate upon E_a/RT means that the rate of a reaction will be extremely sensitive to E_a and to T. For example, at 100°C, only one collision in 10^{15} will have 25 kcal/mole of kinetic energy, but one in 10^9 will have 15 kcal/mole, and one in 10^3 will have 5 kcal/mole. Thus, all other factors being equal, at 100°C a reaction with $E_a = 5$ kcal/mole will proceed a million times faster than a reaction with $E_a = 15$ kcal/mole, which will be a million times faster than a reaction with $E_a = 25$ kcal/mole. The dependence of rate upon temperature is also pronounced. The common rule of thumb is that a 10°C temperature rise will double the rate.* In other words, a 10°C temperature difference gives a 100% difference in rate, or a 1°C temperature variation could give a 10% discrepancy in a measured rate. Thus, we see the great importance of maintaining a constant, well-calibrated temperature (usually ±0.02°C) when taking kinetic data.

Transition State Theory

A somewhat more sophisticated treatment of reaction rates has been developed that provides an expanded interpretation of the pre-exponential factor and a different view of the physical processes involved in a reaction. This theory pays particular attention to the *transition state*—the highest-energy state of the reactants along the reaction pathway. The specific structure of the reactants at the transition state is called the activated complex. A simple potential energy versus reaction coordinate diagram may be constructed incorporating these features (Figure 3.1).

The significance of the "reaction coordinate" parameter is frequently confusing. Basically, it is intended to represent the progress of the reaction along some smooth transition between reactants and products. In a unimolecular reaction, it is usually easy to find an appropriate reaction parameter—for example, the C_1–C_4 bond distance in 1,3-butadiene as it closes to cyclobutene would make a reasonable parameter to measure the

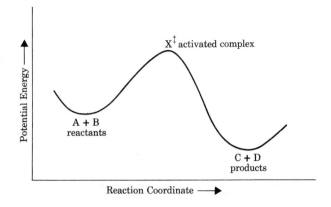

Fig. 3.1 *A typical potential energy reaction diagram.*

* This is strictly true around room temperature for a reaction with $E_a \sim 35$ kcal/mole. The effect is even greater with a larger E_a or at lower temperatures.

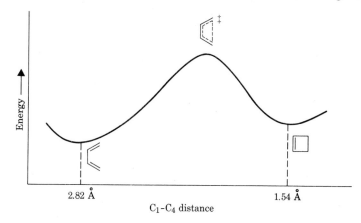

Fig. 3.2 *A potential energy diagram for the electrocyclic conversion of butadiene to cyclobutene.*

progress of reaction (Figure 3.2). (In addition, the orbitals on C_1 and C_4 should simultaneously be undergoing a conrotatory motion as the C_1–C_4 distance changes.)

For a bimolecular reaction, such as the substitution of hydroxide for iodide in methyl iodide, we may usually choose several possible reaction coordinates. For example, either the decreasing C–O bond length or the increasing C–I bond length would be appropriate. The ideal potential energy diagram would show potential energy as a function of all possible reaction variables. In our specific case, a three-dimensional graph of potential energy versus both C–O and C–I bond lengths would be interesting and valuable, since it would indicate whether the lowest energy pathway involves initial C–I bond cleavage, initial C–O bond formation, or simultaneous bond cleavage and formation. Theoretical chemists with access to large computers are beginning to tackle problems of this type. However, the complexity of such an analysis becomes overwhelming when there are several reaction parameters to keep track of (not to mention the difficulty in displaying such an n-dimensional graph). In all cases, we shall write potential energy diagrams two-dimensionally, as though they were a cross-section cut through the n-dimensional diagram. In most cases, it is not critical to know exactly what the reaction coordinate represents, as long as it is recognized that the potential energy curve is taken to be the lowest-energy pathway, and the reaction coordinate measures progress along that curve.

Along the potential energy curve, the transition state represents the point highest in energy and, therefore, most difficult to reach. Thus, the rate of formation of the activated complex will determine the overall rate of the reaction. Absolute rate theory (transition state theory) is based upon two rather unusual postulates: (*a*) *The reactants establish an equilibrium with the activated complex.* This is unusual in that the activated complex has only a transistory existence, since it lies at an energy maximum. (*b*) *All activated complexes proceed on to products at a fixed rate.* This remarkable assumption may be derived from statistical mechanics, which fixes this universal rate constant as $\mathbf{k}T/h$, where $\mathbf{k}$ and h are the Boltzmann and Planck constants, respectively. At room temperature, $\mathbf{k}T/h = 6 \times 10^{12}\ \text{sec}^{-1}$. If one insists upon a physical feeling for this postulate, it might be considered that an activated complex, perched atop an energy maximum, will fall equally well back to products or on to reactants, and at a rate that is essentially independent of the nature of the transition state and how it got there.

From these two postulates, we derive a rate law that is simply the universal rate constant times the concentration of activated complex:

$$\text{Rate} = \left(\frac{kT}{h}\right)[X^\ddagger]$$

The concentration of activated complex may be determined by consideration of the equilibrium between it and the reactants. A normal equilibrium gives rise to the following equations:

$$K = \frac{[\text{products}]}{[\text{reactants}]}$$

$$K = e^{-\Delta G^\circ/RT}$$

$$\Delta G^\circ = G^\circ \text{ (products)} - G^\circ \text{ (reactants)}$$

$$\Delta G^\circ = \Delta H^\circ - T \Delta S^\circ$$

For the special activated complex equilibrium, using the double dagger superscript to represent the activated complex and other features of the transition state, we obtain:

$$A + B \rightleftharpoons X^\ddagger$$

$$K^\ddagger = \frac{[X^\ddagger]}{[A][B]}$$

$$[X^\ddagger] = K^\ddagger[A][B]$$

Thus,

$$\text{Rate} = \left(\frac{kT}{h}\right)K^\ddagger[A][B]$$

Expanding further on $K^\ddagger$,

$$K^\ddagger = e^{-\Delta G^\ddagger/RT}$$

$$\Delta G^\ddagger = \Delta H^\ddagger - T \Delta S^\ddagger$$

$$K^\ddagger = e^{\Delta S^\ddagger/R}e^{-\Delta H^\ddagger/RT}$$

$$\text{Rate} = \left(\frac{kT}{h}\right)e^{\Delta S^\ddagger/R}e^{-\Delta H^\ddagger/RT}[A][B]$$

From statistical mechanics,

$$\Delta H^\ddagger = E_a - nRT$$

where n is the order of the reaction. Thus,

$$\text{Rate} = \left(\frac{kT}{h}\right)e^n e^{\Delta S^\ddagger/R}e^{-E_a/RT}[A][B]$$

This final equation does in fact take the form of the Arrhenius equation,

$$k = Ae^{-E_a/RT}$$

where

$$A = \left(\frac{kT}{h}\right)e^n e^{\Delta S^\ddagger/R}$$

The Arrhenius pre-exponential factor is indeed observed to be linearly dependent upon temperature, but it is easy to see how this might have been masked by the overwhelming exponential dependence.

Thus, we have an absolute rate equation from which we can derive $\Delta H^{\ddagger}$ (or E_a) and $\Delta S^{\ddagger}$ from the temperature dependence of a rate constant. Mechanistically, the values of $\Delta H^{\ddagger}$ and $\Delta S^{\ddagger}$ can be extremely useful evidence. Since

$$\Delta H^{\ddagger} = H°(X^{\ddagger}) - H° \text{ (reactants)}$$

we obtain from $\Delta H^{\ddagger}$ the additional enthalpy content of the activated complex relative to the reactants. This may be compared with known bond energies or other thermochemical data in order to help deduce the nature of the transition state.

Furthermore, since

$$\Delta S^{\ddagger} = S°(X^{\ddagger}) - S° \text{ (reactants)}$$

we obtain from $\Delta S^{\ddagger}$ a knowledge of the relative entropy of the activated complex compared to that of the reactants. This corresponds to what we have called the orientation probability in the simple collision rate theory.

Entropy of activation $\Delta S^{\ddagger}$ can give particularly useful insight into the nature of the activated complex. A negative $\Delta S^{\ddagger}$ indicates that the activated complex has a lower entropy than the reactants; since entropy is a measure of disorder, the reactants must become more ordered as they pass through the transition state. This is the situation encountered with most bimolecular reactions, since two molecules, initially in random motion, must come together with a consequent loss of entropy. If there are further spatial requirements, as with the specific backside attack of hydroxide ion on methyl iodide, $\Delta S^{\ddagger}$ will be strongly negative. A unimolecular reaction will frequently show a negative $\Delta S^{\ddagger}$, especially if it is a concerted reaction. For example, the transition state for closure of 1,3-butadiene to cyclobutene involves a substantial restriction of molecular freedom, particularly with regard to the relative orientation of C_1 and C_4. The orbital symmetry requirement of a conrotatory closure further lowers the entropy of the activated complex. A positive value of $\Delta S^{\ddagger}$ is usually associated with dissociation or fragmentation reactions or other processes in which there is greater disorganization in the transition state than had existed in the reactants.

In dealing with $\Delta S^{\ddagger}$ and $\Delta H^{\ddagger}$, it may be well to remind ourselves that these values only relate the activated complex to the reactants and are not relevant to the products. Reactants and products are related by the traditional $\Delta H°$ and $\Delta S°$, obtained from equilibrium data or thermochemical calculations. For example, the reactants in a typical displacement reaction $(A + B\text{--}C \rightarrow A\text{--}B + C)$ might be expected to contain nearly the same entropy and enthalpy as the products $(\Delta H°, \Delta S° \sim 0)$. Nevertheless, the activated complex for such a reaction will have a substantially higher enthalpy and substantially lower entropy than either $(\Delta H^{\ddagger} > 0, \Delta S^{\ddagger} < 0)$.

Experimentally, the temperature dependence of rate constants is obtained in the form of the Arrhenius equation, and pre-exponential factors are frequently reported rather than $\Delta S^{\ddagger}$ values. Table 3.1 lists typical values for $\Delta S^{\ddagger}$ and for Arrhenius pre-exponential factors for some common types of organic reactions. Roughly, a log A value greater than 13 corresponds to a positive $\Delta S^{\ddagger}$, and log A below 13 reflects a negative $\Delta S^{\ddagger}$.

Table 3.1 *Typical Frequency Factors for Different Reaction Types*

Reaction Type (example)	$\log A$ (A in sec^{-1} or M^{-1} sec^{-1})	$\Delta S^{\ddagger}$ (cal mole^{-1} deg^{-1})
Unimolecular dissociation A—B → A + B	15 to 17	+8 to +17
Unimolecular elimination	12.5 to 14	−3 to +4
Unimolecular rearrangement (Cope reaction)	9 to 13	−20 to 0
Bimolecular combination A + B → A—B	9 to 10.5	−20 to −15
Bimolecular displacement A + B—C → A—B + C	7 to 11	−30 to −10
Bimolecular multicenter (Diels–Alder)	5 to 9	−40 to −20

Kinetic Isotope Effects

We have already discussed the utility of isotopic labeling in elucidating product studies. Additionally, an isotopic substitution may affect the observed kinetics and thereby provide mechanistic information. Normally, kinetic isotope effects are determined with deuterium or occasionally tritium as a replacement for hydrogen. Isotopes of other elements usually have small effects upon rates but are used occasionally.

Chemically, the only significant difference between deuterium and hydrogen lies in their masses. Since deuterium contains a neutron in its nucleus, it has an atomic mass double that of the hydrogen atom. Electronic and bonding properties remain virtually the same except for the relative zero-point vibrational energy. The characteristic C–H stretching frequency lies in the range 2900 to 3100 cm^{-1}, as shown by infrared spectroscopy. Because of the increased mass of deuterium, C–D stretching occurs with a substantially lower frequency, and hence lower energy, around 2050 to 2200 cm^{-1}. Thus, for any reaction in which a C–H bond is broken, or even weakened, we would expect that a C–H bond would be more reactive than a C–D bond. In essence, the C–H bond has a head start of about 1.2 kcal/mole over the C–D bond, due to the difference in zero-point vibrational energy (Figure 3.3).

Thus, C–H bond breaking has a lower effective activation energy than C–D bond breaking, and therefore the rate of C–H reaction would be greater than the rate of C–D reaction. A kinetic isotope effect is defined on this basis:

$$\text{kinetic isotope effect} = \frac{k_{\text{H}}}{k_{\text{D}}}$$

A primary isotope effect is defined as the effect of isotopic substitution at a bond that is broken during the reaction. A secondary isotope effect is the

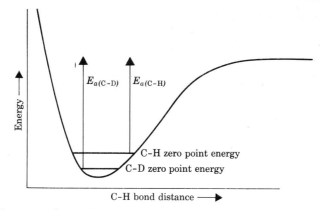

Fig. 3.3 *Zero-point vibrational energy as the source of kinetic deuterium isotope effects.*

effect of isotopic substitution at a bond that is not broken during the reaction (but may be weakened or rehybridized or subject to other effects). The magnitude of deuterium kinetic isotope effects will depend upon the extent of bond breaking in the transition state. If the C–H (or C–D) bond remains intact throughout the reaction, there will be no primary kinetic isotope effect, since the zero-point vibrational energy will remain in the C–H bond and will be unavailable as a "head start" towards the activation energy. Reactions that do involve C–H bond breaking (or even stretching or bending) in the transition state reflect the extent of bond breaking directly in the magnitude of the kinetic isotope effect. The base-catalyzed elimination from 1-bromo-2-phenylethane shows a strong isotope effect ($k_H/k_D = 7.1$), indicating that the C–H bond is nearly completely broken in the transition state (see problem 8 at the end of this chapter):

$$\frac{k_H}{k_D} = 7.1$$

The absence of a primary kinetic isotope effect in most electrophilic aromatic substitution reactions provides a clear indication that the hydrogen is not lost in the rate-determining step, thereby demonstrating that at least one intermediate must be involved in the reaction:

$$\frac{k_H}{k_D} = 1.0$$

Experimentally, kinetic isotope effects may be obtained by either intramolecular or intermolecular competition between hydrogen and deuterium. The isotope effect of 2.0 for abstraction of benzylic hydrogen by a chlorine atom could be calculated from any of the experiments shown below; it is important to recognize that an appropriate statistical factor is needed to compensate for the relative abundance of H and D:

Intermolecular competition $(k_H/k_D = 2)$

Intramolecular competition $(k_H/k_D = 2)$

Substituent Effects

Isotopic effects may be considered to be a special case of substituent effects—the effects upon a reaction caused by changing one part of the molecule. Many studies involving systematic variation of substituents have been extremely successful in providing mechanistic information. An interpretation of substituent effects depends upon a knowledge of the nature of each substituent and its capabilities for interactions with the reaction under consideration.

Basically, a substituent may act upon a neighboring atom to which it is bonded in two different ways: (*a*) inductive effects—electron donation or withdrawal through σ bonds and (*b*) resonance effects—electron donation or withdrawal through π bonds, which can be represented specifically by drawing resonance forms. Other kinds of substituent effects are the field effect, which operates through space rather than through bonds, and steric effects, in which special strain is introduced due to the size or shape of the substituent.

A classic example of inductive substituent effects is the increased acidity of acetic acid as α-hydrogens are replaced by chlorines. Inductive electron withdrawal by the chlorine atoms makes the carboxylic acid more likely to lose a proton because the resulting carboxylate anion can be better stabilized (Table 3.2).

The importance of resonance effects can be seen from a comparison of the acidities of *m*- and *p*-nitrophenol. The *para* isomer is more than ten times

Table 3.2 *Inductive Effects upon Acid Strength*

	pK_a
CH_3COOH	4.76
$ClCH_2COOH$	2.86
$Cl_2CHCOOH$	1.29
Cl_3CCOOH	0.65

stronger as an acid than the *meta* isomer, because the *para* anion can be stabilized by conjugative delocalization of the negative charge onto the nitro group. This delocalization is indicated by an additional stable resonance form which is not possible for the *meta* anion. Both isomers are substantially more acidic than phenol itself because of the strong inductive effect of the nitro substituents (Table 3.3).

The key to the substituent effects noted above lies in the observation that both Cl and NO_2 are electron-withdrawing substituents relative to the hydrogen that they replaced, and this electron withdrawal affects the reaction in a predictable way. In fact, we can make a quantitative scale of the relative electron donating or withdrawing ability of such substituents by simply measuring the substituent effect upon some quantity such as pK_a. This type of analysis was first done by Louis P. Hammett (of Columbia University) using benzoic acid ionization in aqueous solution at 25°C as the standard reaction and then measuring the effects of substituents upon the observed pK_a:

Table 3.3 *Inductive and Resonance Effects upon Phenol Acidities*

	pK_a
	10.0
	8.3
	7.1

Thus, Hammett was able to obtain quantitative substituent effects σ_X by the effect of a given substituent on benzoic acid ionization:

$$\log\left(\frac{K_X}{K_H}\right)_{\text{benzoic acid}} = \sigma_X$$

In order to extend this treatment to other reactions, the assumption was made that the effect of a substituent on a reaction is proportional to its effect upon benzoic acid ionization:

$$\log\left(\frac{K_X}{K_H}\right)_{\text{any rxn}} = \rho \cdot \log\left(\frac{K_X}{K_H}\right)_{\text{benzoic acid}}$$

or in general,

$$\log\left(\frac{K_X}{K_H}\right) = \rho \cdot \sigma_X \qquad \text{(Hammett equation)}$$

Note that this key assumption can also be regarded with respect to free energy differences. It amounts to an assumption that the change in free energy associated with a substituent effect on one reaction $(\Delta G_X^\circ - \Delta G_H^\circ)$ is proportional to the corresponding effect on another reaction. Hence, such a treatment is called a linear free energy relationship:

$$\log K_X = -\frac{\Delta G_X^\circ}{2.3RT}$$

$$\log K_H = -\frac{\Delta G_H^\circ}{2.3RT}$$

$$\log\left(\frac{K_X}{K_H}\right) = -\frac{(\Delta G_X^\circ - \Delta G_H^\circ)}{2.3RT}$$

The same kind of treatment may also be applied to rate constants, in which case the linear free energy assumption is made for $\Delta G_X^\ddagger$ and $\Delta G_H^\ddagger$.

$$\log\left(\frac{k_X}{k_H}\right) = \rho \cdot \sigma_X$$

Probably the best justification for the linear free energy assumption is the fact that these equations work very well. The Hammett equation and other linear free energy relationships have been highly successful at correlating substituent effects in a wide variety of reactions, and they provide one of the most useful tools available for mechanistic organic chemistry.

σ CONSTANTS

The experimentally obtained σ constants thus provide a quantitative measure of the ability of a substituent to donate or withdraw electrons. An electron-withdrawing substituent would have a positive σ value, since it enhances the ionization of benzoic acid, and an electron-donating substituent would have a negative σ value. Hydrogen is defined as the standard with a σ value of zero.

Some representative σ constants are shown in Table 3.4.

ρ VALUES

The Hammett ρ value is a proportionality constant that measures the extent to which a reaction depends upon the electron donating or withdrawing properties of the substituents. A reaction showing no effect of substituents

Table 3.4 *Hammett σ Constants*

Substituent	σ_{meta}		σ_{para}
$(CH_3)_2N$	−0.21		−0.83
NH_2	−0.16		−0.66
$(CH_3)_3C$	−0.10		−0.20
CH_3	−0.07		−0.17
H	0.00	(by definition)	0.00
OCH_3	+0.11		−0.27
OH	+0.12		−0.37
C_6H_5	+0.22		+0.01
F	+0.34		+0.06
I	+0.35		+0.18
Cl	+0.37		+0.23
$COOC_2H_5$	+0.37		+0.45
Br	+0.39		+0.23
CF_3	+0.42		+0.54
CN	+0.56		+0.66
NO_2	+0.71		+0.78
N_2^+	+1.7		+1.8

would have a ρ value of zero. A reaction that is facilitated by electron withdrawal, such as benzoic acid ionization, will show a positive ρ value. A negative ρ value indicates a reaction that is enhanced by electron-donating substituents. The magnitude of a ρ value obtained from equilibrium substituent effects reflects the difference between the reactants' and the products' demand for stabilization by electron withdrawal or donation. In terms of elucidation of a mechanism, a ρ value obtained from rate constant substituent effects can be quite useful, since it will describe the electron demand of the transition state relative to the reactants. Some typical ρ values are shown in Table 3.5.

OTHER SUBSTITUENT EFFECTS—THE FIELD EFFECT

Besides electronic effects through bonds—the inductive and resonance effects—substituents may exert their influence in other ways. Field effects

Table 3.5 *Hammett ρ Values*

Reaction	ρ
$ArNH_3^+ \rightleftharpoons ArNH_2 + H^+$ (H_2O)	+2.77
$ArOH \rightleftharpoons ArO^- + H^+$ (H_2O)	+2.11
$ArCOOH \rightleftharpoons ArCOO^- + H^+$ (C_2H_5OH)	+1.96
$ArCOOH \rightleftharpoons ArCOO^- + H^+$ (H_2O)	+1.00 (by definition)
$ArCH_2COOH \rightleftharpoons ArCH_2COO^- + H^+$ (H_2O)	+0.49
$ArCH_3 + Cl\cdot \rightleftharpoons ArCH_2\cdot + HCl$ (CCl_4)	−0.66
$Ar_3CCl \rightleftharpoons ArC^+ + Cl^-$ (SO_2)	−3.97
$ArH + HNO_3 \rightarrow ArNO_2 + H_2O$ (Ac_2O)	−7.29
$ArCH_2\cdot \rightarrow ArCH_2^+ + e^-$ (vapor phase)	−20

are electronic effects exerted through space rather than through bonds. The separation of field and inductive effects presents a special challenge because for most molecules both effects operate simultaneously and in the same direction. Commonly, field and inductive effects are considered together as polar effects. Construction of appropriate molecules, such as the one shown below, was necessary to show that a field effect can be significant and can work in opposition to inductive effects.

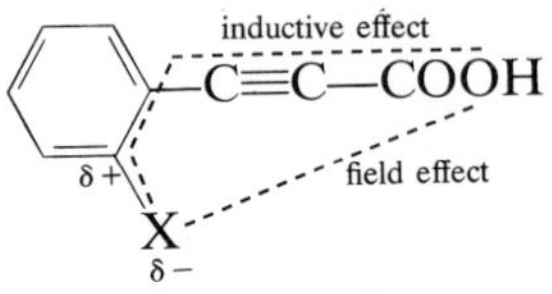

For X as a halogen, the inductive effect would be an acid-strengthening effect, while the field effect of the C–X dipole would be acid-weakening. The finding that the *ortho*-haloacids are weaker acids than the corresponding *meta* or *para* isomers indicates a significant field effect.[3]

STERIC EFFECTS

Steric effects are substituent effects caused by the particular size or shape of the substituent. The study of steric effects upon molecular structure comprises the field of conformational analysis. While we cannot cover this important area in any depth, a few examples should serve to review some of the important aspects.

Virtually all substituents are more stable in an equatorial rather than an axial position of cyclohexane. This is due to steric repulsions between 1,3 axial positions (nonbonded interactions). For a methyl substituent, this difference amounts to about 1.8 kcal/mole:

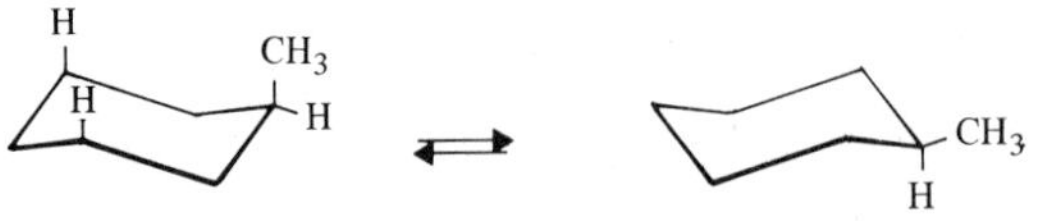

Axial methyl   Equatorial methyl

The preference for equatorial over axial positions is so great for a *t*-butyl substituent that a *t*-butyl group is frequently used to lock a cyclohexane ring into one conformation. For example, the two compounds below would allow the different properties of axial versus equatorial X to be studied:

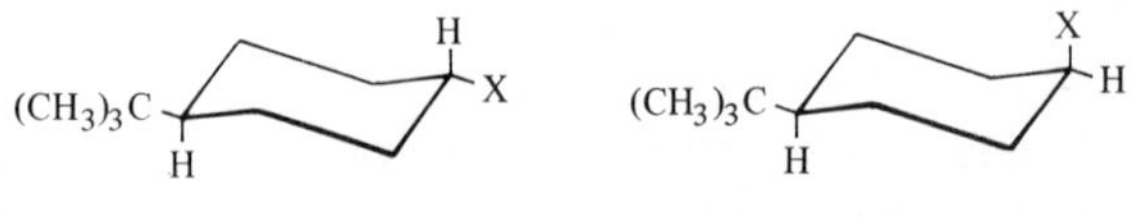

trans, equatorial X   *cis*, axial X

Steric hindrance to rotation or other molecular motions can also lead to unusual isomerism, as in the case of *ortho* substituted biphenyls. The two mirror image isomers (enantiomers)

are isolable because the rotation of the large bromine atoms past the neighboring aromatic ring is severely hindered sterically. The activation energy for interconversion of the enantiomers is 19 kcal/mole.[4]

In terms of steric effects upon reactivity, steric hindrance is the most common observation, resulting when a bulky substituent simply gets in the way of the reaction processes. For example, backside S_N2 attack on an alkyl halide by a nucleophile is very sensitive to the size of the alkyl groups attached to the backside of the central carbon atom. Considering only primary alkyl bromides with different β substituents, in order to minimize inductive effects, the data in Table 3.6 show that increasing alkyl bulk greatly decreases the rate of reaction.

Steric hindrance can also be observed in acid–base reactions. Such effects are usually minimal when a proton is the acid because of its small size, but substantial effects may be noted with other Lewis acids. For example, in nonaqueous solvents with proton acids, the order of amine basicity is that predicted by the inductive effects of alkyl groups:

$$R_3N > R_2NH > RNH_2 > NH_3$$

Relative to a large Lewis acid such as trialkylboranes, this order can be completely reversed because of steric repulsions that render the bond formation more difficult. This type of strain, where a covalent bond is strained by face-to-face steric repulsion, has been called F-strain, for face or frontal strain:

F − strain

Table 3.6 *Relative Ethanolysis Rates of Primary Alkyl Bromides*[5]

R	Relative Rate
CH_3	1.0
CH_3CH_2	0.28
$(CH_3)_2CH$	0.030
$(CH_3)_3C$	0.0000042

Steric effects do not always hinder reactivity. In those cases where the reaction leads to relief of strain, larger substituents can enhance reactivity. The most common example is back strain, or B-strain, as applied to the ionization of tertiary alkyl halides. As the tetrahedral (sp^3) alkyl halide ionizes to a planar (sp^2) carbocation, strain is relieved with larger substituents showing greater reactivity (see Table 3.7, also Chapter 4 for more details):

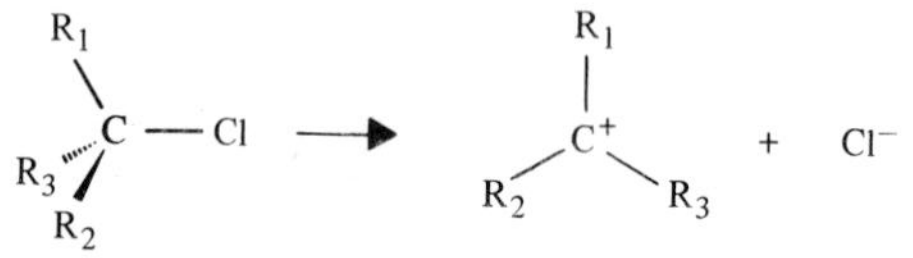

B — strain

Table 3.7 *Relative Ionization Rates of Tertiary Alkyl Chlorides*[6]

R_1	R_2	R_3	Relative Rate
Me	Me	Me	1.0
Me	Me	Et	1.7
Me	Et	Et	2.6
Et	Et	Et	3.0

The special kind of strain inherent in small rings has been called I-strain (internal strain). Its effect upon reactivity may be assessed by considering the bond angles involved; hence, it is also called angle strain. The ionization of cyclopropyl derivatives is substantially slower than their acyclic analogs. Given the fixed cyclopropane bond angle of 60°, ionization to sp^2 hybridization (with a desired bond angle of 120°) would produce even greater strain than already exists with sp^3 hybridization (desired bond angle 109.5°):

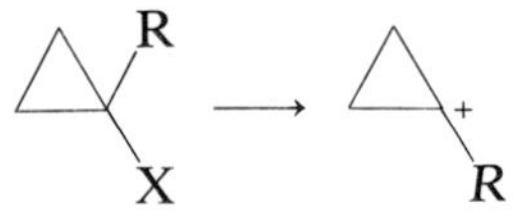

(prefers 109.5°)   (prefers 120°)
sp^3 sp^2

angle strain

Steric effects may also work indirectly via other substituent effects. For example, because of resonance effects, benzoic acids are weaker acids than aliphatic carboxylic acids. However, *o*-*t*-butylbenzoic acid is about ten times stronger than the *para* isomer because the large *t*-butyl group forces the carboxylic acid out of the plane of the benzene ring and thus diminishes the acid-weakening resonance effects.

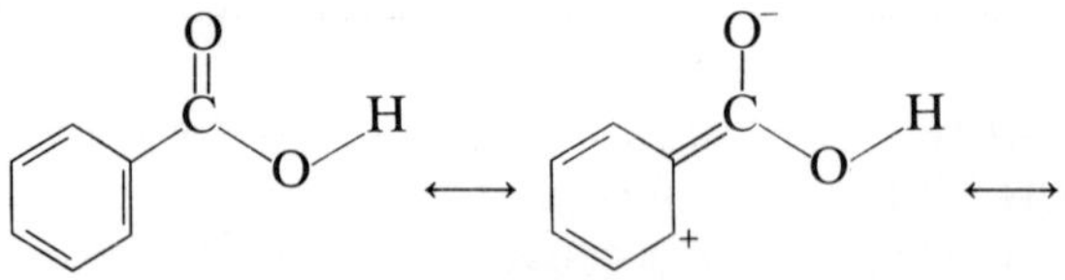

Acid-weakening resonance effects

Nonplanar–smaller resonance effect

Solvent Effects

Besides the Hammett and other relationships describing substituent effects, several other linear free energy relationships have been developed describing solvent effects (the Grunwald–Winstein equation), catalysis (the Bronsted catalysis law), and substituent effects in specific reactions such as the Swain equation for nucleophilic substitution and the Brown equation for aromatic substitution. Furthermore, there have been modifications of the Hammett equation for different types of reactions. Some of these will be dealt with in Chapter 4.

In evaluating the effect of some variation upon a reaction, we are really evaluating the change in $\Delta G^{\ddagger}$, since free energy governs reaction rate. Therefore, we must consider the effect of this variation upon both the reactants and the transition state, since $\Delta G^{\ddagger} = G^{\ddagger} - G_{(reactants)}$. The necessity of dealing with changes in $\Delta G^{\ddagger}$ (sometimes called $\Delta\Delta G^{\ddagger}$) is probably clearest in the area of solvent effects. A change from one solvent to another can strongly affect not only the stability of the transition state, but also the stability of the reactants.

We will refer mainly to solvent polarity as measured by dielectric constant to determine relative solvating ability. It should be kept in mind, however, that hydrogen bonding, acid–base properties, and other specific interactions with solvent molecules often play a very important role in solvent effects. Solvent polarity simply describes the solvent as a medium, specifically with respect to its ability to support ionic charges. Thus a polar solvent, such as water or ethanol, can effectively solvate and stabilize ions and therefore will facilitate ionization reactions by lowering the activation energy. In nonpolar solvents, such as benzene or hexane, ions are not well solvated and hence an ionization reaction would have a large activation energy.

The effect of solvent polarity may be summarized for the two limiting cases: (*a*) increased solvent polarity will enhance those reactions for which the transition state involves an increase in charge separation (relative to reactants); (*b*) increased solvent polarity will depress those reactions for which the transition state involves a decrease in charge separation (relative to reactants). These two effects are probably more readily understandable graphically, as shown in Figures 3.4 through 3.6. If attainment of the transition state involves a separation of charge, then a more polar solvent will lower the energy of the transition state, and thereby increase the rate. On the other hand, if attainment of the transition state requires a bringing together of charged species, then the major effect of an increased solvent polarity will be to stabilize the reactants more than the transition state. There will be many intermediate cases in which charge is rearranged in the transition state rather than clearly created or destroyed, and these cases must be treated individually (see Chapter 4 for further analysis of solvent effects).

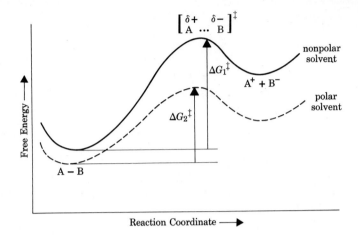

Fig. 3.4 *Rate enhancement by solvent polarity where the transition-state effect is dominant.*

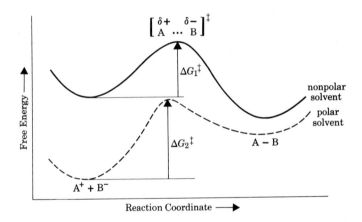

Fig. 3.5 *Rate depression by solvent polarity where the ground-state effect is dominant.*

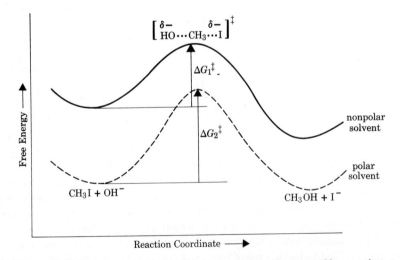

Fig. 3.6 *Slight rate depression with increased solvent polarity with comparable ground-state and transition-state effects.*

3.5 Reactive Intermediates

The remainder of this book will deal with the subject of reactive intermediates. Any time a reaction involves more than one step, at least one
intermediate must be involved. To call an intermediate a reactive intermediate implies some amount of reactivity or instability. We will simply
consider that a reactive intermediate is any intermediate molecule involved
in a reaction mechanism that is not stable enough to be isolated with other
products of the reaction.

Furthermore, there is a minimum requirement of stability in order to
classify a molecule as a reactive intermediate. An intermediate must lie at a
relative energy minimum (potential energy well) on a potential energy
reaction diagram. Thus, a transition state is not an intermediate but a
transitory structure that exists at a relative energy maximum.

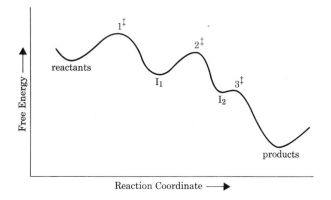

Fig. 3.7 *A three-step reaction with two intermediates.*

Frequently, distinguishing an intermediate from a transition state (as I_2
and $3^{\ddagger}$ in Figure 3.7) can be extremely difficult and ultimately indetermina-
ble and, therefore, meaningless for mechanistic implications. Nevertheless,
some very distinguished chemists have occasionally embroiled themselves in
lengthy and even heated debates over the existence of a small dip in a
potential energy diagram.

Reactive intermediates are relatively short-lived species, derived from the
reactants, and from which the final products are derived. A mechanism is
then an accounting of the formation and destruction of these intermediates.
Obtaining evidence for such intermediates, thus, becomes a major focus of
organic reaction mechanism studies. We will present some of the more
common methods for direct and indirect detection of intermediates.

Isolation of Intermediates

The best evidence for the existence of an intermediate, of course, is the
actual isolation and characterization of the intermediate. This can quite
readily be done in many cases by simply running the reaction at a lower
temperature or otherwise controlled conditions so that the intermediate
does not proceed on to product as usual. This technique is not absolutely

conclusive, however. It then becomes necessary to demonstrate that the isolated intermediate is not simply a side product and that it will go on to the normal product under the usual conditions. For example, alkyl fluorides dissolved with antimony pentafluoride produce alkyl cations that may be unequivocally identified by their nmr spectra. Upon addition of water, the spectrum of the corresponding alcohol is observed. However, this is not conclusive proof that cations are intermediates in the usual hydrolysis reaction, since the conditions have been severely altered:

$$R\text{--}F \xrightarrow{\text{SbF}_5} R^+ + SbF_6^-$$

$$R^+ \xrightarrow{\text{H}_2\text{O}} ROH + H^+$$

Trapping of Intermediates

If a careful product study under controlled experimental conditions does not provide evidence for any intermediates, other types of product studies are occasionally quite successful. The most popular of these is the technique of "trapping" an intermediate by adding some molecule with which it will react readily. The trapping molecule will divert the intermediate from its normal pathway and create a new, characteristic adduct with the intermediate. For example, the intermediacy of benzyne in the reaction of sodium amide with iodobenzene was supported by the observation that reactive dienes gave Diels–Alder adducts as additional products:

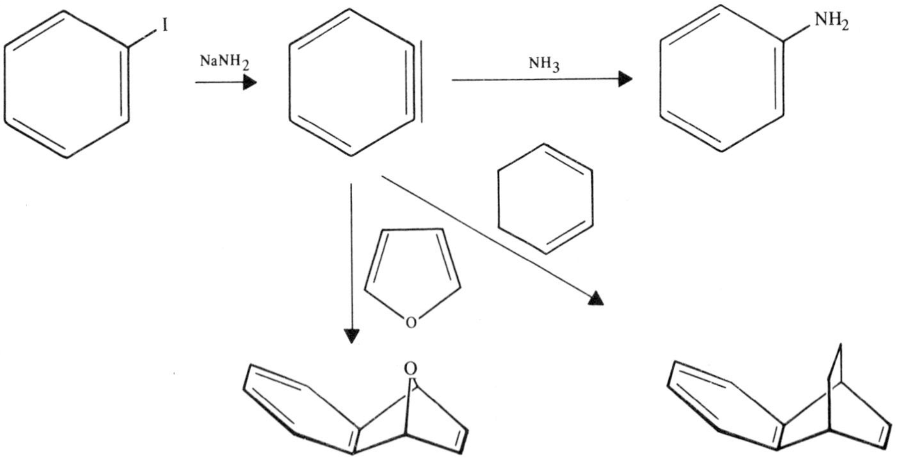

Alternative Sources of Intermediates

The existence of the molecule benzyne was further substantiated by the many different methods developed to create it. Benzyne results from pyrolysis of any of the compounds below, as well as others:

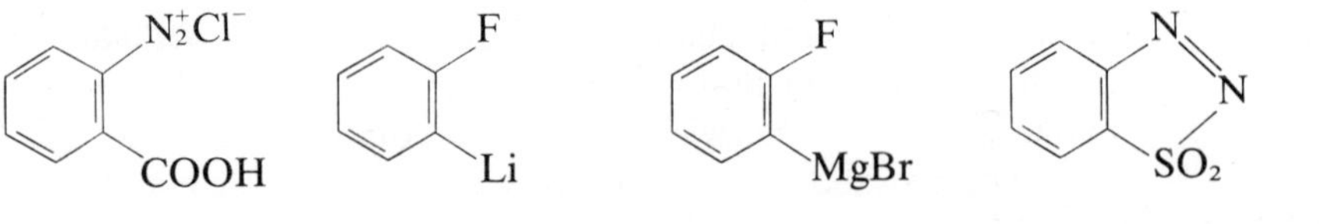

The compound produced in all the foregoing pyrolyses is concluded to be the same intermediate (free benzyne), since the relative reactivity towards furan and 1,3-cyclohexadiene is constant regardless of source.

Kinetic Evidence for Reactive Intermediates

The first evidence for the necessity of an intermediate often comes from a comparison of the stoichiometry of the overall reaction and the observed rate law. A concerted reaction must show a correlation between the kinetic orders observed in the rate law and the stoichiometry of the balanced equation. If this correlation does not exist, the reaction cannot be concerted, and at least one intermediate is necessary. Beyond this, it becomes a matter of developing a multistep reaction mechanism consistent with the observed rate law.

There are several available techniques utilizing kinetics to test for possible intermediates. For example, the addition of a specific "trap" for an intermediate, while designed to add to the product information, may also alter the kinetics. If the addition of the trap does not affect the kinetics, trapping occurs after the rate-determining step. If the trap competes with the normal reaction pathway in the rate-determining step, a rate increase would be observed. In the latter case, where both product and kinetic information are obtained, they may be checked against one another for consistency:

$$A \underset{}{\overset{fast}{\rightleftharpoons}} B \xrightarrow[k_1]{slow} C$$
$$\Big\downarrow k_2[\text{Trap}]$$
$$D$$

If sufficient trap is added to double the rate, for example, then the trapped adduct should account for 50% of the products.

If $k_2[\text{Trap}] = k_1$

then $[\text{D}] = [\text{C}]$

Finally, information may be obtained about the nature of an intermediate from the selectivity that it displays. Addition of a trapping molecule allows the molecule a choice of pathways. An intermediate is said to be selective if it shows a wide range of rates of reaction toward different molecules. This selectivity is usually associated with a relatively stable intermediate; the intermediate has a sufficiently long lifetime to choose carefully among its alternatives. This corresponds to a relatively deep energy well on a potential energy diagram. A less stable intermediate reacts more indiscriminately with the available molecules and shows lower selectivity. Such an unstable intermediate lies in a shallow energy minimum and has rapid, low-activation-energy pathways open to it.

With respect to the reactivity of halogen atoms in hydrogen abstraction reactions, bromine atoms are found to be substantially less reactive (and more selective) than chlorine atoms. The key to this behavior lies in the relative energetics of the reactions. Bromine abstractions of alkyl hydrogens are endothermic reactions, while abstractions by chlorine are exothermic. Thus, the nature of the alkyl group has a much greater effect upon the transition state for the bromine reaction than for the chlorine reaction.

A useful principle that correlates these concepts of reactivity and selectivity is Hammond's postulate (Figure 3.8):

> *An exothermic reaction involves an early transition state that resembles reactants; an endothermic reaction involves a late transition state that resembles products.*

In the exothermic abstraction of hydrogen by chlorine, the transition state has achieved relatively little alkyl radical character. Hence the difference between alkyl groups is minimal, and chlorine is not very selective. In contrast, the endothermic abstraction by bromine involves substantial alkyl

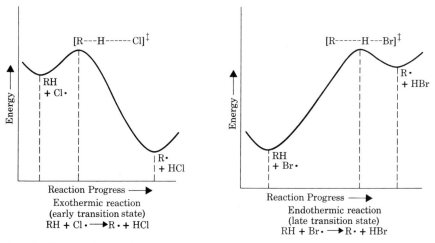

Fig. 3.8 *Hammond's postulate.*

radical character. Hence the transition state stability will be a direct reflection of the alkyl radical stability, and bromine abstracts very selectively.

While Hammond's postulate provides a convenient and straightforward method for correlation of reactivity and selectivity, it should be pointed out that exceptions have been noted recently,[7,8] and the generality of the widely accepted reactivity–selectivity correlation has been questioned.

Some Common Types of Reactive Intermediates

In the remaining five chapters we shall deal in detail with five of the most common classes of reactive intermediates. Except for excited states, all derive their reactivity from a lack of the normal four covalent bonds or eight electrons about carbon. These criteria may be used to catalog four major types of reactive intermediates, as in Table 3.8.

Of course, many reactive intermediates do not fall into this simple classification scheme. Excited states derive their special reactivity from the displacement of one electron from its normal position in the molecular orbitals of the molecule. Benzyne is a unique type of reactive intermediate

Table 3.8 *General Classification of Some Reactive Intermediates*

Reactive Intermediate	General Structure	Covalent Bonds	Valence Electrons
Carbanions	—C:⊖	3	8
Free Radicals	—C·	3	7
Carbocations	—C⊕	3	6
Carbenes	—C:	2	6

that could be considered a diradical, a zwitterion, or just an extremely strained ring:

Strained rings represent an important class of reactive intermediates, as do valence tautomers, unstable oxidation states, radical ions, and protonated or deprotonated molecules. Some of these will be mentioned in more detail in the following chapters.

References

1. H. J. Reich, M. Jautelat, M. T. Messe, F. J. Weigert, and J. D. Roberts, *J. Amer. Chem. Soc.*, **91,** 7445 (1969).
2. J. D. Roberts, D. A. Semenow, H. E. Simmons, and L. A. Carlsmith, *J. Amer. Chem. Soc.*, **78,** 601 (1956).
3. J. D. Roberts and R. A. Carboni, *J. Amer. Chem. Soc.*, **77,** 5554 (1955).
4. M. M. Harris, *Proc. Chem. Soc.*, 367 (1959).
5. E. D. Hughes, *Quart. Rev.*, **2,** 107 (1948).
6. H. C. Brown and R. S. Fletcher, *J. Amer. Chem. Soc.*, **71,** 1845 (1949).
7. T. J. Gilbert and C. D. Johnson, *J. Amer. Chem. Soc.*, **96,** 5846 (1974).
8. D. Farcasiu, *J. Chem. Educ.*, **52,** 76 (1975).

Bibliography

Brief General Texts

R. Breslow, "Organic Reaction Mechanisms," 2nd ed. Benjamin, New York, 1969.

R. A. Jackson, "Mechanism: An Introduction to the Study of Organic Reactions," Clarendon, Oxford, 1972.

O. T. Benfey, "Introduction to Organic Reaction Mechanisms," McGraw-Hill, New York, 1970.

P. Sykes, "A Guidebook to Mechanism in Organic Chemistry," 3rd ed., Longman, London, 1970.

P. Sykes, "The Search for Organic Reaction Pathways," Halsted, New York, 1972.

Mechanistic Reference Books

J. March, "Advanced Organic Chemistry: Reactions, Mechanisms, and Structure," McGraw-Hill, New York, 1968.

C. K. Ingold, "Structure and Mechanism in Organic Chemistry," 2nd ed., Cornell Univ. Press, Ithaca, N.Y., 1969.

Annual Review Periodicals

V. Gold, Ed., "Advances in Physical Organic Chemistry," Academic Press, London, 1963 –

A. Streitwieser, Jr., and R. W. Taft, Eds., "Progress in Physical Organic Chemistry," Wiley-Interscience, New York, 1963 –

C. W. Rees, M. J. Perkins, and B. Capon, Eds., "Organic Reaction Mechanisms," Wiley-Interscience, New York, 1966 –

Kinetics

A. A. Frost and R. G. Pearson, "Kinetics and Mechanism," 2nd ed., Wiley, New York, 1961.

R. Huisgen, *Angew. Chem. Int. Ed. Engl.*, **9**, 751 (1970).

Linear Free Energy Relationships

P. R. Wells, "Linear Free Energy Relationships," Academic Press, New York, 1968.

J. E. Leffler and E. Grunwald, "Rates and Equilibria of Organic Reactions," Wiley, New York, 1963.

J. Hine, "Structural Effects on Equilibria in Organic Chemistry," Wiley-Interscience, New York, 1975.

Isotope Effects

C. J. Collins and N. S. Bowman, Eds., "Isotope Effects in Chemical Reactions," Van Nostrand Reinhold, New York, 1970.

Reactive Intermediates

S. P. McManus, Ed., "Reactive Intermediates," Academic Press, New York, 1973.

Stereochemistry

K. Mislow, "Introduction to Stereochemistry," Benjamin, New York, 1966.

Problems

---

1. Comment upon the reasonableness of the following reactions as elementary mechanistic steps:

(a) $CH_4 + Cl\cdot \longrightarrow CH_3Cl + H\cdot$

(b) $4 \text{ HC} \equiv \text{CH} \xrightarrow{\text{Ni(CN)}_2}$

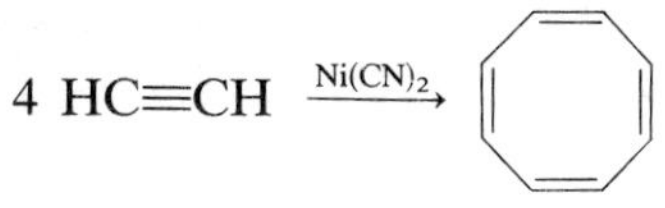

(c) $\xrightarrow{\text{D}_2\text{SO}_4}$ $+ \text{H}^+$

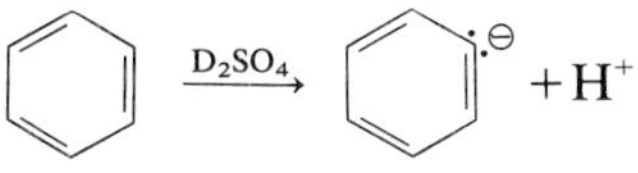

(d) $\text{Br}_2 +$ $\longrightarrow$ (D,L)-2,3-dibromobutane

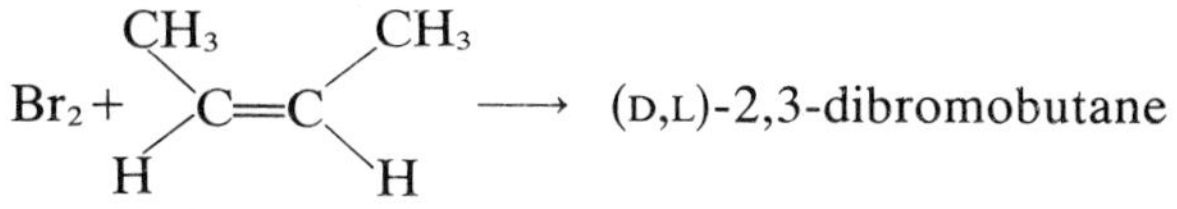

(e) $\text{CH}_4 + \text{Cl}\cdot \xrightarrow[\text{gas phase}]{} \text{CH}_4^{\oplus} + \text{Cl}^{\ominus}$ ($\longrightarrow \text{CH}_3\cdot + \text{HCl}$)

2. Bromine adds to ethylene in solution to give 1,2-dibromoethane. In the presence of dissolved NaCl, some 1-bromo-2-chloroethane is formed. Explain.

3. If the nitration of toluene is carried out with nitric acid in nitromethane solvent, the rate of reaction is exactly the same as the nitration of benzene under the same conditions.

   Nevertheless, if an equimolar mixture of benzene and toluene is nitrated under these conditions, the toluene is nitrated preferentially.

   Explain the apparent discrepancy in determining relative reactivity by these two methods. Refer to a mechanism, the rate law, and an appropriate potential energy diagram.

4. Explain how the kinetic orders (a, b, and c) and the rate constant (k) can be obtained experimentally from the following general rate law. Assume that the rate of reaction can be followed only by monitoring the disappearance of A:

$$\text{Rate} = k[\text{A}]^a[\text{B}]^b[\text{C}]^c$$

5. The acid-catalyzed chlorination of acetone is zero-order in Cl_2, except at very low concentrations of Cl_2, in which case the rate law is first-order with respect to Cl_2. Explain this with a mechanism and a general rate law that can be applied to both circumstances.

6. Suggest a reasonable mechanism for the following reaction if the observed rate law is first-order in compound A only.

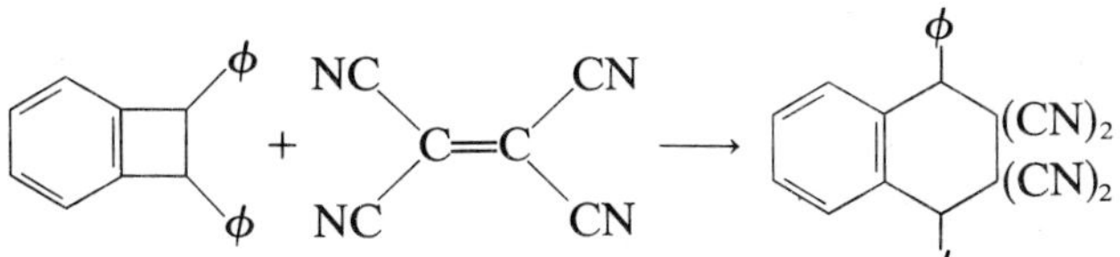

7. Devise an intramolecular and an intermolecular experiment that would allow determination of the kinetic isotope effect for the bromination of acetone.

8. If the difference between the zero-point vibrational energy of a C–H and a C–D bond is 1.2 kcal/mole, calculate the theoretical maximum kinetic isotope effect at room temperature.

9. Which of the following two equilibria would you predict to lie farther to the right? Explain.

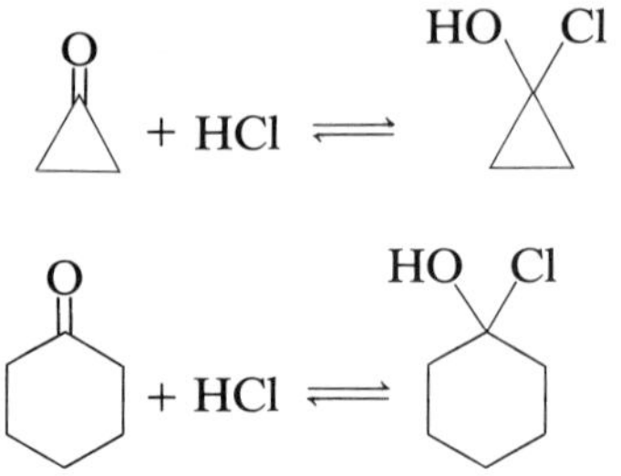

10. Hydrolysis of [^{13}C] o-phthalamic acid in water enriched in ^{18}O gives phthalic acid with the ^{18}O label equally located on each carboxyl group. Explain with a reasonable mechanism. What control experiment should be done with the phthalic acid product?

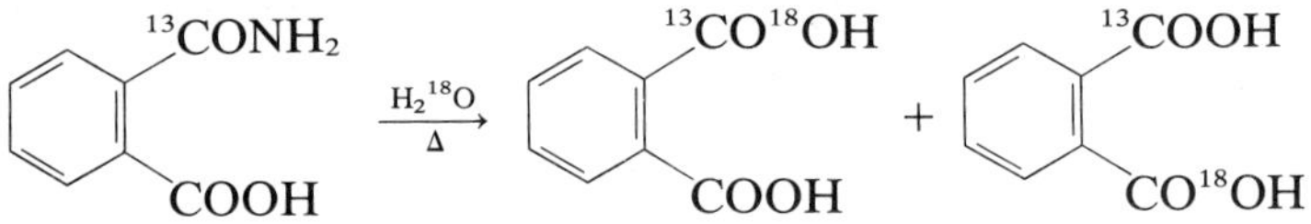

11. The decomposition of phenyldiazene (C$_6$H$_5$N=NH $\rightarrow$ C$_6$H$_6$ + N$_2$) is second-order in phenyldiazene and shows the following activation parameters: $\Delta H^{\ddagger} = +9$ kcal/mole and $\Delta S^{\ddagger} = -23$ e.u. Propose a consistent mechanism.

12. The reaction of anilines with ethyl chloroformate involves the two steps shown below. The Hammett plot shows two distinct linear regions: $\rho = -5.5$ for electron-donating substituents and $\rho = -1.6$ for electron-withdrawing substituents. Explain.

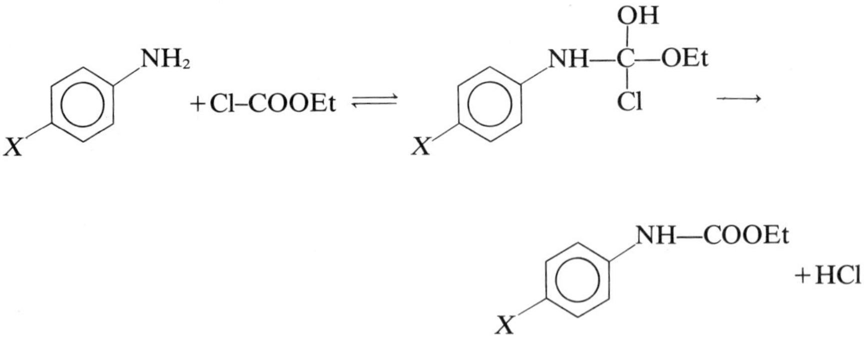

4

Carbocations

4.1 Introduction

Carbocations are common organic intermediates in which carbon is positively charged. These species can be generated in several different ways of which the more common types will be considered here: ionization leading to nucleophilic aliphatic substitution and elimination (eq. 1), electrophilic aromatic substitution (eq. 2), and electrophilic addition to alkenes (eq. 3).

The study of carbocations has been one of the cornerstones of physical organic chemistry. A large percentage of organic reactions proceed by carbocation formation, and these reactions were among the first to be subjected to mechanistic study.[1] Many of the fundamental principles governing interactions within and between organic molecules came initially from the study of carbocations. Additionally, many of the common techniques for the study of organic reaction mechanisms were developed during examination of carbocation reactions.

$$(CH_3)_3C—Cl + H_2O \underset{-Cl^-}{\rightleftharpoons} (CH_3)_3C^+ + H_2O \quad\begin{cases} \xrightarrow{sub} (CH_3)_3C—\overset{+}{O}H_2 \\[2em] \xrightarrow{elim} CH_2{=}C\overset{CH_3}{\underset{CH_3}{\diagdown}} + H_3O^+ \end{cases} \quad (1)$$

$$\bigcirc + HNO_3 \xrightarrow{H_2SO_4} \left(\overset{H\quad NO_2}{\underset{+}{\bigcirc}}\right) \xrightarrow{-H^+} \overset{NO_2}{\bigcirc} \quad (2)$$

$$CH_3—CH{=}CH_2 + HCl \longrightarrow CH_3—\overset{+}{C}H—CH_3 \overset{Cl^-}{\longrightarrow} CH_3—\overset{Cl}{\underset{|}{C}H}—CH_3 \quad (3)$$

In developing and presenting current knowledge of carbocation reactions, we will first treat some elementary aspects of these intermediates and the reactions by which they are generated, and then move briefly into specific areas of active research interest. Specific research topics are discussed in part to point out that there is much about these intensively studied reactions that remains poorly understood.

Nomenclature

For many years, trivalent organic cations were referred to as carbonium ions. However, in recent years there has been some debate regarding the use of this term.[2-4]

$$\overset{|}{\underset{\diagup\,\diagdown}{C}}{}^+$$

It has been pointed out that the term carbonium ion should be used similarly to the carbinol nomenclature for alcohols; thus, we have methylcarbinol and the methylcarbonium ion:

$$CH_3CH_2OH \qquad CH_3CH_2^+$$

Other workers prefer the description of $CH_3CH_2^+$ as the ethyl carbonium ion (note the space between the ethyl and carbonium).[3,4] Others have suggested that the "onium" suffix should be reserved for the highest valence state of the particular element, which for carbon would be the pentavalent not the trivalent species;[2] thus, there are oxonium, ammonium, and carbonium ions:

$$R_3O^+ \qquad R_4N^+ \qquad R_5C^+$$

We will avoid this controversy and accompanying ambiguity by use of a generally accepted cation nomenclature. The entire class of positive carbon intermediates will be referred to as carbocations and the individual species will be referred to as cations; for example,

CH_3^+ methyl cation $C_2H_5^+$ ethyl cation

C_6H_5—CH_2^+ benzyl cation $(CH_3)_3C^+$ *tert*-butyl cation

Structure

Two possible structures of a trivalent carbocation are the pyramidal, sp^3-hybridized form and the planar or trigonal sp^2-hybridized form:

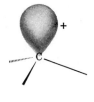
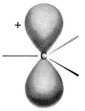

sp^3 hybridization sp^2 hybridization

The planar form has been shown to be the preferred form.[5–10] This result is probably due both to steric and to electronic effects. In the planar form, the groups attached to the cationic center and the electrons in the bonds are further apart. Additionally, electrons in the sp^2 orbitals of the planar form will be more stable than if they were in sp^3 orbitals, because sp^2 orbitals, having a greater percentage of s character, are closer to the carbon nucleus. We shall discuss experiments that demonstrate this general preference for the planar form.

Stability

SOLVENT EFFECTS

Only in a few unusual instances are carbocations sufficiently stable to be isolated. More commonly, carbocations are studied as transient reaction intermediates *in solution*. The role of the solvent in carbocation-generating reactions is no minor one; indeed, few of these reactions would proceed in the absence of solvent. For example, approximately 20 kcal/mole are required to ionize *tert*-butyl bromide to the *tert*-butyl cation and the bromide anion in aqueous solution, yet in the gas phase approximately 200 kcal/mole are required for ionization.[11] Obviously, the solvent plays an important role in facilitating the process of ionization. This role is in large part due to the dipolar character of certain solvents and the consequent ability to stabilize ions. Solvent effects will be discussed in more detail in the following sections.

ELECTRONIC EFFECTS

Carbocations are electron deficient. Obviously then any structural change that increases electron density at the positive center will stabilize the carbocation. An alternative way of visualizing stabilization by electron donation is to consider the positive charge becoming delocalized; donation of electrons to one center must give an electron deficiency at another center.

Such electronic effects can be treated in terms of three "effects": inductive, field, and resonance effects. An inductive effect is electron withdrawal or donation by polarization of a bond. For example, attachment of an electronegative fluorine to the ethyl cation would result in a destabilization of the cation by an inductive withdrawal of electrons through the carbon–carbon σ bond:

$$CH_3\text{—}CH_2^+ \qquad \overset{-\delta}{F}\text{—}\overset{+\delta}{CH_2}\text{—}CH_2^+$$

Similarly, carbocation stability increases with the degree of substitution

$$(CH_3)_3C^+ > (CH_3)_2CH^+ > CH_3CH_2^+ > CH_3^+$$

because of inductive electron donation by alkyl groups.[12]

The field effect is a simple through-space electrostatic interaction. This effect is difficult to separate from the inductive effect, and the two are frequently referred to jointly as the inductive-field or polar effect. The greater stability of the parent 1-adamantyl cation relative to the 3-cyano derivative is believed to be primarily due to a field effect:[13]

The resonance effect is a shift in electron density by conjugation through a π system. For example, the benzyl cation is substantially more stable than the methyl cation because of electron donation by resonance. A similiar effect explains the increase in stability upon substitution of a methoxy group on the methyl cation:

$$CH_3\text{—}\overset{..}{\underset{..}{O}}\text{—}CH_2^+ \longleftrightarrow CH_3\text{—}\overset{+}{\underset{..}{O}}\text{=}CH_2$$

Interestingly, there is an increasing amount of evidence[14] that the inductive effect of alkyl groups may be in part the result of hyperconjugation: conjugation or resonance because of the contribution of a resonance form in which there is no covalent bond to one atom:

$$\overset{H}{\underset{CH_2\text{—}CH_2^+}{\diagdown}} \longleftrightarrow \overset{H^+}{} \quad CH_2\text{=}CH_2$$

As stated previously, the resonance effect operates by conjugation or delocalization through a π system:

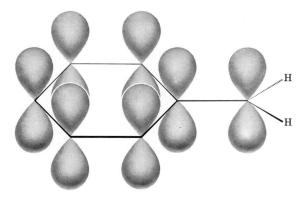

Although we normally treat σ bonds and σ systems as localized, this is only an approximation (see Chapter 1). Hyperconjugation is nothing more than conjugation between σ and π systems:

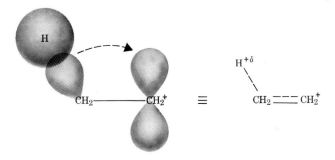

STERIC EFFECTS

Carbocations prefer a trigonal configuration, that is, sp^2 hybridization in which the three bonds lie in one plane with 120° between them:

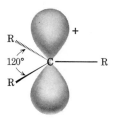

Actually, a trigonal carbocation is obtained only when the three groups attached to the central carbon are identical. For example, the *tert*-butyl cation will be trigonal but the isopropyl cation will have an angle of greater than 120° between the methyl groups because of steric repulsion between the two larger groups:

In studying reactions and intermediates, we frequently utilize reaction rates which, of course, depend on the difference in energy between reactants and transition states. The production of a reactive carbocation is an endothermic process, and for such a process, it is usually reasonable to assume that the transition state resembles the product (in this case the carbocation) more than the reactant (see Chapter 3 for a discussion of Hammond's postulate). In other words, we can assume a transition state for carbocation production resembles a carbocation and treat rates of reactions yielding carbocations by comparison of reactant and carbocation. It is in such reactant–carbocation comparisons that steric effects can be most easily seen.[6] In the ionization of *tert*-butyl chloride, for example, the cation produced is relatively strain-free but the reactant alkyl chloride is not:

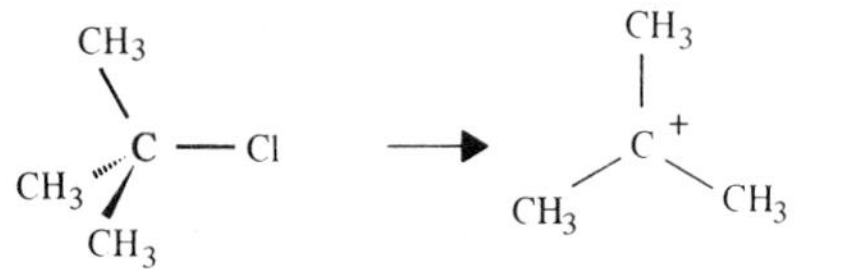

Strain in the alkyl chloride is due to the fairly large chlorine atom forcing the three methyl groups closer together than they would prefer to be. Thus, ionization results in a release of strain with bond angles increasing from 109° to 120°.

For certain geometrically constrained molecules, ionization gives an increase in strain rather than a decrease. For example, what happens when ionization cannot be accompanied by formation of a planar carbocation with approximately 120° between substituents? Methylcyclopropyl chloride is a strained molecule, primarily because of the 60° bond angles in the three-membered ring when 109° is preferred for carbon–carbon bond angles. Ionization, however, produces a carbocation in which bond angles of 120° are desired, yet the cyclopropane remains restricted to 60°:

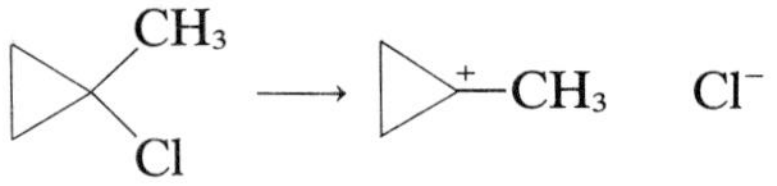

Ionization is accompanied by an increase in strain that will be reflected in an especially low rate.[15]

A similar situation occurs for the ionization of bridgehead derivatives, those in which the three alkyl groups attached to carbon are members of a ring:

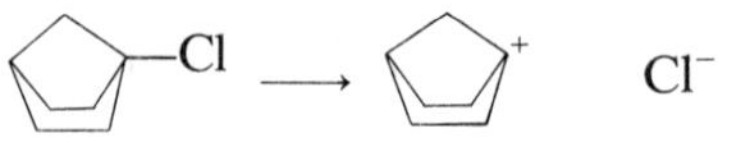

In these cases, ionization is retarded because the resultant carbocation cannot achieve planarity.[16]

4.2 Nucleophilic Aliphatic Substitution

Reactions involving substitution of a nucleophile for a leaving group on an aliphatic system constitute a large percentage of known organic reactions including those done in the laboratory and those occurring in living things. Understandably these reactions have been extensively investigated.[1,5–10,17]

$$R - X + Nucl \rightarrow R - Nucl + X$$

Nucleophilic aliphatic substitutions have long been treated in terms of the S_N1 (substitution, nucleophilic, unimolecular) and S_N2 (substitution, nucleophilic, bimolecular) mechanisms developed by E. D. Hughes and Sir Christopher Ingold.[1]

$$\text{(S}_N\text{1)}$$

Intermediate

$$\text{(S}_N\text{2)}$$

Transition state

Note that only the S_N1 mechanism involves formation of a carbocation intermediate. Reaction by the S_N2 mechanism does proceed through a transition state in which there is a partial positive charge on carbon, and this type of reaction will depend in diminished degree on many of the factors governing carbocation properties. The unimolecular and bimolecular labels for the two mechanisms derive from the molecularity of the rate-determining steps: unimolecular for S_N1 and bimolecular for S_N2.

$$\text{Rate} = k_1[R_3CX] \qquad \text{(S}_N\text{1)}$$
$$\text{Rate} = k_2[\text{Nucl}][R_3CX] \qquad \text{(S}_N\text{2)}$$

Utilization of these two mechanisms can provide a rationalization of many of the properties of substitution reactions. In the first part of this study, we will consider the effects of system variation (such structural effects as alkyl group, leaving group, nucleophile, and solvent) on reaction by the S_N2 and S_N1 mechanisms. There are certain aspects of nucleophilic substitution reactions that require modification of these classic mechanisms, however. We will consider two such phenomena: ion pairing (anion–cation association effects) and borderline reactions (reactions that exhibit properties intermediate between S_N1 and S_N2 properties). Additionally, we will examine the mechanisms of carbocation rearrangements.

Structural Effects on S_N1 Reactions

Reaction by an S_N1 mechanism commonly occurs for tertiary and resonance-stabilized secondary alkyl halides and esters in polar solvents:[6]

$$(CH_3)_3C—Cl \xrightarrow[\substack{\text{acetone} \\ —Cl^-}]{H_2O} (CH_3)_3C^+ \xrightarrow{H_2O} (CH_3)_3C—\overset{+}{O}H_2 \xrightarrow{—H^+} (CH_3)_3C—OH$$

$$(C_6H_5)_2CH—O_2CC_6H_5 \xrightarrow[—C_6H_5CO_2^-]{C_2H_5OH} (C_6H_5)_2CH^+ \xrightarrow{C_2H_5OH} (C_6H_5)_2CH—\overset{\overset{\displaystyle H^+}{|}}{O}C_2H_5$$

$$\downarrow —H^+$$

$$(C_6H_5)_2CH—OC_2H_5$$

Determination that a reaction is following an S_N1 mechanism is frequently made on the basis of kinetic and stereochemical information. Carbocations are indicated to be planar. A planar species necessarily has a plane of symmetry and, therefore, must be achiral and capable of giving only a racemic product mixture. In other words, a planar carbocation can be attacked by a nucleophile from top or bottom to give equal amounts of

enantiomeric products. Substitution by an S_N1 mechanism should then occur to give a racemic product mixture. Many such examples have been observed:[6]

Racemic mixture

Reaction by an S_N1 mechanism must also be kinetically first-order. This fact is frequently of no value, however, because in many cases the substituting nucleophile is a solvent molecule. In such solvolysis reactions, as they are referred to, it is not possible by the usual kinetic techniques to detect or rule out involvement of a solvent molecule because the change in solvent concentration is too small relative to the total solvent concentration. These reactions in which substitution is by solvent are said to exhibit pseudo-first-order kinetics; they may or may not be unimolecular. It is usually possible to deduce the molecularity of the rate-determining step of a solvolysis reaction by qualitative comparison with the rate response of model substrates to changes in alkyl group, leaving group, and solvent.

Reaction by an S_N1 mechanism can also be verified if a rate independence of nucleophile concentration (nonsolvolytic) and a product independence of leaving group are observed. For example, benzhydryl chloride reacts at approximately the same rate in aqueous acetone containing 0.02 M NaN_3 or 0.06 M NaN_3. Such a result is consistent with reaction involving nucleophile

after a slow ionization step. Similarly, benzhydryl chloride and benzhydryl bromide react at different rates in aqueous acetone containing NaN_3, but both halides give the same product ratio. This again is a result consistent with rate determining ionization to a carbocation followed by fast nucleophilic attack:[1]

$$(C_6H_5)_2CH-Cl \xrightarrow[\substack{acetone \\ -Cl^-}]{H_2O} (C_6H_5)_2CH^+ \xrightarrow{N_3^-} (C_6H_5)_2CH-N_3$$

$$(C_6H_5)_2CH^+ \xrightarrow{H_2O} (C_6H_5)_2CH-OH$$

$$\xleftarrow[\substack{-Br^-}]{\substack{H_2O \\ acetone}} $$

$$(C_6H_5)_2CH-Br$$

ALKYL GROUP

The effects of alkyl group variation on reaction by an S_N1 mechanism can be understood well in terms of carbocation stability. Any structural change that stabilizes the intermediate carbocation also stabilizes the transition state leading to the intermediate and thus facilitates ionization:

$$R_3C-X \rightarrow [R_3\overset{+\delta}{C}\cdots \overset{-\delta}{X}]^{\ddagger} \rightarrow R_3C^+ + X^-$$
$$\text{Transition state}$$

In Table 4.1, some relative carbocation stabilities derived from experimental, gas-phase heats of formation are presented.[12] The observed increase in stability with degree of substitution is due to the electron-donating inductive effect of alkyl groups and correlates well with observed increases in solvolytic reactivity under S_N1 conditions (Table 4.1). S_N1 conditions can be defined as reaction in a weakly nucleophilic, strongly ionizing medium; trifluoracetic acid is a good example of such a medium.

Primary carbocations are much less stable than secondary carbocations, and it is doubtful that they are generated in solvolysis reactions. More probably, primary derivatives undergo substitution by the S_N2 mechanism, and the relative rates in Table 4.1 in all likelihood do not reflect actual differences in S_N1 reactivity. Secondary alkyl tosylates, however are believed to react by an S_N1 mechanism in trifluoracetic acid.

According to Table 4.1, tertiary carbocations are approximately 15 kcal/mole more stable than secondary carbocations. If all of this energy difference is reflected in the transition states for ionization of a secondary

Table 4.1 *Carbocation Stabilities Relative to the Ethyl Cation, and Relative Trifluoroacetolysis Rates of the Corresponding Alkyl Tosylates*[12,18]

Cation	Relative Stabilities (kcal/mole)	R-OTs	k_{rel} (25°C)
$CH_3CH_2^+$	0	Et-OTs	1.0
$CH_3CH_2CH_2^+$	6	n-Pr-OTs	5.3
$(CH_3)_2CH^+$	22	i-Pr-OTs	2.6×10^4
$CH_3CH_2-CH^+-CH_3$	26	2-Bu-OTs	1.5×10^5
$(CH_3)_3C^+$	40	t-Bu-OTs	6.3×10^{13}

and a tertiary derivative, it can be shown that the tertiary derivatives will react approximately 10^{11} times faster than the secondary:

$$k = Ae^{-E_a/RT}$$

$$\log k_{rel} = \frac{\Delta E_a}{2.303\,RT}$$

$$\log k_{rel} = \frac{15}{2.303\,(0.6)}$$

$$\log k_{rel} = 10.8$$

A rate difference of about 10^8 is observed, and it can be assumed that the transition states for these reactions have a great deal of carbocation character.

A linear free energy relationship similar in form to the Hammett relationship (see Chapter 3) has been derived by R. W. Taft for the prediction of inductive effects on reactivity in saturated systems.[20]

$$\log \left(\frac{k}{k_0}\right) = (\textstyle\sum \sigma^*)\rho^* \tag{4}$$

These substituent constants (σ^*—Table 4.2) are derived from substituent effects on rates of ester hydrolysis, and accurately reflect the ability of a substituent to inductively stabilize a carbocation. The reaction constant ρ^* measures the response of a particular reaction type to inductive effects. As an example of the application of this equation, consider the reaction of a series of tertiary alkyl chlorides in 80% aqueous ethanol:

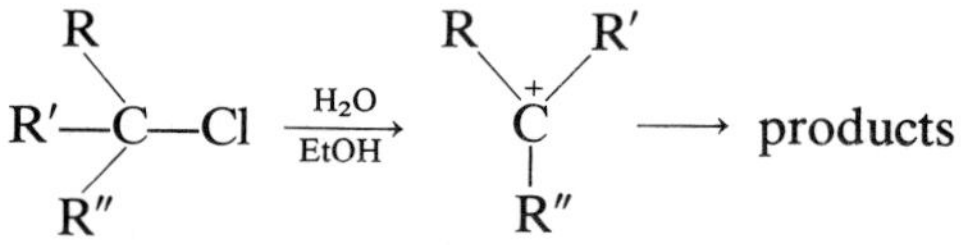

For this series, the parent compound is *tert*-butyl chloride; the rate constant for its reaction is substituted for k_0 of eq. 4, and $\sum \sigma^*$ for this compound is zero (σ^* for methyl equals zero—Table 4.2). For other members of the series, the rate constant is substituted for k of eq. 4, and the sum of σ^* is entered. The Taft plot for this reaction series is given in Figure 4.1.

Table 4.2 *Taft Polar Substituent Constants (σ^*) for* $R—\overset{\shortmid}{\underset{\shortmid}{C}}—X$[20]

R	σ^*	R	σ^*
Cl_3C	2.65	$C_6H_5CH_2CH_2$	0.08
CH_3OOC	2.00	CH_3	0
NO_2CH_2	1.40	C_2H_5	−0.10
$ClCH_2$	1.05	$n\text{-}C_3H_7$	−0.115
CF_3CH_2	0.92	$i\text{-}C_4H_9$	−0.125
C_6H_5	0.60	$n\text{-}C_4H_9$	−0.130
CH_3OCH_2	0.52	$(CH_3)_3CCH_2$	−0.165
H	0.49	$i\text{-}C_3H_7$	−0.190
$C_6H_5CH_2$	0.26	$t\text{-}C_4H_9$	−0.30

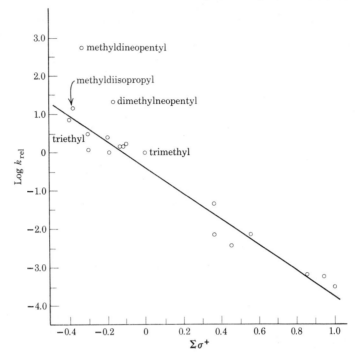

Fig. 4.1 *Correlations of rates of solvolysis of tertiary carbinyl chlorides in 80% ethanol with $\Sigma\sigma^*$* $(\rho^* = -3.29)$.[6]

Interestingly, relief of strain upon ionization does not appear to be as important as might possibly have been expected. Reference to Figure 4.1 shows that even relatively crowded tertiary chlorides such as methyldiiso-propylcarbinyl chloride fall close to the correlation line and can, therefore, be treated solely in terms of inductive effects. More highly branched substrates, such as methyldineopentylcarbinyl chloride, do react more rapidly than predicted by σ^* values.

Carbocation stabilities can also be determined by study of carbocation–alcohol equilibria in strong acids such as sulfuric acid.[21] These studies have been conducted for a large series of triarylmethyl and diarylmethyl cations.

$$R^+ + H_2O \rightleftharpoons R\!\!-\!\!OH + H^+$$

Some of these data, in terms of free energy relative to the triarylmethyl cation, are given in Table 4.3. The data clearly show the effects of conjugative stabilization of carbocations. For example, the stability order tri-*p*-anisylmethyl > triphenylmethyl > diphenylmethyl is due to the ability of methoxy and phenyl groups to donate electron density by resonance. Inductive effects are shown by the greater stability of di-*p*-tolylmethyl cation and lesser stability of the di-(*p*-chlorophenyl)methyl cation relative to the diphenylmethyl (or benzhydryl) cation.

The triphenylmethyl cation is indicated in Table 4.3 to be 9.3 kcal/mole more stable than the diphenylmethyl cation. Examination of solvolysis rates in 40% ethanol–60% ethyl ether shows that triphenylmethyl chloride reacts about 10,000 times faster than diphenylmethyl chloride. If the entire 9.3 kcal/mole stability difference were reflected in the transition states for

Table 4.3 *Relative Carbocation Stabilities from Carbocation–Alcohol Equilibria in Sulfuric Acid*[21]

Cation	ΔG_R^+ (kcal/mole)[a]
Tri-(*p*-dimethylaminophenyl) methyl	21.8
Tri-*p*-anisylmethyl	10.1
Tri-*p*-tolylmethyl	4.1
Di-*p*-anisylmethyl	1.3
Triphenylmethyl	0.0
Tri-(*p*-chlorophenyl) methyl	−1.3
Di-*p*-tolylmethyl	−5.2
Diphenylmethyl	−9.3
Di-(*p*-chlorophenyl) methyl	−10.0
Tri-(*p*-nitrophenyl) methyl	−13.3

[a] $\Delta G_R^+ = -RT \ln K$ relative to triphenylmethyl.

S_N1 solvolysis a rate difference of approximately 10^7 should have been observed. Again we see that the theory of transition states *resembling* the first intermediate for an endothermic process is a good one.

The Hammett σ constants are useful for correlating the effects of benzene ring substituents on reactivity except when there is direct conjugation between the substituent and the reaction center:

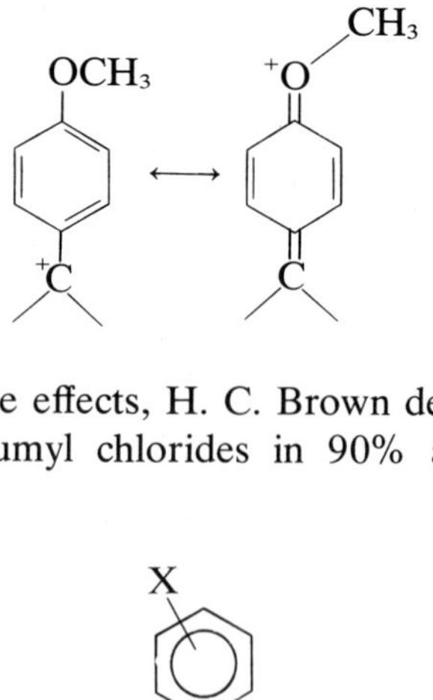

To treat direct conjugative effects, H. C. Brown defined σ^+ constants from the solvolysis rates of cumyl chlorides in 90% aqueous acetone (Table 4.4):[22]

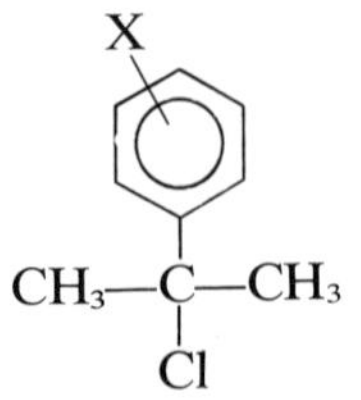

$$\log \frac{k}{k_0} = \sigma^+ \rho \qquad (5)$$

All rates are relative to the parent molecule (k_0 for X=H), and ρ in 90% aqueous acetone (−4.62) is chosen to make σ^+ for *meta* substituents (for which conjugation is not possible) as close as possible to Hammett σ values. With ρ so chosen, σ^+ values for the substituents can then be derived from

Table 4.4 *Some σ^+ and σ Values*[22]

Substituent	σ_p^+	σ_m^+	σ_p	σ_m
CH₃O	−0.764	0.047	−0.268	0.115
CH₃	−0.306	−0.069	−0.170	−0.069
F	−0.071	0.337	0.062	0.337
H	0	0	0	0
Cl	0.112	0.373	0.227	0.373
I	0.132	0.352	0.18	0.352
Br	0.148	0.391	0.232	0.391
NO₂	0.777	0.662	0.778	0.710

solvolysis rates of the cumyl chlorides. Significantly, excellent linear relationships are observed between σ^+ and ΔG_R^+ (see Figure 4.2). Also, it is
interesting to note that the slope of the plots is steeper for the diarylmethyl
series than for the triarylmethyl series; such a result is consistent with there
being more need or demand for stabilization in the secondary series and a
consequent larger response on the part of the substituents.

A molecular orbital method for correlating solvolytic reactivity of arylmethyl derivatives is given in Chapter 2.

The examples of S_N1 reactions given have all involved saturated or sp^3
hybridized carbon derivatives. Of unsaturated esters and halides, only the
vinyl derivatives have been observed to undergo ionization and then only

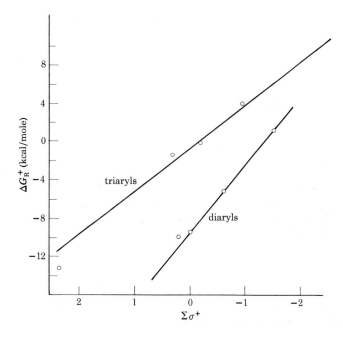

Fig. 4.2   *A plot of ΔG_R^+ against $\sum \sigma^+$ for a series of diarylmethyl and triarylmethyl cations.*[21]

when stabilized by other groups or when very good leaving groups are used:[23]

This unreactivity is due in part to the stronger bonds formed between sp^2 hybridized carbon and the leaving group. Also as the degree of carbon unsaturation increases, the carbon becomes more electronegative because increasing the amount of s character of a hybrid orbital brings the orbital physically closer to the nucleus. Thus, there is also an electronic destabilization of the carbocation resulting from ionization of an unsaturated derivative.

SOLVENT

Most S_N1 reactions involve ionization of a neutral reactant and, as expected for a process in which charge is generated, are facilitated by a polar solvent:

$$R—X \xrightarrow{\text{slow}} R^+ + X^-$$

Table 4.5 *Y Values from tert-Butyl chloride Solvolysis*[24]

Solvent	Y	Solvent	Y
100% C_2H_5OH	−2.03	90% Acetone	−1.86
80% C_2H_5OH	0.00	50% Acetone	1.40
60% C_2H_5OH	1.12	CH_3COOH	−1.64
40% C_2H_5OH	2.20	HCOOH	2.05
H_2O	3.49	CF_3COOH	4.6

One measure of solvent polarity is the dielectric constant, and a rough correlation between solvolysis rate and dielectric constant is observed.[24] A much more satisfactory measure of solvent polarity or solvent ionizing power comes from a linear free energy relationship developed by Winstein and Grunwald, which is based on the reaction of *tert*-butyl chloride as a model S_N1 substrate.[24] These workers presented the following relationship:

$$\log \frac{k}{k_0} = mY$$

where k is the solvolysis rate of *tert*-butyl chloride in the solvent in question, k_0 is the rate in 80% aqueous ethanol, m is a measure of substrate sensitivity to solvent ionizing power and is assigned a value of unity for *tert*-butyl chloride, and Y is a measure of solvent ionizing power. With m equal to unity, Y is thus determined by the solvolysis rates of *tert*-butyl chloride. Some of the Y values determined in this manner are given in Table 4.5. A positive Y value means a solvent has a stronger ionizing power than 80% ethanol.

Measurement of solvolysis rates of other compounds in solvents of known Y gives m for these compounds (Table 4.6). Many secondary and primary substrates solvolyze with significant *nucleophilic* solvent participation, that is, by an S_N2, not an S_N1, mechanism. Consequently, m is in many cases also a function of solvent nucleophilicity and not a simple measure of suscepti- bility to solvent ionizing power. A more accurate measure of susceptibility to solvent ionizing power is made possible by the observation that acetic and formic acids are approximately equally nucleophilic (see p. 151); the m value for reaction in these two solvents (m_{AF}) will, therefore, not be a function of solvent nucleophilicity.

When discussing solvent effects on reactions such as those proceeding by an S_N1 mechanism in which charge is generated in the transition state, it is quite commonly assumed that the effects can be explained on the basis of the ability of the solvent to solvate the polar transition state.[1] The effect of a solvent change on the stability of the nonpolar reactant molecule would seem to be relatively unimportant. However, thermodynamic studies have shown that this is certainly not the case.[24] In fact, solvent effects on reactant stability can actually be the dominant factor in certain instances. In the solvolysis of *tert*-butyl chloride, for example, a decrease of 6.25 kcal/mole in $\Delta G^{\ddagger}$ results upon going from methanol to water, but the major contribution (4.09 kcal/mole) to this decrease results from destabilization of the relatively nonpolar ground state, not stabilization of the polar transition state.[26]

Table 4.6 *Effect of Solvent Ionizing Power in Terms of m values in Aqueous Ethanol and m_{AF} Values in Acetic and Formic Acids*[24,25]

Compound	m	m_{AF}
Methyl tosylate	0.23	0.30
Ethyl tosylate	0.25	0.44
Isopropyl tosylate	0.42	0.72
Benzyl chloride	0.43	0.66
Cyclopentyl tosylate	0.49	0.73
Cyclohexyl tosylate	0.44	0.79
3,3-Dimethyl-2-butyl brosylate	0.71	0.83
2-Adamantyl tosylate	0.91	0.89
tert-Butyl chloride	1.00	1.00
1-Bromobicyclo[2.2.2]octane	1.03	0.92

(tosylate = *p*-toluenesulfonate, brosylate = *p*-bromobenzenesulfonate)

The relative contribution of changes in reactant solvation $\Delta\Delta G^{\text{react}}$ and transition state solvation $\Delta\Delta G^{\text{t.s.}}$ to changes in free energy of activation for reaction in different solvents can be determined by measurement of relative rates to give the change in free energies of activation $\Delta\Delta G^{\ddagger}$ and by calorimetric determination of reactant solvation energies:

$$\Delta\Delta G^{\ddagger} = \Delta\Delta G^{\text{t.s.}} - \Delta\Delta G^{\text{react}}$$

An increasing number of such determinations are being done, and the theories of solvation are undergoing dramatic changes as a consequence of the increased knowledge provided.

LEAVING GROUP

In the transition state for reaction by an S_N1 mechanism, the leaving group has acquired almost a full negative charge:

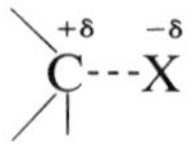

Thus, an increase in the ability of the leaving group to accept this negative charge should give a better leaving group and more rapid reaction. That this is the case can be seen from the solvolysis rates of 1-adamantyl arenesulfonates (Table 4.7); as the electron-withdrawing ability of X increases, the rate

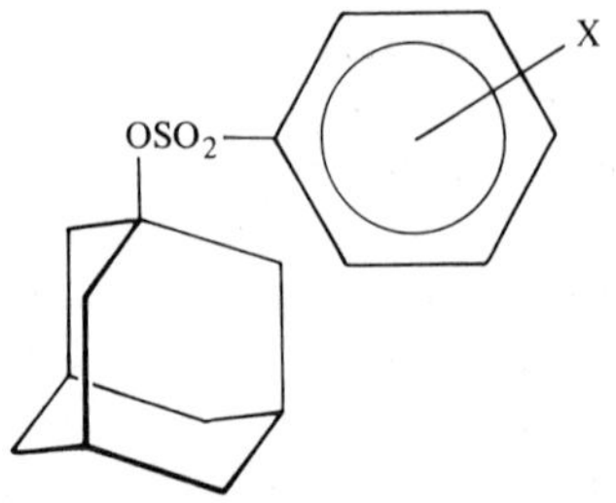

Table 4.7 *Relative Leaving Group Abilities for 1-Adamantyl Arenesulfonates in Ethanol and tert-Butyl halides in 80% Ethanol*[6,27]

1-AdmOSO$_2$ —⟨⟩—X X =	k_{rel}	tert-butyl-X X =	k_{rel}
p–OCH$_3$	1.0	F	10^{-5}
p–CH$_3$	1.6	Cl	1.0
p–H	3.1	Br	39
p–F	5.2	I	99
p–Cl	8.4	OTs	$>10^5$
p–Br	9.0		
p–NO$_2$	55		
m–NO$_2$.	78		

increases. As would be expected, a plot of log k for 1-adamantyl arenesulfonate ethanolysis against Hammett σ constants is linear.[27]

S_N1 reactivity, of course, depends on the free energy difference between reactants and transition state. Reactant stability can also influence leaving group ability, and the reactivity order I > Br > CI > F for *tert*-butyl halide solvolysis is largely due to increases in C–X bond strength for the order I < Br < CI < F (Table 4.7).[6]

Structural Effects on S$_N$2 Reactions

Reaction by an S_N2 mechanism commonly occurs for primary and secondary derivatives:[28]

$$CH_3CH_2{-}Br + C_2H_5OH \rightarrow \left[\begin{array}{c} H \quad H \\ \overset{+\delta}{C_2H_5{-}O}{\cdots}\overset{+\delta}{C}{\cdots}\overset{-\delta}{Br} \\ H \quad CH_3 \end{array} \right]^{\ddagger} \rightarrow \overset{+}{C_2H_5{-}O{-}CH_2CH_3}\ Br^- \\ \qquad H$$

$$CH_3{-}\underset{\underset{CH_3}{|}}{CH}{-}OTs + N_3^- \rightarrow \left[\begin{array}{c} H \quad CH_3 \\ \overset{-\delta}{N_3}{\cdots}\overset{+\delta}{C}{\cdots}\overset{-\delta}{OTs} \\ CH_3 \end{array} \right]^{\ddagger} \rightarrow N_3{-}\underset{\underset{CH_3}{|}}{CHCH_3} + OTs^-$$

OTs = p–toluenesulfonate

Reaction by an S_N2 mechanism can be determined from kinetics if the nucleophile concentration can be monitored, as is the case for all nucleophiles except solvent:

$$R{-}X + Nucl \rightarrow R{-}Nucl + X$$

$$Rate = k_2[R{-}X][Nucl]$$

But if Nucl is a solvent

$$[solvent] = constant$$

and

$$Rate = k_1[R{-}X] \quad (pseudo\text{-}first\text{-}order)$$

Substitution by an S_N2 mechanism must occur with inversion of stereochemistry and such examples have been observed.[6,28] Observation of inversion upon substitution by solvent is probably the most direct method for demonstrating solvolysis by an S_N2 mechanism:

$$HCOOH + \underset{C_3H_7}{\overset{H}{\underset{|}{\overset{D}{\backslash}}C}} - OBs \longrightarrow H - \overset{O}{\overset{||}{C}} - O - \underset{C_3H_7}{\overset{D}{\underset{\backslash}{\overset{/}{C}}}}H + HOBs$$

— OBs = p − bromobenzenesulfonate

$$CH_3OH + \underset{C_6H_{13}}{\overset{H}{\underset{|}{\overset{CH_3}{\backslash}}C}} - OBs \longrightarrow CH_3O - \underset{C_6H_{13}}{\overset{H}{\underset{\backslash}{\overset{/CH_3}{C}}}} + HOBs$$

Qualitative probes that prove useful in demonstrating an S_N2 mechanism are the degree of kinetic response to nucleophilicity, solvent, and leaving group changes. Additionally, the S_N2 mechanism, unlike the S_N1 mechanism, requires that a product dependence on leaving group be observed:

$$CH_3CH_2Cl \underset{\underset{H_2O}{\overset{N_3^-}{\Big\langle}}}{} \quad \begin{array}{l} CH_3CH_2-N_3 \overset{N_3^-}{\longleftarrow} \\ CH_3CH_2-OH \underset{H_2O}{\longleftarrow} \end{array} \Big\rangle CH_3CH_2Br$$

$$\left[\frac{R-N_3}{R-OH}\right]_{Cl} \neq \left[\frac{R-N_3}{R-OH}\right]_{Br}$$

ALKYL GROUP

As the degree of carbocation substitution decreases (tertiary to secondary to primary), there is a decrease in carbocation stability (about 15 kcal/mole for each degree).[12] Unimolecular heterolysis of an alkyl derivative (S_N1), therefore, becomes less probable as the degree of substitution is decreased. Additionally, primary and, to a lesser extent, secondary derivatives are more open sterically toward backside nucleophilic attack. Thus, it is not surprising that many primary and secondary esters and halides follow the S_N2 mechanism:[28]

$$CH_3-\underset{\underset{CH_3}{|}}{CH}-OTs + N_3^- \longrightarrow CH_3-\underset{\underset{CH_3}{|}}{CH}-N_3 + OTs^-$$

$$Rate = k_2 [N_3^-][2-Pr-OTs]$$

A simple S_N2 displacement on a tertiary system has never been observed.

The major effect of alkyl group variation on S_N2 substitution within a primary or secondary series appears to be a steric effect. For example, a plot of log k against Taft's σ^* constant for reaction of primary alkyl bromides with sodium iodide in acetone does not give a linear relationship but rather a plot in which the more highly branched derivatives react more slowly than predicted on the basis of inductive effects (Figure 4.3). This result is

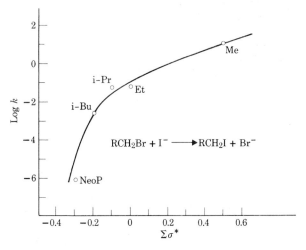

Fig. 4.3 *A Taft σ^* plot for reaction of primary alkyl bromides with NaI in acetone.*[28]

consistent with an increase in hindrance to nucleophile approach with an increase in branching (see also Table 3.6).

Also consistent with the major importance of steric effects in S_N2 reactions is the observation that sterically hindered nucleophiles are poor nucleophiles. For example, 2-*tert*-butylpyridine reacts with methyl iodide 11,900 times slower than does 3-*tert*-butylpyridine:[29]

In the pentacovalent transition state for S_N2 displacement, bond angles between the three groups attached to the central carbon will be approximately 120°:

Therefore, angle strain effects will be quite similar for S_N1 and S_N2 reactions. Consistent with this interpretation is the following rate order observed for bimolecular displacement on cyclic derivatives: cyclopentyl > cyclobutyl > cyclopropyl; strain increases on passing through the 120° configuration are greatest for the smaller rings.[28]

Although the primary effects of alkyl group variation on S_N2 reactivity are due to steric effects, electronic effects can also be detected. Substitution of a β fluorine on ethyl bromide slows the rate of thiophenoxide displacement by a factor of eight;[6] more dramatically, 2,2,2-trifluoroethyl triflate (trifluoromethane sulfonate) undergoes aqueous ethanolysis 10^6 more slowly than methyl triflate.[30] An S_N2 transition state with a partial positive charge on carbon is indicated.

Interestingly, it also appears that conjugative effects may be important for S_N2 displacements despite formation of a pentavalent transition state. For example, the following reactivity order is observed for reaction of alkyl chlorides with sodium iodide in acetone (an S_N2 reaction):

$$C_6H_5-CH_2-Cl > CH_2=CH-CH_2-Cl > CH_3CH_2-Cl$$

$k_{rel} = 93 \qquad\qquad 33 \qquad\qquad\qquad 1.0$

These results have been explained by assuming that there is conjugative orbital overlap in the transition state, as shown for allyl:[6]

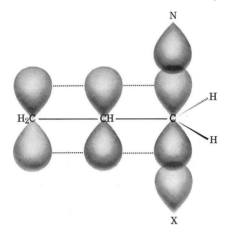

NUCLEOPHILE

The facility with which a molecule or anion displaces a leaving group from carbon is referred to as nucleophilicity. Nucleophilicity is a function of the ability to donate electrons and also of the energy required to desolvate the nucleophile.

As would be expected, substitution of electron-donating substituents onto a nucleophile produces a better nucleophile; thus, the following order of nucleophilicity is observed:[6]

There is an obvious connection between nucleophilicity and basicity, since basicity is determined by the tendency to bond to a proton and nucleophilicity is determined by the tendency to bond to carbon. Basicity data are readily available and attempts have been made to predict nucleophilicities from these data. Among the cases in which attachment to a common atom is compared (e.g., all oxygen nucleophiles or all nitrogen nucleophiles) the agreement between basicity and nucleophilicity is good. For example, both properties decrease in the following series: $C_2H_5O^- > C_6H_5O^- > CH_3COO^- > NO_3^-$.[6]

When the attacking atom is changed, the correlation between basicity and nucleophilicity becomes a poor one in many cases. Ethoxide ion is a better base than iodide ion, but iodide is a better nucleophile. Butyl mercaptide ion is about as basic as phenoxide ion but is 10^3 times as nucleophilic. An explanation of these results that is frequently given is that polarizability (the ability of electron clouds to distort) is important for nucleophilicity but not for basicity. Thus, the larger, more polarizable species exhibit enhanced nucleophilicity. The greater importance of polarizability for attack on carbon relative to attack on hydrogen can be rationalized in terms of positions on the reaction coordinate of the two transition states. The reaction of a proton with bases such as the halide ions is extremely rapid. The transition state for this process will be an early one with relatively little electron transfer. The reaction of these same bases with saturated carbon is much less rapid. This reaction takes place by a late transition state in which electron transfer from nucleophile to carbon is extensive. Consequently, the ability of the nucleophile to give up its electrons (*i.e.*, polarizability) is of greater importance.

Increasing the size of a nucleophile makes it more difficult for the nucleophile to closely approach the backside of a carbon and transfer its available electrons to carbon. As would be expected on this basis, any structural variation that leads to crowding about a nucleophilic center should also lead to a decrease in rate of S_N2 displacement; this steric factor is shown in the following relative nucleophilicities:[31]

As the degree of substitution of the substrate under attack increases, the rate-retarding effect of nucleophile crowding should become increasingly important; this effect is shown in Table 4.8. The relative rates of the 3-substituted pyridines serve to separate electronic factors from steric factors.

Before a nucleophile can attack carbon, the shell or sheath of solvent molecules about the nucleophile must be removed. The importance of this desolvation process is clearly illustrated by comparing relative nucleophilicities in protic solvents (e.g., methanol or water) and in dipolar aprotic solvents (e.g., dimethylsulfoxide, DMSO, or dimethylformamide, DMF). In methanol the order of nucleophilicities is $I^- > Br^- > Cl^-$, but in dimethylformamide the reverse order is observed, $Cl^- > Br^- > I^-$. Also, azide

Table 4.8 *Relative Displacement Rates for Reaction of Pyridines with Alkyl Iodides (rates relative within each series only)*[31]

R =	CH_3I	CH_3CH_2I	$(CH_3)_2CHI$
H	1.0	1.0	1.0
2-Me	0.47	0.23	0.054
2-Et	0.22	0.11	—
2-*i*-Pr	0.071	0.030	—
3-Me	2.1	2.2	1.8
3-Et	2.2	2.2	1.9
3-*i*-Pr	2.4	2.2	1.8

ion reacts with methyl chloride 2000 times faster in DMF than in water.[32] Analysis of solvent effects on reactants and on transition states shows that these results are due to the absence of strong hydrogen bonds to reactant anions in the dipolar aprotic solvent. Hydrogen bonding to anions is most important for small anions with concentrated charge; therefore, the nucleophilicity order in methanol of $I^- > Br^- > Cl^-$ is seen to be due to the ease of desolvating the larger anions. In an aprotic solvent, anionic desolvation becomes less important, and the order $Cl^- > Br^- > I^-$ is observed. Similarly, the enhanced nucleophilicity of N_3^- in DMF relative to water is due to weaker solvation of the nucleophile in DMF; in essence, this is nothing more than rate enhancement by reactant destabilization.

At first glance, it might appear that the desolvation argument invalidates the polarizability argument used above to explain the enhanced nucleophilicities of large nucleophiles; is not this enhancement just a desolvation effect? It should be recalled, however, that the enhanced nucleophilicity was based on predictions from basicity, and basicity is also a function of desolvation of the base.

Linear free energy relationships have been used to derive empirical measures of nucleophilicity. One of these is the Swain–Scott equation based on displacement rates on methyl bromide:[33]

$$\log \frac{k}{k_{H_2O}} = sn$$

In this equation, k is the rate of attack by some nucleophile, k_{H_2O} is the rate of attack by water, s is the sensitivity to nucleophilic attack (assigned a value of unity for CH_3Br), and n is a measure of nucleophilicity. An n value of greater than zero indicates a nucleophile stronger than water. Some of the available n values are given in Table 4.9.

To measure the nucleophilicity of solvent molecules, the effect of solvent ionizing power on the model reaction must first be subtracted. This can be done by use of Y values as a measure of solvent ionizing power, m values as a measure of sensitivity to solvent ionizing power, and

$$\log \frac{k}{k_0} = sN + mY \tag{6}$$

Using methyl tosylate as a model substrate and s of unity, m of 0.30, and previously determined Y values, the N values given in Table 4.10 were derived.[34]

Table 4.9 *Swain–Scott Nucleophilicity Values*[33]

Nucl	n	Nucl	n
$S_2O_3^{-2}$	6.64	OH^-	4.27
CN^-	5.45	Cl^-	2.83
I^-	4.92	F^-	1.67
SCN^-	4.87	H_2O	0
Br^-	4.45		

Table 4.10 *Solvent Nucleophilicities*[34]

Solvent	N	Solvent	N
Ethanol	0.09	56% Acetone	−0.47
Methanol	0.01	Acetic acid	−2.05
80% Ethanol	0	Formic acid	−2.05
50% Ethanol	−0.20	Trifluoroacetic acid	−5.55
Water	−0.26	Fluorosulfonic acid	−5.5

SOLVENT

A detailed examination of solvent effects on S_N2 reactions must include thermodynamic separation of the effect of solvent on reactants and transition states (e.g., see discussion of nucleophilicities in aprotic solvents). It is possible, however, to make qualitatively accurate predictions of solvent effects simply by considering the change in charge type as reactants form products.[1] There are four combinations of reactant charge types:

Nucl	R–X
Neutral	Neutral
Neutral	Cation
Anion	Neutral
Anion	Cation

For the neutral–neutral combination, charge is created in the transition state. An increase in solvent polarity would be expected to aid this charge development and give an increase in reaction rate:

$$H_2O + (CH_3)_2CH\!-\!Br \longrightarrow \left[\begin{array}{c} CH_3 \quad\; CH_3 \\ \overset{+\delta}{H_2O}\text{---}\overset{|}{C}\text{---}\overset{-\delta}{Br} \\ | \\ H \end{array} \right]^{\ddagger} \longrightarrow H_2\overset{+}{O}CH(CH_3)_2 + Br^-$$

Charge development

An anion–cation combination results in charge destruction, a process retarded by an increase in solvent polarity:

$$OH^- + CH_3\!-\!\overset{+}{S}\!\!\overset{CH_3}{\underset{CH_3}{\big\langle}} \longrightarrow \left[\begin{array}{c} H \qquad H \\ \overset{-\delta}{HO}\text{---}\overset{|}{C}\text{---}\overset{+\delta}{S}(CH_3)_2 \\ | \\ H \end{array} \right]^{\ddagger} \longrightarrow HOCH_3 + S(CH_3)_2$$

Charge destruction

The other two combinations, neutral–cation and anion–neutral, result in a dispersal of charge in the transition state and should be slightly inhibited by an increase in solvent polarity. Verification of these predictions is given in

Table 4.11 *Effects of Solvent Polarity on S_N2 Reactions of Various Charge Types*[1]

	Relative Rates			
Reaction	EtOH	60% EtOH	H_2O	Charge
$H_2O + 2\text{-PrBr}$	1.0	39	—	Created
$(CH_3)_3N + {}^+S(CH_3)_3$	1.0	—	0.097	Dispersed
$OH^- + 2\text{-PrBr}$	1.0	0.5	—	Dispersed
$OH^- + {}^+S(CH_3)_3$	1.0	0.0021	0.000051	Destroyed

Table 4.11, where reaction rates are contrasted in ethanol and the more polar 60% aqueous ethanol and water:

$$(CH_3)_3N + CH_3 - \overset{+}{S}\overset{CH_3}{\underset{CH_3}{\diagdown}} \longrightarrow \left[\underset{H}{\overset{CH_3}{\underset{|}{(CH_3)_3\overset{+\delta}{N}\text{---}C\text{---}\overset{+\delta}{S}(CH_3)_2}}} \right]^{\ddagger} \longrightarrow (CH_3)_3\overset{+}{N} - CH_3 + S(CH_3)_2$$

Charge dispersal

$$OH^- + (CH_3)_2CHBr \longrightarrow \left[\underset{H}{\overset{H \qquad H}{\underset{|}{HO\overset{-\delta}{\text{---}}C\overset{-\delta}{\text{---}}Br}}} \right]^{\ddagger} \longrightarrow HO\text{---}CH(CH_3)_2 + Br^-$$

Charge dispersal

LEAVING GROUP

The same factors that affect S_N1 leaving group ability affect S_N2 leaving group ability. Primarily, this means that increasing the ability of the leaving group to bear a negative charge increases reaction rate. Nonbonded interactions between bulky leaving groups and the alkyl group are more important for S_N1 reactions because S_N1 reactions usually occur with bulky tertiary alkyl groups. This factor, in large part, accounts for a tosylate to bromide ratio of 9750 for 1-adamantyl solvolysis (in 80% ethanol) and a tosylate to bromide ratio of only eleven for displacement by water or ethanol on a methyl derivative.[35] For less sterically demanding leaving groups, S_N1 and S_N2 reactions vary in about the same manner upon a leaving group change (Table 4.12).

Table 4.12 *Leaving Group Effects on S_N1 and S_N2 Reactions*

	k_{rel}	
Leaving Group (X)	S_N2 $N_3^- + CH_3X\ (CH_3OH)$[32]	S_N1 $t\text{-Bu-X} + H_2O/EtOH$[36]
Cl^-	1.0	1.0
Br^-	63	39
I^-	100	99
$CH_3\text{---}\langle\bigcirc\rangle\text{---}SO_2O^-$	630	150,000

Ion Pairing[28,37,38]

The S_N1–S_N2 scheme was used for over 20 years with excellent success. Beginning in the early 1950s, however, it was recognized that this simple mechanistic formulation was not adequate for the treatment of certain experimental data. A more detailed formulation, the Winstein ion-pair scheme, is now seen to provide the simplest treatment of nucleophilic aliphatic substitution reactions consistent with the available experimental evidence. Species **I** is neutral substrate, **II** is a tight or intimate ion pair, **III** is a solvent-separated ion pair (frequently represented as $R^+\|X^-$), and **IV** is a free cation.

THE WINSTEIN ION-PAIR SCHEME[37,38]

$$R-X \underset{k_{-1}}{\overset{k_1}{\rightleftharpoons}} R^+X^- \underset{k_{-2}}{\overset{k_2}{\rightleftharpoons}} R^+---Solvent---X^- \underset{k_{-3}}{\overset{k_3}{\rightleftharpoons}} R^+ + X^-$$

I	II	III	IV
$\downarrow$ Nucl	$\downarrow$ Nucl	$\downarrow$ Nucl	$\downarrow$ Nucl
R–Nucl	R–Nucl	R–Nucl	R–Nucl

The discussion, in prior sections, of S_N1 and S_N2 reactions is not invalidated by the new formulation. Rather, S_N1 and S_N2 mechanistic types are included in modified form in the ion-pair scheme. An S_N1 reaction can be seen as one in which the rate-limiting step is unimolecular formation of **II, III,** or **IV** (i.e., k_1, k_2, or k_3). An S_N2 reaction is now seen to be a reaction in which the rate-limiting step is bimolecular nucleophilic attack on **I, II, III,** or **IV.**

Several factors influence whether **I, II, III,** or **IV** will be involved in a particular reaction. An increase in carbocation or leaving group stability promotes reaction by a more dissociated intermediate. An increase in solvent ionizing power also promotes dissociation, but an increase in nucleophilicity has the opposite effect, promoting reaction at an earlier, less dissociated stage.

There is an increasing amount of evidence that this Winstein ion-pair scheme may itself be too simple; for example there is evidence for the existence of nucleophilically-solvated ion pairs and solvent-separated ion pairs with two insulating solvent molecules.[28]

$$Nucl^----R^+---X^- \qquad R^+---solvent---solvent---X^-$$

It is because of these recent developments that we earlier referred to the Winstein ion-pair scheme as "the simplest treatment of nucleophilic aliphatic substitution reactions." The S_N1-S_N2 scheme was shown to be too simple; the more detailed Winstein ion-pair scheme has been adequate for the treatment of almost all available data to this point, but it is becoming increasingly difficult to fit certain observations to this scheme, and an even more complicated formulation may eventually be required. The determination of ion pair involvement and rates for the various dissociation, return, and nucleophilic attack steps is one of the major goals of research on nucleophilic aliphatic substitution reactions.

In the following sections we will consider five types of experiments which have led to the development of the Winstein ion-pair scheme.

COMMON ION RATE DEPRESSION[1,6]

The addition of common ion salts to a solvolysis reaction (e.g., sodium chloride to reaction of an alkyl chloride) sometimes results in a decrease in reaction rate, referred to as a common ion rate depression. This rate depression has been explained as due to the occurrence of return (designated external ion return) from a free carbocation; adding X^- increases the

$$R\!-\!X \underset{k_{-1}}{\overset{k_1}{\rightleftharpoons}} R^+ + X^- \underset{SOH}{\overset{k_2}{\longrightarrow}} ROS + HX$$

amount of return and decreases the rate of product formation. Application of the steady-state approximation to this reaction scheme gives

$$\frac{-d[RX]}{dt} = \frac{k_1 k_2 [RX]}{k_2 + k_{-1}[X]}$$

and it can be seen that addition of X^- to a reaction will give a rate retardation if $k_{-1}[X^-]$ is significant compared to k_2; that is, if return is important. If k_2 is much greater than $k_{-1}[X^-]$, the expression reduces to that for the original S_N1 mechanism.

Increasing $[X^-]$ also can produce a rate enhancement because of an increase in ionic strength of the medium (a "normal" salt effect) that acts in opposition to the common ion rate depression. This salt effect can most easily be seen by examination of the effects of noncommon-ion salts (Table 4.13). Depending on the system, either the common ion rate depression or the normal salt effect may dominate. Detection of the common ion rate depression is made possible even when the normal salt effect is dominant, by holding constant the total salt concentration and contrasting the effects of common- and noncommon-ion salts.

It has been observed that substrates that give a common-ion rate depression also undergo racemization and anion exchange:

Optically active　　　　　　　　Both racemic

Table 4.13 *Effects of Added Salts on Solvolysis of Benzhydryl Halides in 80% Aqueous Acetone*[39]

	k_{rel}	
Added Salt (0.1 M)	$(C_6H_5)_2CHCl$	$(C_6H_5)_2CHBr$
None	1.0	1.0
LiBr	1.16	0.87
LiCl	0.87	1.27

This result is also consistent with formation of a planar, free carbocation that reacts with anion to give return or reacts with solvent to give product.

The common-ion rate depression required only a slight modification of the S_N1–S_N2 scheme; however, this modification proved to be only a forerunner of other changes that were not so slight.

ALLYLIC RETURN[1,28,37]

Allyl cations are conjugated and subject to attack at two positions:

$$C_2H_5\overset{H}{\underset{H}{\text{C}}}\overset{}{\underset{+}{\text{C}}}\overset{CH_3}{\underset{H}{\text{C}}} \xrightarrow{\text{Nucl}} C_2H_5\text{—}\underset{\text{Nucl}}{\text{CH}}\text{—CH}=\text{CH—CH}_3 + C_2H_5\text{—CH}=\text{CH—}\underset{\text{Nucl}}{\text{CH}}\text{—CH}_3$$

If external ion return were to occur for an allylic cation, rearranged substrate should sometimes occur, and this has been observed. More significantly, some 3,3-dimethylallyl chloride results from hydrolysis of 1,1-dimethylallyl chloride, even though, in this case, the reaction was *not* subject to common-ion rate depression:

$$CH_3\text{—}\underset{Cl}{\overset{CH_3}{\text{C}}}\text{—CH}=\text{CH}_2 \xrightarrow[\text{H}_2\text{O}]{\text{90\% acetone}}$$

$$CH_3\text{—}\underset{OH}{\overset{CH_3}{\text{C}}}\text{—CH}=\text{CH}_2 + CH_3\text{—}\overset{CH_3}{\text{C}}=\text{CH—}\underset{OH}{\underset{|}{\text{CH}_2}} + \overset{CH_3}{\text{C}}=\text{CH—}\underset{Cl}{\text{CH}_2}$$

The rearranged substrate must have come from a route other than external ion return. This other route is return (designated internal return) from an ion pair intermediate:

$$CH_3\text{—}\underset{Cl}{\overset{CH_3}{\text{C}}}\text{—CH}=\text{CH}_2 \rightleftharpoons \underset{CH_3}{\overset{CH_3}{\text{C}}}\text{⋯}\overset{+}{\text{CH}}\text{⋯CH}_2 \longrightarrow \underset{CH_3}{\overset{CH_3}{\text{C}}}=\text{CH—}\underset{Cl}{\text{CH}_2}$$
$$Cl^-$$

The ion-pair concentration cannot be affected by addition of common ion, so the amount of internal return and the observed rate are independent of common ion concentration (except for minor ionic strength effects).

A similar study can be performed by examination of optically active allylic compounds:

$$\underset{CH}{\overset{R}{\diagdown}}\text{CH}\overset{CH}{\diagup}\underset{X}{\overset{H}{\text{C}}}\text{R} \longrightarrow \underset{H}{\overset{R}{\diagdown}}\overset{+}{\diagup}\underset{H}{\overset{R}{\diagdown}}$$

Ionization of symmetrically substituted, optically active allylic derivatives gives an intermediate with a plane of symmetry, and return must, therefore, give racemic substrate. The solvolysis in aqueous acetone of optically active *trans*-α,γ-dimethylallyl p-nitrobenzoate (OPNB) was shown to give racemic ester four times as fast as solvolysis product:

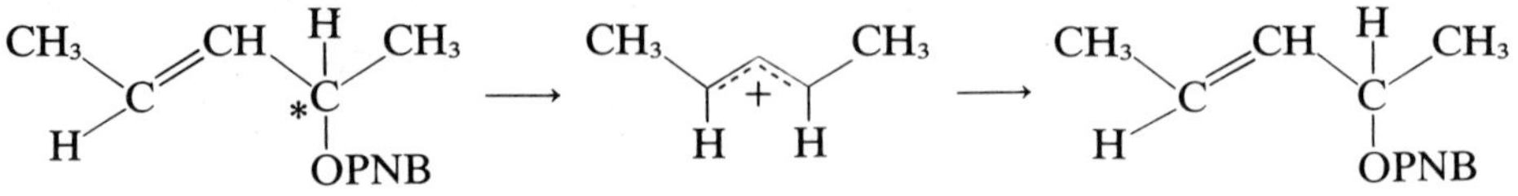

However, reaction of labeled ester (^{18}O) was observed to occur *without* exchange with unlabeled acid. Again, return is occurring that does not involve free carbocation; external ion return from a free carbocation should give anion exchange. The explanation, of course, is that return is from an ion pair that can give racemic ester by internal return but that cannot undergo anion exchange.

It is not necessary to isolate starting material to detect racemization of starting material. Reaction via an achiral intermediate must give racemic starting material or product. The rate of loss of optical activity, thus, is a measure of ionization rate, and if the rate of ionization equals the rate of product formation, then there can be no return to give racemic substrate. On the other hand, if the rate of loss of optical activity exceeds the rate of product formation, then there must also be return to give racemic ester, a process that would not be detected by monitoring the rate of product formation. Reference to these processes will frequently be made in the following pages, and the following notation will be used: rate of loss of optical activity k_α, rate of product formation k_t, and rate of formation of racemic starting material k_{rac}. Reference to the equation below shows that the sum of k_{rac} and k_t must be k_α (the steady-state condition for R$^+$):

$$R\text{—}X \underset{\text{racemization}}{\overset{\text{ionization}}{\rightleftharpoons}} R^+ \xrightarrow{\overset{\text{product}}{\text{formation}}} R\text{—Nucl}$$

$$k_\alpha = k_{rac} + k_t$$

SPECIAL SALT EFFECT[28,37,38]

Several other types of chiral alkyl derivatives ionize to give achiral intermediates. Threo-p-anisyl-2-butyl brosylate* is such a substrate. It undergoes ionization to give a "bridged" ion with a plane of symmetry (bridged ions are discussed in more detail in the section "Neighboring Group Participation"):

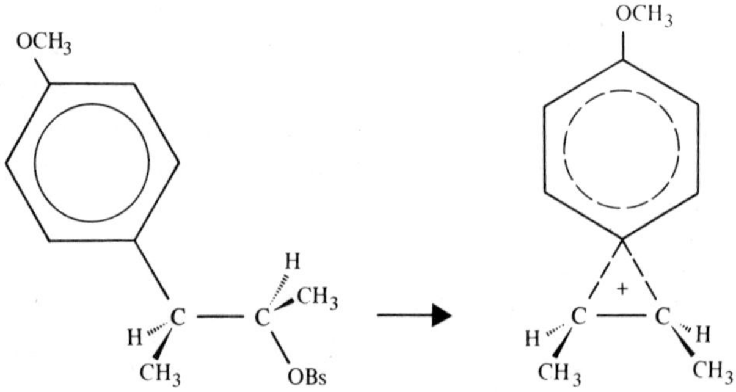

* Refer to a beginning organic chemistry text to review threo and erythro stereochemistry.

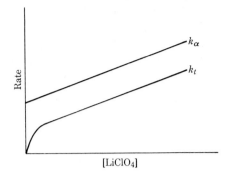

Fig. 4.4 *Effect of* [LiClO₄] *on rate of loss of optical activity* (k_α) *and rate of product formation* (k_t) *for acetolysis of threo-p-anisyl-2-butyl brosylate.*[38]

The acetolysis of this compound has been examined by Winstein and his co-workers.[37] As with the allylic compounds discussed above, internal return was demonstrated by observation of $k_\alpha > k_t$ and the absence of a common-ion rate depression. Addition of nonnucleophilic salts to the reaction medium did produce an increase in ionization rate (k_α) as can be seen in Figure 4.4. The linear relationship between salt concentration and rate is that expected from an increase in ionic strength of the ionization medium.

$$\frac{k}{k^\circ} = 1 + b[\text{salt}]$$

A normal salt effect on the rate of product formation (k_t) was also found except at low salt concentrations. Addition of small amounts of lithium perchlorate gave a rapid increase in rate as shown by the steep portion of Figure 4.4. The addition of small amounts of salt increases the rate of product formation to an extent which is greater than that expected from an ionic strength effect (the effect of ionic strength can be seen on k_α). This effect was labeled the special salt effect; to account for it, Winstein found it necessary to postulate the existence of two types of ion-pair intermediates, a tight ion pair (R⁺X⁻) and a solvent-separated ion pair (R⁺‖X⁻).[38]

It was suggested that perchlorate ion reacts with the solvent-separated ion pair *in competition* with return from solvent-separated ion pair to tight ion pair (k_{-2}). The perchlorate ion pair goes directly to solvolysis product (acetate) either because any alkyl perchlorate that is formed is unstable or because perchlorate anion is so weakly nucleophilic that attack by solvent dominates:

$$\text{R—OBs} \underset{k_{-1}}{\overset{k_1}{\rightleftarrows}} \text{R}^+\text{OBs} \underset{k_{-2}}{\overset{k_2}{\rightleftarrows}} \text{R}^+\|\text{OBs}^- \longrightarrow \text{ROAc} + \text{HOBs}$$

$$\downarrow \text{ClO}_4^-$$

$$\text{R}^+\|\text{ClO}_4^- \longrightarrow \text{R—ClO}_4$$

Thus, the initial addition of perchlorate competes with k_{-2}. When all k_{-2} is eliminated, the slope of the rate versus [ClO₄⁻] plot flattens out.

The following critical point provides evidence for the existence of two ion-pair intermediates: even when one return step (k_{-2}) has been eliminated, another return step continues to operate as shown by $k_\alpha > k_t$. This step, of course, is k_{-1}, return from the tight ion pair. Return from the tight ion pair is subject only to a normal salt effect (an ionic strength effect) as shown by the linear slope for k_α. The k_α and k_t slopes are parallel after the initial steep slope for k_t because any increase in ionization rate must also give a corresponding increase in rate of product formation.

PRODUCT DEPENDENCE ON LEAVING GROUP[28,40]

It would seem reasonable that solvent-separated ion pairs should exist. Between the stages of tight ion pair and free carbocation, there must be a stage where only one solvent molecule lies between anion and cation; whether this species is a stable intermediate is, of course, another matter. More direct evidence for the existence of a solvent-separated ion pair comes from the study of product dependencies on leaving group identity.

According to the original S_N1 mechanism, ionization gives a free cation that should give the same products even if the leaving group is changed:

$$R\!-\!X \xrightarrow[-X^-]{} R^+ \begin{array}{c} \xrightarrow{\text{Nucl}_A} R\!-\!\text{Nucl}_A \\ \xrightarrow{\text{Nucl}_B} R\!-\!\text{Nucl}_B \end{array}$$

However, a unimolecular rate-determining step may lead to an ion pair, and product formation from an ion pair will certainly depend on leaving group identity. A test for ion-pair involvement can be made by demonstrating a product dependence on leaving group identity for a reaction shown to be unimolecular; S_N2 reactions must be ruled out.

Application of this test to the aqueous ethanolysis of 1-adamantyl halides shows an increase in the 1-Adm-OEt to 1-Adm-OH ratio with a change in leaving group from chloride to bromide to iodide. Such a result is consistent with increasingly favored attack by the more nucleophilic ethanol on an increasingly stable (and therefore increasingly selective) ion pair as X^- is varied from chloride to bromide to iodide.

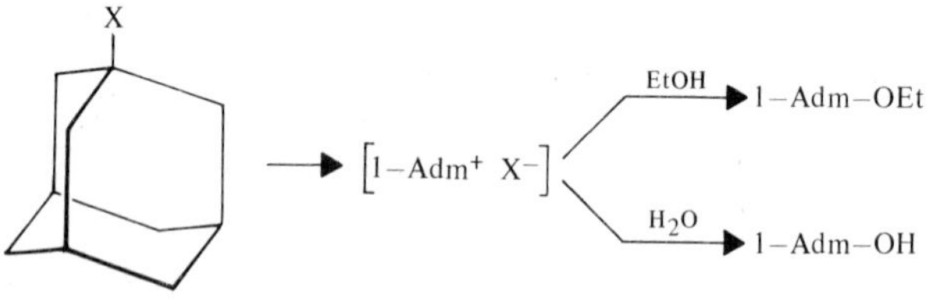

X = Cl, Br, and I

Performance of the same experiment for 2-adamantyl arenesulfonates gives an interesting result: increasing the stability of the anionic fragment gives a decrease in the 2-Adm-OEt to 2-Adm-OH ratio:

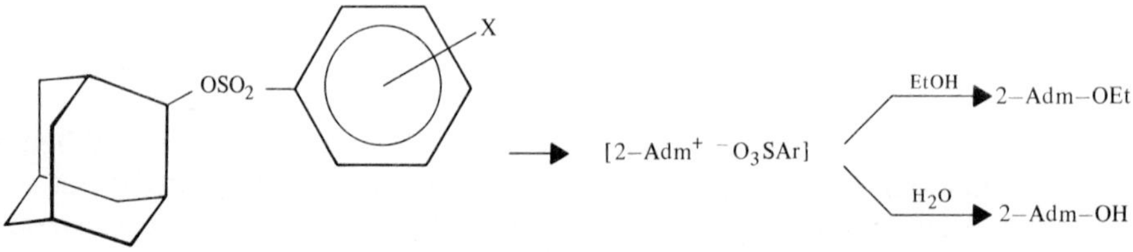

How can this result be explained when ethanol is known to be more nucleophilic than water? The answer may be that products are coming from the solvent-separated ion pairs **1** and **2**:

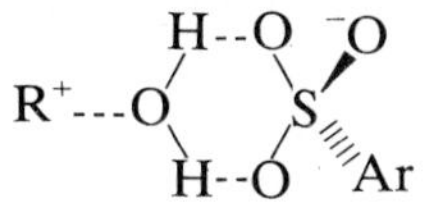

1	**2**

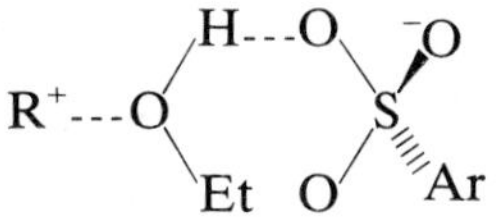

Ion pair **1** has two hydrogen bonds and should be more stable than ion pair **2**. Consequently, as anion stability increases, **1** should be increasingly favored over **2** since the more stable anion should be more selective. A solvent-separated ion pair can give products by reaction between the cation and the insulating solvent molecule. Thus, increasing anion stability increases the relative concentration of **1** and also decreases the ROEt/ROH ratio because of reaction between cation and insulating solvent. This interpretation is supported by the observation of excess retention of configuration for 2-adamantyl tosylate solvolysis.

OXYGEN EQUILIBRATION IN ESTERS[37,38]

Esters labeled in one position with ^{18}O have been shown to give ^{18}O-scrambled or ^{18}O-equilibrated esters as a result of return processes. The study of ester ^{18}O equilibration (extensively utilized by H. L. Goering of the University of Wisconsin) is one of the more powerful methods for elucidating ion pairing phenomena and has also been used to demonstrate the involvement of two ion pairs in a reaction. Two examples of such equilibration processes follow:

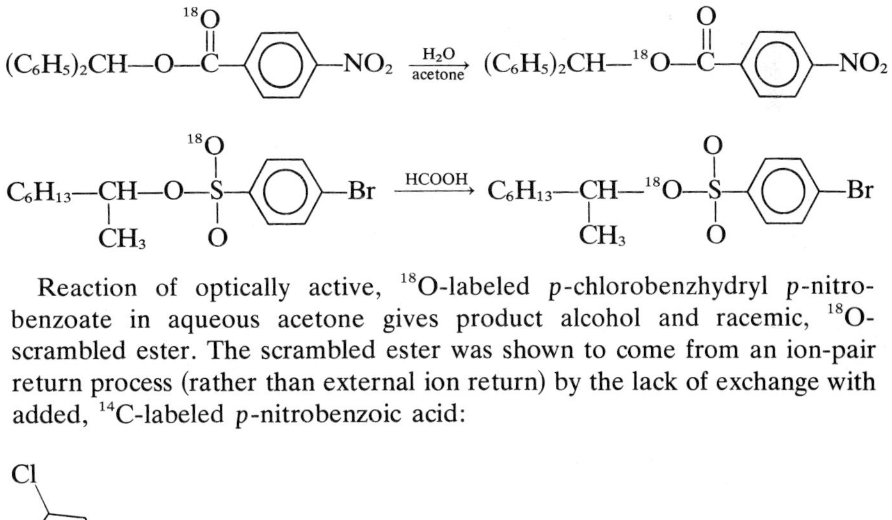

Reaction of optically active, ^{18}O-labeled p-chlorobenzhydryl p-nitrobenzoate in aqueous acetone gives product alcohol and racemic, ^{18}O-scrambled ester. The scrambled ester was shown to come from an ion-pair return process (rather than external ion return) by the lack of exchange with added, ^{14}C-labeled p-nitrobenzoic acid:

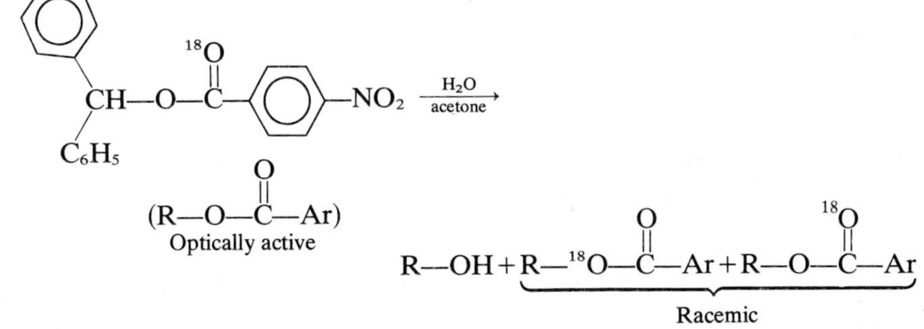

Azide ion is a strong nucleophile, and it has been demonstrated that the presence of a strong nucleophile favors product formation over return. Addition of sodium azide to the solvolysis of *p*-chlorobenzhydryl *p*-nitrobenzoate gave a very interesting result: ion-pair return was decreased as evidenced by the elimination of formation of racemic ester; however, the rate of formation of ^{18}O-scrambled ester remained essentially unchanged. This result can be explained if two ion pairs are again assumed to be involved and if it is also assumed that N_3^- can trap only the solvent-separated ion pair:

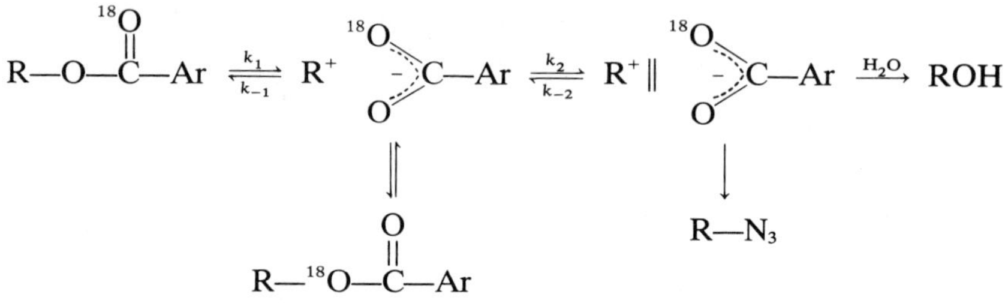

If this is the case, ^{18}O equilibration can be seen to result from return from tight ion pair, and racemization from return from solvent-separated ion pair. Only the latter process is eliminated by azide attack.

Rearrangement and Neighboring Group Participation

Rearrangements during nucleophilic aliphatic substitution reactions are common. These can occur by two mechanisms, a sequential ionization-rearrangement mechanism:

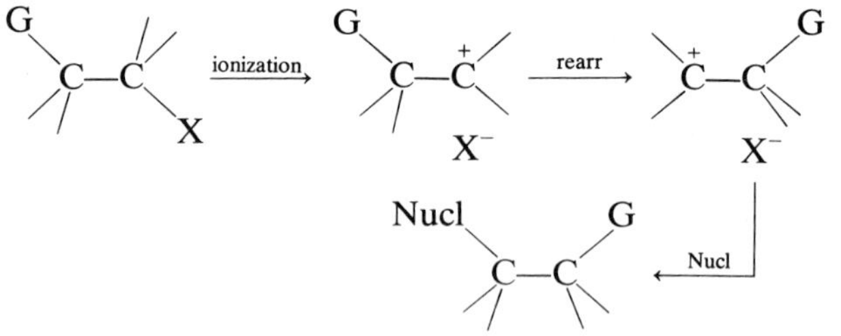

or a concerted ionization-rearrangement mechanism:

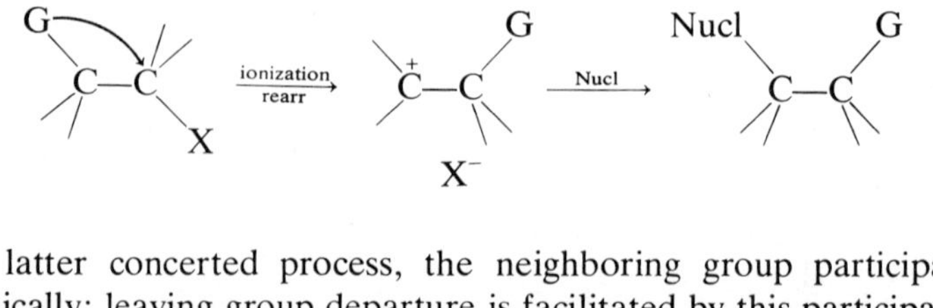

In the latter concerted process, the neighboring group participates nucleophilically; leaving group departure is facilitated by this participation and the reaction is said to be anchimerically assisted or neighboring group assisted. Additionally, neighboring group participation may occur to form an *intermediate* "bridged" ion in which the neighboring group is bonded to both the carbon of the migration origin and the migration terminus:

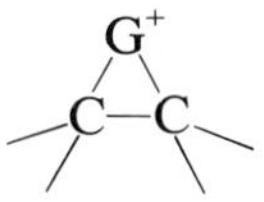

Bridged ion formation is well established for neighboring groups with lone-pair and π electrons:[41,42]

These ions are frequently drawn with dotted lines to the neighboring group to represent the potential contributions of all possible resonance forms:

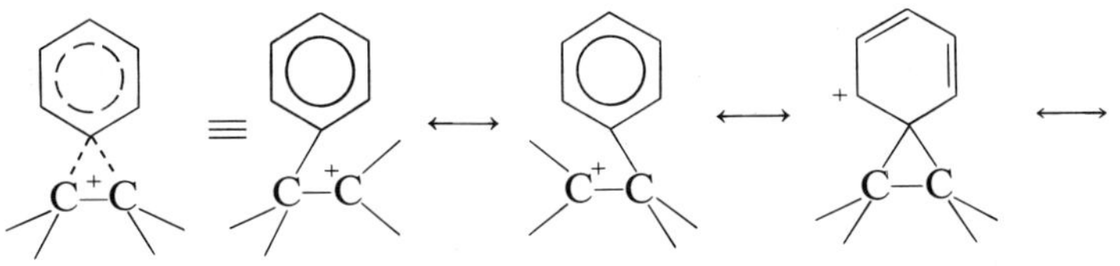

Bridged ions have also been postulated as being formed for neighboring groups without lone-pair or π electrons:

The last intermediates are referred to as nonclassical ions because there are insufficient electrons (two electrons, three centers) to form the bonds to the neighboring group; the dotted lines are necessary in such cases to denote bonds other than two-center, two-electron bonds. The possible existence of nonclassical carbocations and the importance of neighboring group assistance have been the subject of much controversy. We will discuss the matter further in the following sections.

PRIMARY AND TERTIARY SYSTEMS

The mechanism of rearrangements in primary and tertiary systems is fairly well understood: for primary systems ionization and rearrangement are concerted; for tertiary systems ionization and rearrangement are sequential.[28] Primary derivatives form relatively unstable carbocations and are sterically open to nucleophilic attack and therefore readily undergo S_N2 displacement by solvent nucleophiles. Thus, if rearrangement is to occur during nucleophilic substitution on a primary derivative, it must do so with neighboring group assistance (denoted k_Δ) in competition with a facile nucleophilic displacement process (denoted k_s); carbocation intermediates are not formed:

$$CH_3-\underset{\underset{CH_3}{|}}{\overset{\overset{H}{|}}{C}}-CH_2-OH$$

$$CH_3-\underset{\underset{CH_3}{|}}{C}-CH_2-OTs \quad \xrightarrow{k_s}$$

$$\xrightarrow{k_\Delta} \quad \underset{\underset{CH_3}{|}}{\overset{\overset{CH_3}{|}}{\overset{+}{C}}}-CH_3 \quad \xrightarrow{H_2O} \quad (CH_3)_3C-OH$$

$$OTs^-$$

When rearrangement yields a stable carbocation, it is this stability gain (relative to simple ionization) that provides the "driving force" for neighboring group participation. There are also instances of rearrangement in primary systems in which rearrangement does not yield a stable carbocation but is degenerate; that is, the same product is obtained from direct displacement by external nucleophile or from rearrangement:[41]

$$CH_2-CH_2-OTs \quad \xrightarrow[H_2O]{k_s} \quad CH_2-CH_2 \quad \underset{OH}{|}$$

$$\xrightarrow{k_\Delta} \quad CH_2-CH_2 \quad \underset{HO}{|}$$

As has been elegantly demonstrated for rearrangement during acetolysis of 2-phenylethyl tosylate, the driving force for such rearrangements is probably formation of a relatively stable bridged ion (see p. 163).[41] Note that according to the k_s versus k_Δ mechanism, the k_s process should give inversion of diastereomeric configuration (*threo* to *erythro*), and the k_Δ process should give retention of diastereomeric configuration (*threo*) and racemization (the products are enantiomers). This has been observed. Also, the amount of racemization was found to be twice the amount of rearranged product formed; the bridged ion (a phenonium ion) gives 50% rearranged and 50% unrearranged product.

Tertiary derivatives undergo nucleophilic substitution by an S_N1 mechanism because tertiary carbocations are relatively stable and because of steric hindrance to nucleophilic attack.[28] These factors also make neighboring group assistance a relatively unfavorable process in tertiary systems, and

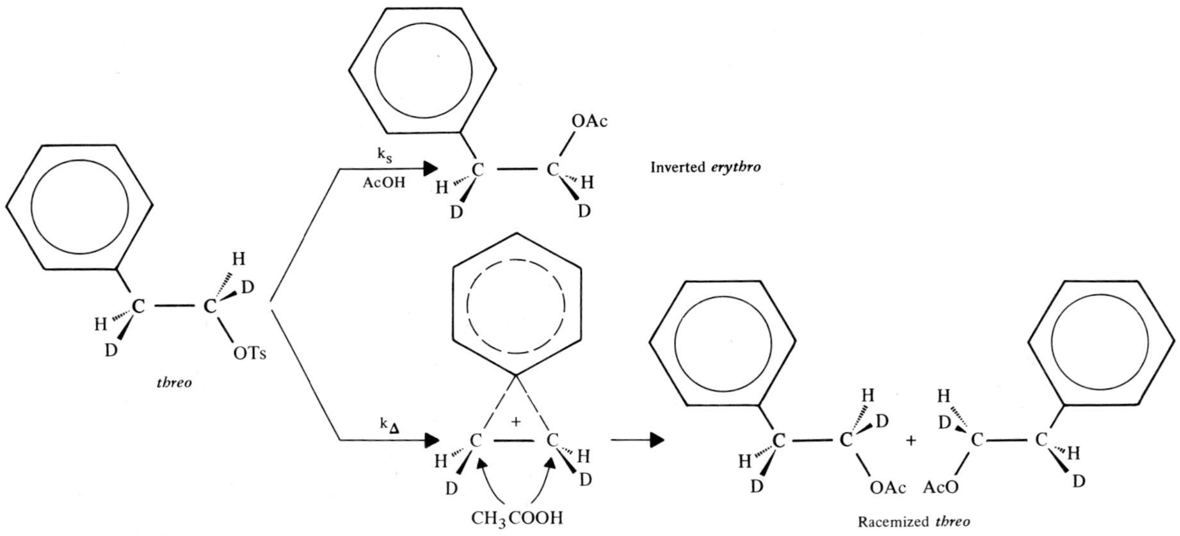
threo
OTs
k_S
AcOH
OAc
Inverted erythro
k_Δ
CH_3COOH
OAc
AcO
Racemized threo

rearrangement in these systems commonly occurs subsequent to rate-determining ionization:[44,45]

However, strong neighboring groups (i.e., π or n, not σ) that are properly oriented at the backside of the C–X bond apparently do act in displacing the leaving group. For example, the 6-methoxybenzonorbornenyl compound below undergoes hydrolysis 22 times faster than the 7-methoxy compound.[46] Such a rate order is consistent with π-electron displacement of the leaving group; a 6-methoxy group, but not a 7-methoxy group, can aid such electron movement as shown by resonance forms of the carbocation intermediate from the 6-methoxy derivative. If ionization were unassisted, the two methoxy compounds would be expected to have virtually identical reaction rates.

6–Methoxy   7–Methoxy   A carbocation resonance form

There is little evidence for the formation of tertiary nonclassical ions. Degenerate rearrangements of tertiary alkyl cations probably occur simply as a result of thermal activation.[28]

SECONDARY SYSTEMS

Secondary carbocations are, of course, intermediate in stability between primary and tertiary carbocations. Also, steric hindrance to nucleophile approach to the backside of a secondary derivative is moderate but not sufficient to prevent attack. Thus, although strong nucleophiles such as azide ion are known to substitute secondary derivatives by an S_N2 mechanism, it is not surprising that there has been much debate about the mechanism of substitution by weak nucleophiles such as solvent; both S_N1 and S_N2 mechanisms have been proposed.[28] It is precisely this point that is also critical to the question of neighboring group assistance in secondary systems: if nucleophilic solvent attack is a facile process, such that secondary carbocations (or ion pairs) are not permitted to form, then if neighboring group assistance is to occur it must do so in competition with this strong nucleophilic solvent participation. If solvolytic substitution of secondary systems takes place by an S_N2 mechanism, then the occurrence of rearrangement can only be taken as evidence for neighboring group assistance in displacement of the leaving group; if the neighboring group is a poor nucleophile, then it simply cannot compete with strong solvent nucleophile, and rearrangement will not occur. Alternatively, if solvolytic substitution of secondary derivatives is an S_N1 process, then rearrangement could be either a sequential or a concerted process.

It would seem then that determination of the rearrangement mechanism for secondary derivatives depends on determination of the mechanism of solvolytic substitution in secondary derivatives, and this is the case for those secondary derivatives that are open to backside nucleophilic attack. However, there are a large number of secondary derivatives that are sterically hindered toward backside nucleophilic approach; these include cyclic and acyclic molecules:

For these hindered molecules, it is possible that ionization or dissociation may take place in a slow step that is followed by rearrangement of neighboring group and nucleophilic attack, even if solvolytic substitution of simple secondary systems is an S_N2 process. In other words, rearrangement in a hindered system does not have to occur in competition with a strong solvent displacement process, and we can make no conclusions as to the relative strength of the neighboring group as a nucleophile. The neighboring group may be such a poor nucleophile that ionization occurs without neighboring group assistance and rearrangement follows; we just cannot tell in the absence of solvent displacement as a calibration point.

In the following sections, we consider examples of specific secondary systems, first nonhindered and then hindered, that illustrate the current extent of our understanding of rearrangements during nucleophilic substitution of secondary derivatives.

The 3-aryl-2-butyl system.[41]   The acetolysis of ^{14}C-labeled, optically active, *threo*-3-phenyl-2-butyl tosylate gives as the predominant substitution product a racemic mixture of unrearranged L-*threo* acetate and rearranged D-*threo* acetate:[4]

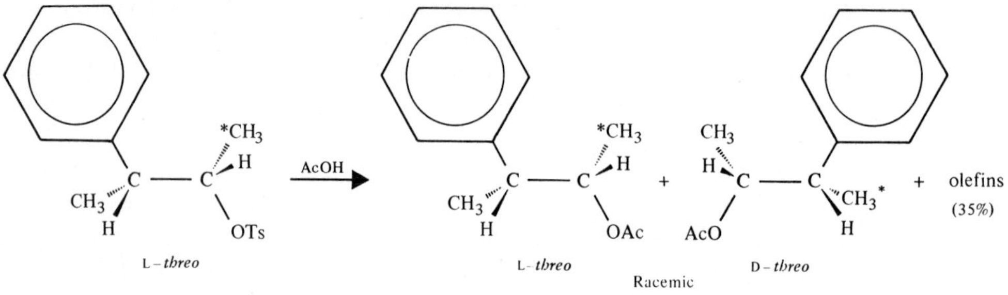

A small amount (6% of substitution) of *erythro* acetate is also formed. Acetolysis of the corresponding *erythro* tosylate gives a substitution product that is 98% optically active *erythro* acetate. This product was also rearranged, however, as shown by the equal distribution of ^{14}C label in C_1 and C_4:

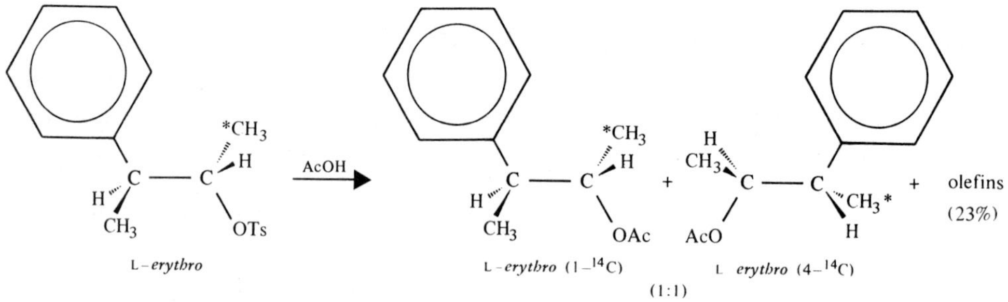

The reactions of these *threo* and *erythro* tosylates have two factors in common: (*a*) diastereomeric configuration is retained in the substitution product and (*b*) the substitution product is 50% rearranged. These results were originally explained in terms of neighboring group participation by phenyl to give a bridged phenonium ion:

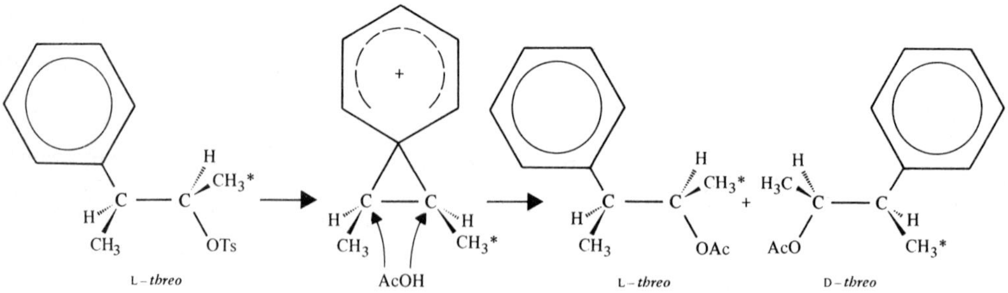

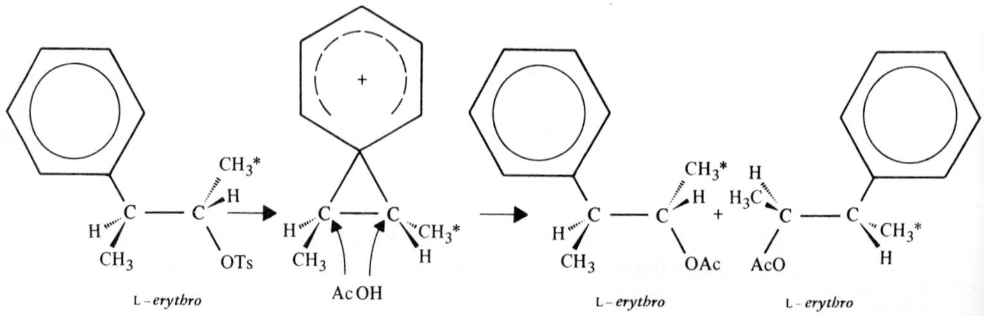

Nucleophilic attack on the phenonium ion is equally probable at either C_2 or C_3 (thus 50% rearrangement), and attack on the phenonium ion can only give retention of diastereomeric configuration. The *threo* tosylate gives a racemic product mixture because the phenonium ion it produces is achiral (has a plane of symmetry) and achiral products only can be formed. The *erythro* tosylate gives an optically active product because *the phenonium ion formed is chiral* and attack at either C_2 or C_3 gives the same enantiomer.

There is an apparent flaw in the phenonium-ion interpretation, however. It is one of the basic tenets of the neighboring-group-assistance theory that participation provides kinetic enhancement (anchimeric assistance) in departure of the leaving group. That is, reaction is faster than in the absence of neighboring group assistance. Yet, *threo*-3-phenyl-2-butyl tosylate undergoes acetolysis only half as rapidly as does 2-butyl tosylate! It is generally difficult to predict what a solvolysis rate should be. For example, steric and polar effects of a phenyl group might be sufficient to greatly decrease the solvolysis rate of 3-phenyl-2-butyl tosylate, in which case the observed rate of one half the rate of 2-butyl tosylate solvolysis would indicate neighboring phenyl assistance. Also ion-pair return is known to be important in this system and reduces the observed rate significantly. We will presently discuss methods for predicting the rate of 3-phenyl-2-butyl tosylate acetolysis in the absence of phenyl participation; for the moment, the fact that 3-phenyl-2-butyl tosylate reacts more slowly than 2-butyl tosylate must be considered damaging to the phenonium-ion interpretation.

An alternative explanation that takes account of the apparent lack of rate acceleration for 3-phenyl-2-butyl tosylate acetolysis has been proposed (primarily by Brown).[47] According to this alternative theory, acetolysis occurs unaided by neighboring phenyl to give a secondary carbocation that equilibrates with its enantiomeric secondary carbocation by phenyl migration—a sequential ionization-rearrangement mechanism:

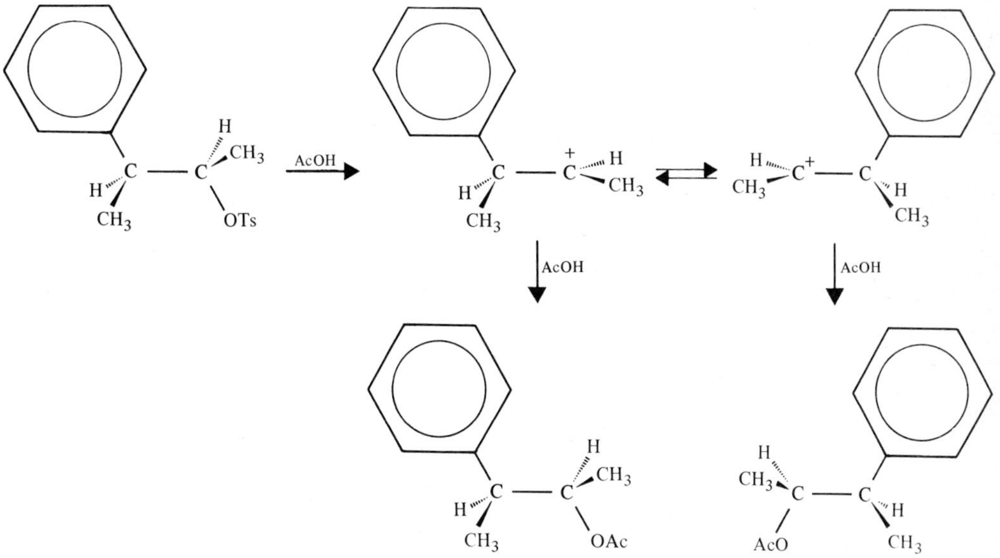

Solvent attack on the pair of equilibrating cations must give a product mixture that is 50% rearranged. Retention of configuration is said to result because of an electrostatic interaction between the electron-rich phenyl ring and the electron-poor carbocation center (a so-called donor-acceptor π

complex), which requires that the nucleophile attack only from the side opposite the phenyl group.

As noted before, the mechanism of rearrangement (concerted or sequential) can be determined for a nonhindered system such as the one under consideration if the mechanism of substitution by solvent on a simple (nonhindered, nonrearranging) secondary system is known. If solvent substitution is such a facile process that secondary carbocation or ion pairs are not formed, then the sequential, equilibrating-cations mechanism can be ruled out.

Several research groups have recently demonstrated that solvent substitution by solvents as nucleophilic as acetic acid on simple secondary systems occurs by an S_N2 mechanism;[28] that is, secondary carbocations or ion pairs are not formed in a unimolecular rate-determining step in these solvents (rapid, reversible ion pair formation cannot be eliminated). Thus, the mechanism of rearrangement in 3-phenyl-2-butyl tosylate acetolysis must be concerted neighboring group participation. The following is a presentation of one of the methods (developed primarily by P. v. R. Schleyer)[41] used to demonstrate the mechanism of solvent substitution in secondary systems.

If the acetolysis of a series of 3-aryl-2-butyl brosylates is considered, it is observed that the deactivated brosylates (those with electron-withdrawing groups) exhibit a reduced tendency to form rearranged, retained products:

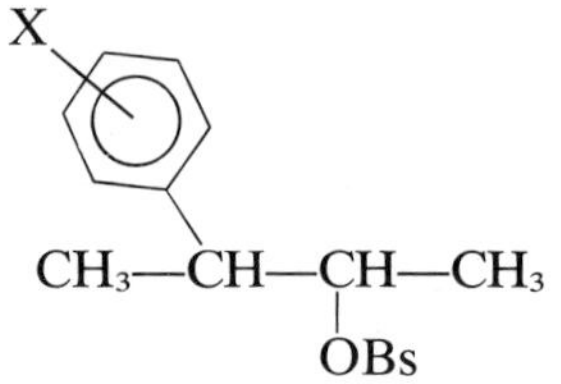

Rather, they form some olefin and inverted substitution products. The activated substrates, as discussed for 3-phenyl-2-butyl tosylate, form rearranged, retained substitution products and reduced amounts of olefins (Table 4.14). These results are consistent with a competition between

Table 4.14 *Product Data and Predicted Percentage of Fk_Δ Values for threo-3-Aryl-2-butyl Brosylate Acetolysis*

	Product(%)			Rate-derived
X	olefins	erythro	threo	% Fk_Δ
p-OCH_3	0.3	0	99.7	99
p-CH_3	12	0	88	87
m-CH_3	31	1	68	73
H	38	3	59	66
p-Cl	53	6	39	37
m-Cl	76	11	12	~0
m-CF_3	76	18	6	~0
p-CF_3	75	14	11	~0
p-NO_2	68	12	1	0
m, m'-$(CF_3)_2$	62	34	1	0

nucleophilic solvent attack, k_s, to give inverted acetate and olefin and neighboring phenyl attack to give retained acetate:

$$CH_3-CH-CH-CH_3 + \text{olefins}$$
Inverted

crossover

$$CH_3-CH-CH-CH_3 \longrightarrow$$
OBs^-

$$CH_3-CH-CH-CH_3$$
OAc
Retained

The formation of olefin during nucleophilic substitution is expected (see the following section on "Elimination Reactions"). Also note that the occurrence of ion-pair return from the phenonium ion is accounted for by the factor F, the fraction of ionization that proceeds to products.

The amount of a particular reaction proceeding by the Fk_Δ pathway can be deduced from the percent retained substitution product *if* there is no crossover between the two pathways (k_s and k_Δ). Crossover is not expected if the k_s pathway is an S_N2 pathway that gives no ion-pair or free carbocation intermediates susceptible to neighboring aryl attack. On the other hand, if solvent substitution takes place by an S_N1 mechanism, crossover would be expected. (See p. 170).

A test for the occurrence of crossover can be achieved by dissecting $\%k_s$ and $\%Fk_\Delta$ on the basis of kinetics and comparing with product-derived values. A kinetic method would depend only on the slow ionization steps k_s and Fk_Δ, and these would be kinetically independent of crossover even if it did occur. Agreement between product-derived (from percent retention) $\%Fk_\Delta$ and rate-derived $\%Fk_\Delta$ would then implicate the S_N2 mechanism of solvent substitution.

A kinetic method for dissecting the rate of 3-aryl-2-butyl brosylate acetolysis into k_s and k_Δ components can be achieved by use of a Hammett plot.[41] The plot of $\log k$ against σ for these brosylates shows that the activated and deactivated substrates define lines of different slopes (Figure 4.5). The deactivated substrates react only by a k_s pathway and ρ_s for these substrates measures the effect of substituent changes on solvent substitution.

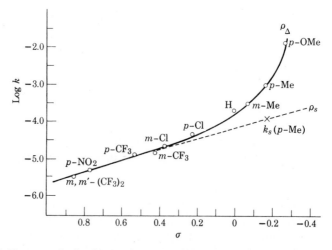

The slope for the activated substrates ρ_Δ measures substituent effects on phenonium-ion formation. As would be expected, ρ_Δ is steeper (more response to substitutent changes) than ρ_s. If ρ_s is extended (dotted line in Figure 4.5), a predicted rate for reaction by the k_s pathway can be obtained for the *activated* substrates. For example, k_s for p-CH$_3$ is indicated by an X in Figure 4.5. Since

$$k_t = k_s + Fk_\Delta$$

knowledge of k_t and k_s yields a rate-derived $\%Fk_\Delta$.

Comparison of rate-derived $\%Fk_\Delta$ and the percent retained substitution product (equivalent to product-derived $\%Fk_\Delta$) (Table 4.14) shows agreement between the two dissection measures. Thus, it must be concluded that (*a*) there is no crossover; (*b*) solvent displacement is an S$_N$2 process; and (*c*)

Fig. 4.5 *A Hammett plot for the acetolysis of 3-aryl-2-butyl brosylates.*[41]

rearrangement occurs by concerted neighboring aryl participation in competition with solvent displacement.

This competitive mechanism also rationalizes the original difficulty with the phenonium-ion explanation of the rearranged, retained product: if neighboring aryl assistance is occurring why is there no rate acceleration? First, the aryl group does have a rate retarding polar effect and possibly a steric effect that makes comparison with 2-butyl solvolysis deceiving. Second, solvent substitution without neighboring group assistance is itself a facile process for secondary derivatives, and rate acceleration (detected from the Hammet plot), no matter how small, is significant and indicative of further assistance. To illustrate how easily neighboring group assistance to ionization could be missed, if 50% of a reaction proceeds by a k_Δ pathway in competition with a k_s pathway, only a twofold rate acceleration will be observed. In other words, the rate acceleration is there, but the nature of the solvent substitution process must be clearly understood if the acceleration is to be detected.

Other studies are also consistent with the $S_N 2$ mechanism of solvent substitution on unhindered secondary derivatives.[28] For example, 2-octyl and cyclohexyl tosylates undergo acetolysis with essentially complete inversion of stereochemistry (the unrearranged product). We can conclude from all this information that *any* rearrangement during nucleophilic substitution on an unhindered secondary system must take place in a concerted fashion in competition with nucleophilic solvent attack. This conclusion assumes that other causes of rearrangement such as elimination-addition and product instability are ruled out. The following rearrangements must, therefore, involve neighboring group assistance:

$$
\underset{\underset{\displaystyle OTs}{|}}{(CH_3)_2 \overset{\displaystyle \overset{H}{|}}{C}-CH-CH_3} \xrightarrow{CH_3COOH} \underset{\underset{}{}}{(CH_3)_2 -\overset{\displaystyle \overset{OAc}{|}}{C}-CH_2CH_3}
$$

$$
(CH_3)_2\ddot{N}-CH_2-\underset{\underset{\displaystyle Cl}{|}}{CH}-CH_3 \xrightarrow[H_2O]{OH^-} HO-CH_2\overset{\displaystyle \overset{\ddot{N}(CH_3)_2}{|}}{C}HCH_3
$$

Involvement of bridged ion intermediates is indicated by formation of equal amounts of rearranged and unrearranged products:[28]

$$
CH_3CH_2-\underset{\underset{\displaystyle OTs}{|}}{CD}-CD_3 \xrightarrow{CF_3COOH} CH_3-\overset{\displaystyle \overset{H}{\overset{\cdots}{|}}}{CH}\overset{+}{-}CD-CD_3
$$

$$
\longrightarrow CH_3CH_2\underset{\underset{\displaystyle OOCCF_3}{|}}{C}DCD_3 + CH_3\underset{\underset{\displaystyle OOCCF_3}{|}}{C}HCHDCD_3
$$

or by formation of stereochemically retained product:[42]

Of course, if rearrangement is not complete (i.e., solvent displacement competes with neighboring group displacement), the percent rearrangement must equal the percent retention; the 3-aryl-2-butyl system discussed previously is a case in which rearrangement and retention correspond.

Neighboring group participation does not necessarily give a bridged ion intermediate. For example, 4,4-dimethylcyclohexyl tosylate gives 13% rearranged substitution product, yet both rearranged and unrearranged substitution product are inverted.[48] (The rearranged product is defined as inverted since it is assumed that the hydrogen trans to the leaving group migrates.) Such a result is inconsistent with formation of a bridged ion since, in this case, the unrearranged product would be retained. Consistent with the foregoing discussion, it must be assumed that the rearranged product derives from neighboring group participation in competition with a k_s process.

The 2-Norbornyl System. As pointed out previously, the mechanism of rearrangement in hindered secondary systems is difficult to ascertain because it is difficult to detect carbocation or ion-pair formation that may precede rearrangement. This mechanism can be ruled out in nonhindered secondary systems because facile solvent attack prevents rate-limiting ionization. One fact from the study of nonhindered systems can be applied to the study of hindered systems: if a neighboring group is sufficiently nucleophilic to compete with solvent nucleophile for attack on a nonhindered system, this same neighboring group will displace the leaving group in a hindered system provided participation is stereoelectronically feasible. In other words, if a neighboring group can compete with nucleophilic solvent attack, it will continue to participate when solvent competition is removed by steric hindrance. For example, double bond participation is known to be important in acyclic systems:[49]

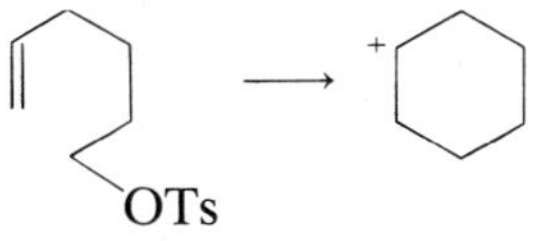

The rearrangement observed in hindered systems where the double bond is properly oriented must also derive from neighboring group participation:

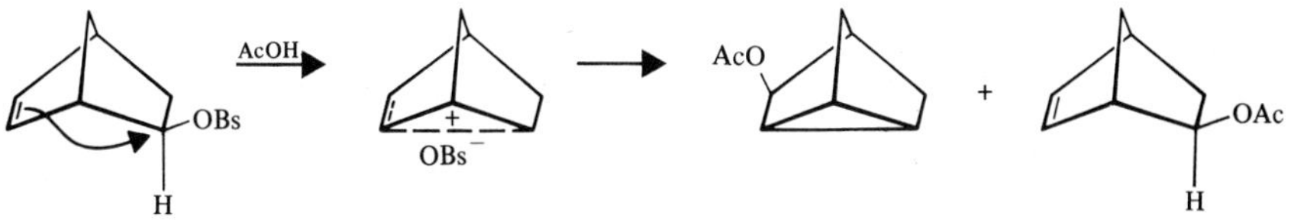

exo−5−norbornenyl−OBs

As evidence for this proposed mechanism for *exo*-5-norbornenyl brosylate, the *exo* derivative reacts approximately 8000 times more rapidly than the *endo* derivative. For *endo*-5-norbornenyl brosylate the double bond does not have the proper stereoelectronic orientation for participation (it cannot attack the backside) and unassisted ionization probably occurs:[50]

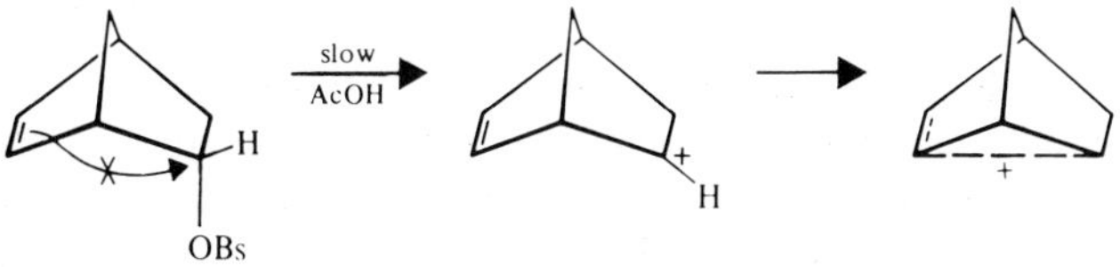

Solvent attack is also hindered in the norbornyl system.

Many of the rearrangements in hindered systems are carbon–carbon single bond rearrangements and neighboring group assistance by carbon–carbon single bonds in nonhindered systems has not been demonstrated:

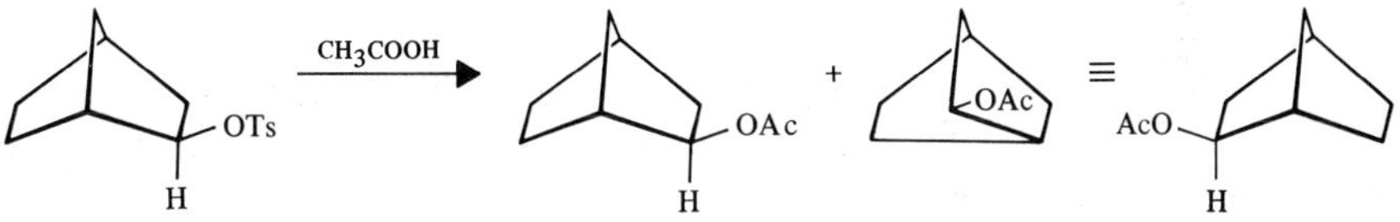

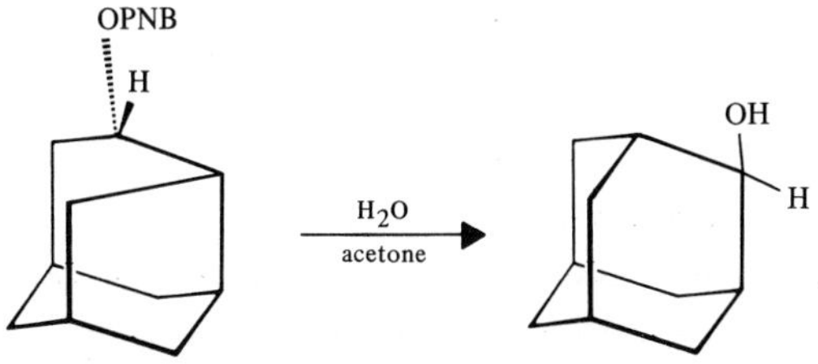

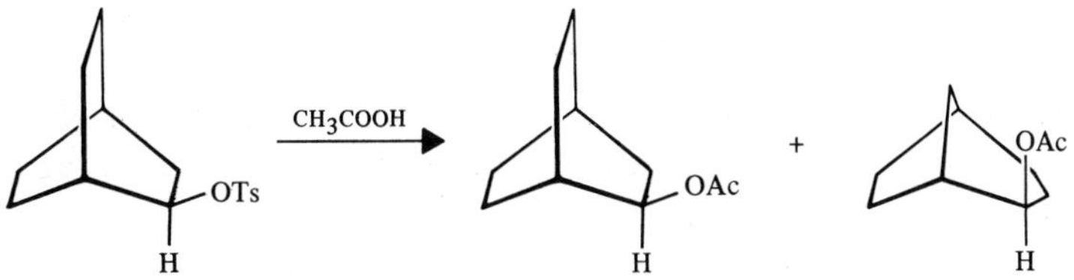

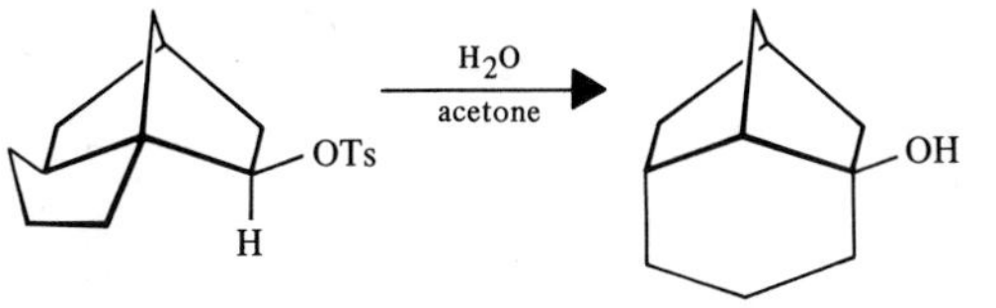

Consequently, the relationship between neighboring group assistance in nonhindered and hindered systems (if it occurs in nonhindered systems it must also occur in hindered systems if stereoelectronically feasible) has limited usefulness. Neighboring alkyl participation in nonhindered secondary systems is difficult to investigate because alkyl rearrangement simply does not occur in these systems; rather, hydrogen migrations and solvent attack dominate:[28]

$$CH_3-\underset{\underset{CH_3}{|}}{\overset{\overset{H}{\diagdown}}{C}}-\underset{\underset{OTs}{|}}{CH}-CH_3 \xrightarrow{CH_3COOH} CH_3-\overset{+}{\underset{\underset{CH_3}{|}}{C}}-CH_2-CH_3 + CH_3-\overset{+}{CH}-\underset{\underset{CH_3}{|}}{CH}-CH_3$$

$$100\% \qquad 0\%$$

Acyclic systems in which alkyl groups do migrate are usually also hindered toward nucleophilic attack by solvent. Pinacolyl tosylate provides an example:

$$CH_3-\underset{\underset{CH_3}{|}}{\overset{\overset{CH_3}{|}}{C}}-\underset{\underset{OTs}{|}}{CH}-CH_3 \xrightarrow{CH_3COOH} CH_3-\overset{+}{\underset{\underset{CH_3}{|}}{C}}-\underset{\overset{CH_3}{|}}{CH}-CH_3$$

To illustrate the techniques used to study rearrangements in hindered systems we will examine the solvolysis reactions of *exo-* and *endo*-2-norbornyl derivatives, two of the most intensively studied molecules in the history of organic chemistry:[51,52]

These compounds are prime examples of rearranging substrates for which nucleophilic solvent attack is hindered and neighboring group assistance difficult to prove or disprove.

Acetolysis of optically active *exo-* and *endo*-norbornyl brosylates gives the following results:[51-53]

1. $k_t^{exo}/k_t^{endo} = 350$;
2. Both *exo* and *endo* give *exo* products only;
3. The products from *exo* solvolysis are racemic;
4. The products from *endo* solvolysis show 8% excess inversion of stereochemistry (i.e., 46% retained, 54% inverted);
5. For *exo*, the rate of loss of optical activity k_α is greater than the solvolysis rate k_t; and for *endo*, $k_\alpha = k_t$.

Furthermore, acetolysis of *exo*-norbornyl brosylate labeled in the 2-position with ^{14}C gives products with essentially equal amounts of ^{14}C at the 1- and 2-positions:

The results for the *exo* derivative (retention of diastereomeric configuration, loss of enantiomeric configuration, 50% rearrangement, ion-pair return, and apparent rate acceleration) are those expected for neighboring group assistance and bridged ion formation; such a mechanism was proposed by Winstein:[53]

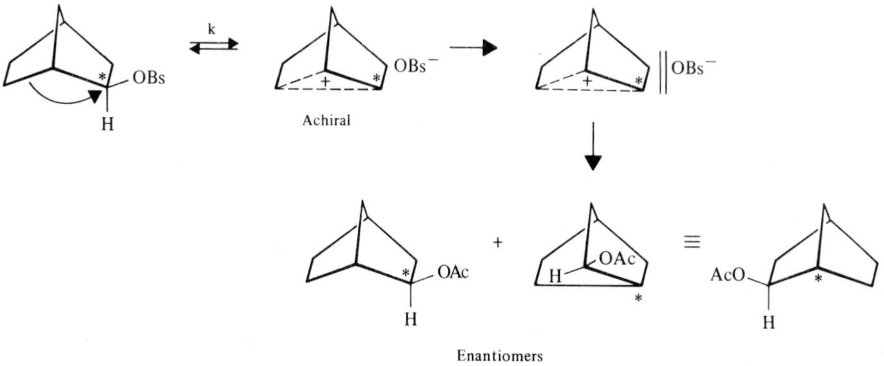

A special salt effect was not observed, so return from a solvent-separated ion pair must not be important. Dissociation at least to the solvent-separated ion pair was said to be necessary because protection of the *endo* side by the bridging electrons and protection of the *exo* side by the leaving group prevent nucleophilic attack by solvent on the tight ion pair.

The C_1 to C_6 bonding electrons cannot act as a neighboring group for displacement of the *endo* leaving group. Winstein proposed that the slow step for the reaction of the *endo* brosylate is formation of a nucleophilically solvated ion pair, followed by partitioning between (*a*) collapse with solvent to give unrearranged, inverted product and (*b*) rearrangement ("leakage") to the nonclassical bridged ion and formation of rearranged, racemic product:

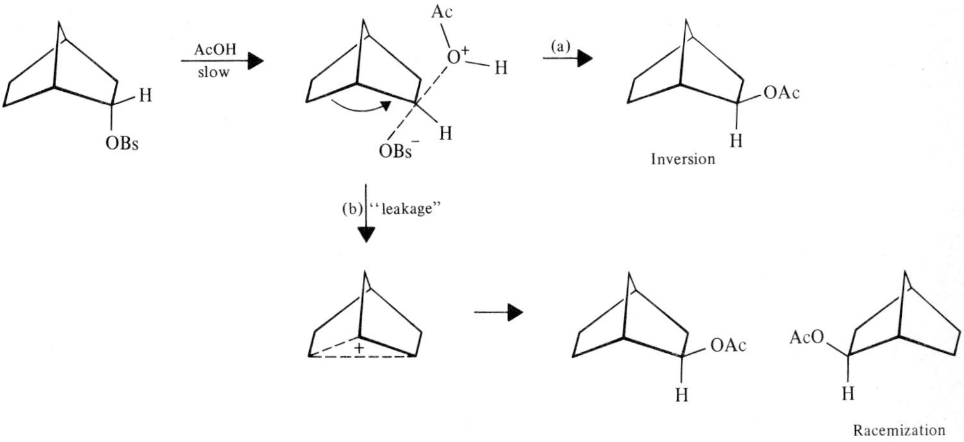

Anchimeric assistance from the neighboring carbon–carbon σ-bond of the *exo* derivative was said to be greater than nucleophilic solvent assistance for the *endo* derivative; that is, *exo* solvolysis is assisted and more rapid than normal for a secondary derivative, and *endo* solvolysis proceeds at a rate that is normal for secondary derivatives. Although we have not discussed nucleophilically solvated ion pairs such as the previous one for the *endo*

derivative, there is a significant amount of evidence that these ion pairs are formed in many S$_N$2 reactions.[28]

An alternative equilibrating-carbocation mechanism for norbornyl solvolysis has been presented. Brown[52] has suggested that the *exo* derivative undergoes unassisted ionization to give a pair of enantiomeric equilibrating cations:

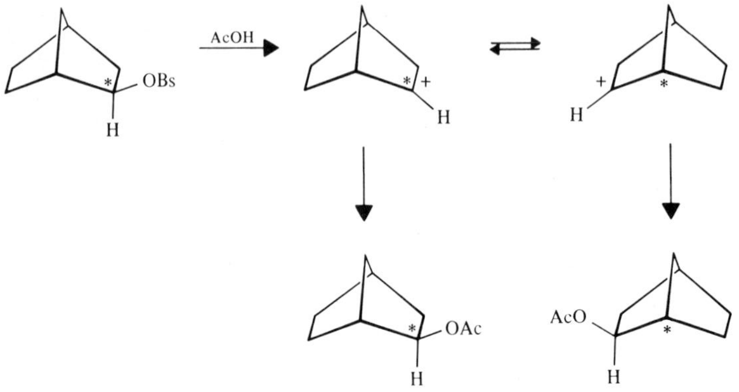

and that the *endo* derivative undergoes sterically retarded ionization, again to give a pair of equilibrating cations:

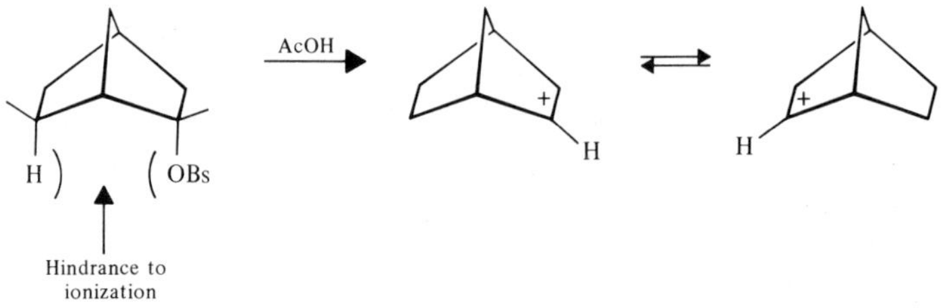

As evidence for this interpretation, nucleophilic attack from the *exo* side was shown to be much more favorable than attack from the *endo* side by examining two-step (nonconcerted) addition reactions to norbornene; addition always occurs from the *exo* side:

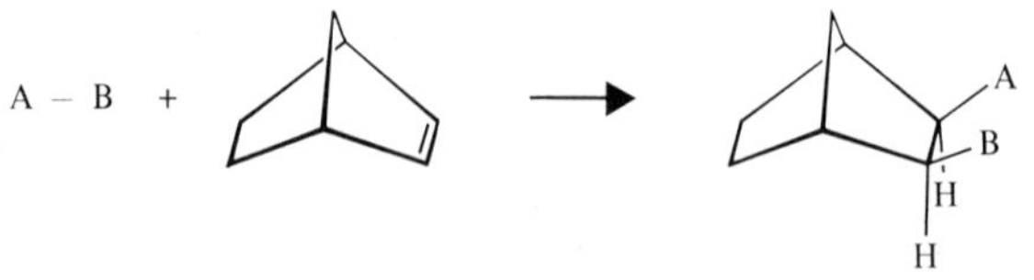

AB = HCl, CH$_3$COOH, SO$_2$Cl$_2$ − peroxides, BH$_3$, and so on

Thus, one of the key pieces of evidence for the bridged-ion theory (bridging protects the backside of the cation and forces *exo* solvent attack) was shown to be nonessential; attack from the *exo* side occurs even when bridging is absent (for the AB additions to norbornene, for example).

The problem of deciding between the two explanations resolves itself into answering two questions: (*a*) Is the *exo* rate fast and *endo* normal, or is *exo* normal and *endo* slow; and (*b*) is the intermediate ion bridged or is there a pair of equilibrating cations? Recall that the equilibrating-ion theory was ruled out for the 3-aryl-2-butyl system previously discussed because solvent attack was too facile to permit formation of such species, but this reasoning cannot be applied to a system such as *exo*-norbornyl in which backside nucleophilic attack cannot occur.

The second question regarding the structure of the intermediate has been answered by spectroscopic examination of the norbornyl cation in strong acid media.[54] The lifetime of carbocations in the usual solvolytic solvents (e.g., aqueous alcohol) is very short ($\ll 1$ sec) and direct spectroscopic examination is impossible. However, carbocations generated in strongly acid, non-nucleophilic mixed solvents (such as SbF_5–SO_2, SbF_5–SO_2ClF, HF–SbF_5, or FSO_3H–SbF_5) are stable and easily studied by spectroscopic means (a technique pioneered by G. A. Olah of Case Western Reserve University). Examination of the norbornyl cation at $-70°C$ by means of ^{13}C nmr spectroscopy shows three absorbances (proceeding downfield): a triplet of area three, a doublet of area one, and a quintet of area three. Since there are only three signals, there must be only three carbon types in the ion. This would be consistent with rapid equilibration of three bridged ions by means of 6,2 hydride shifts (the nmr technique cannot measure processes occurring more rapidly than about $10^3 sec^{-1}$ so an average signal for equilibrating species will be observed):

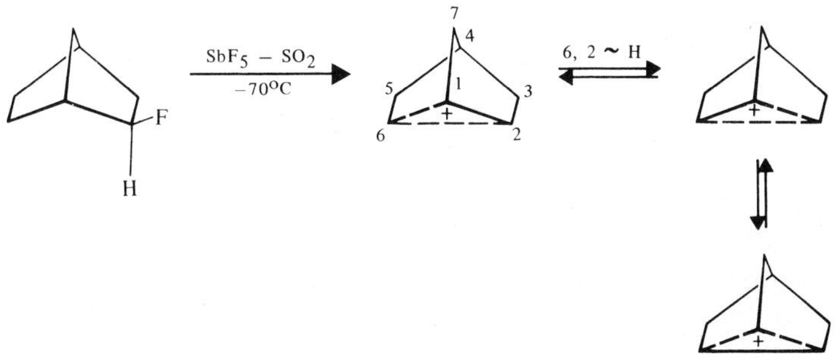

The combination of bridging and cation interconversion by hydride shifts results in nmr equivalency for C_1, C_2, and C_6 and C_3, C_5, and C_7. C_4 is not equivalent to any other carbon. Alternatively, the results could be due to rapid ($>10^3 sec^{-1}$) interconversion of three pairs of equilibrating cations again making C_1, C_2, and C_6 and C_3, C_5, and C_7 equivalent:

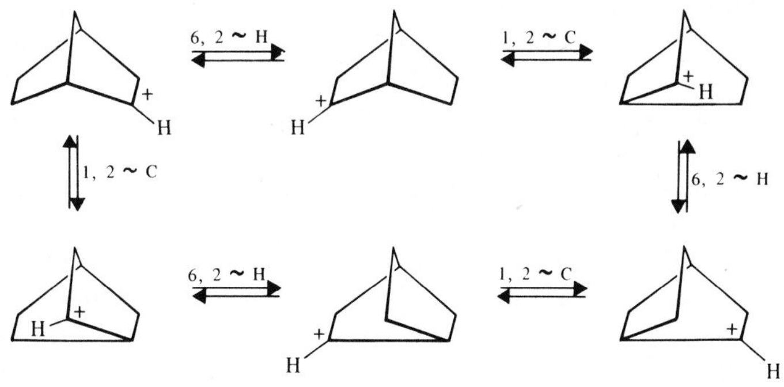

Both of these explanations are consistent with the observed splittings: C_1, C_2, and C_6 are coupled with four equivalent protons; C_3, C_5, and C_7 with two protons, and C_4 with one proton. Thus both the bridged-ion theory and the equilibrating-cation theory can be used to explain these results.

At $-150°C$ the 6,2 hydride shifts are frozen out, and the presence of five different carbons is indicated by nmr; the −70°C low field absorbance (C_1, C_2, and C_6) separates into two signals, one for C_2-C_1 and one for C_6; the absorbance for C_3, C_5, and C_7 also separates into two absorbances, one for C_3-C_7 and one for C_5. These results are consistent with existence of a single bridged cation or a pair of classical cations equilibrating by a 1,2 alkyl shift faster than $10^3 \, sec^{-1}$.

The time-scale limitation of nmr spectroscopy can be avoided by use of photoelectron spectroscopy (PES) or electron spectroscopy for chemical analysis (ESCA) as the technique is referred to in certain instances. In PES the energy required to remove valence shell electrons is determined by irradiating a sample with photons and measuring the kinetic energies of the ejected electrons. The term ESCA refers to PES with an X-ray ionization source that permits study of carbon 1s electron binding energies.

For the present consideration it is important to note that the time scale of the ionization process for PES is 10^{-16} sec. This time scale is fast enough to distinguish between a bridged or nonclassical cation and a rapidly equilibrating pair of classical cations. In the classical norbornyl cation, the distance between C_1 and C_6 should be much less than the distance between C_2 and C_6; the 1,2 bond shift gives the cation in which the distance between C_2 and C_6 is less than that between C_1 and C_6. In other words, there is a bond between C_1 and C_6 in one case and between C_2 and C_6 in the other case. By contrast, in the nonclassical ion, the distance between atoms 1 and 6 and between atoms 2 and 6 is the same. If the bond shift for equilibration of the classical cations occurs so rapidly that the nuclei do not have time to respond, the distances between C_2 and C_6 and between C_1 and C_6 will be identical and the species must be classified as a single nonclassical ion; that is, only electron movement is occurring. Bond vibrations occur on the order of 10^{-12} sec. If the shifting electrons in the norbornyl cation are moving back and forth more rapidly than the time required for a bond vibration, then it seems reasonable to assume that the carbon nuclei have insufficient time to respond to the presence of the electrons and that the species is nonclassical. Electron spectroscopy for chemical analysis provides a 10^{-16} second "picture" of the norbornyl cation, and this time scale is certainly fast enough to capture a classical cation before it equilibrates. If equilibration is occurring on the order of 10^{-16} sec, then the species must be nonclassical since only electron movement is occurring.

The ESCA technique is not capable of high resolution. For example, the spectrum of the *tert*-butyl cation shows two signals only, one of area one for the central positive carbon, and one of area three for the methyl carbons. Additionally, the area-one peak shows a binding energy 3.9 eV higher than the area three peak; electrons are obviously more difficult to remove from the positive carbon. The spectrum of the norbornyl cation at −150°C in strong acid clearly shows two peaks, one of area five and one of area two at 1.5 eV higher binding energy. This spectrum is consistent only with the nonclassical ion structure in which positive charge is simultaneously on C_2 and C_1. The separation of binding energies of only 1.5 eV (compared with

3.9 eV for *tert*-butyl) is consistent with the positive charge being delocalized onto two carbons. If the norbornyl cation were classical, the spectrum should have exhibited a signal of area one at approximately 4 eV higher binding energy than another peak of area six.

It might seem that the determination that the norbornyl cation is nonclassical has settled the question of the solvolysis mechanism of *exo*-norbornyl derivatives; however, it is at least conceptually possible for participation by the C_1 through the C_6 electrons to lag behind ionization:

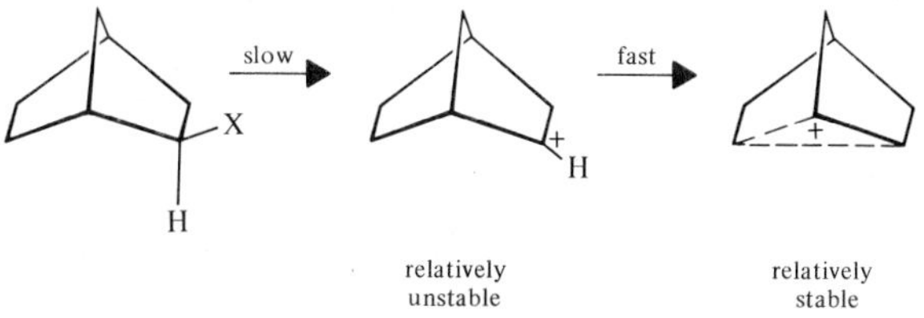

That this is not the case has been shown by demonstrating that reaction of *exo*-2-norbornyl tosylate is much faster than predicted for reaction without assistance (question one, above).[55] The polar and steric effects of secondary cyclic systems can be estimated by examination of solvolysis rates of the corresponding tertiary systems:

Using these estimates of polar and steric effects, the acetolysis rates of secondary tosylates reacting without nucleophilic solvent assistance or neighboring group assistance (e.g., 2-adamantyl and 7-norbornyl) are predicted accurately. *exo*-2-Norbornyl tosylate reacts approximately 10^3 times faster than expected on the nonassistance basis. Clearly, participation is concerted with ionization. Also, *endo*-2-norbornyl tosylate is shown to be assisted; in this instance, the assistance must derive from the nucleophilic solvent attack and the *endo* rate is normal for secondary solvolyses.

It should be noted, however, that the norbornyl problem is an extremely complicated one, and certain experimental results are inconsistent with the above conclusion.[52b] Also the ESCA interpretation has been questioned.[52b] A complete understanding of this controversial problem must await future experimentation.

4.3 β Elimination Reactions[56-58]

Nucleophilic substitution reactions are frequently accompanied by elimination of HX:

$$(CH_3)_3C\!-\!X \xrightarrow[\text{acetone}]{H_2O} \underset{\substack{\text{Substitution}\\\text{product}}}{(CH_3)_3C\!-\!OH} + \underset{\substack{\text{Elimination}\\\text{product}}}{(CH_3)_2C\!=\!CH_2} + HX$$

where X is halide, tosylate, and so on. For certain leaving groups (such as —NR$_3^+$) and for reactions with strong bases, elimination can become dominant:

$$(CH_3)_3C-\overset{+}{N}(CH_3)_3OH^- \longrightarrow (CH_3)_2C=CH_2+(CH_3)_3N+H_2O$$
(no alcohol formed)

These reactions are referred to as β elimination reactions because the hydrogen bond cleaved is β with respect to X. It is important to separate these reactions from α eliminations in which both H and X are eliminated from the same carbon to give a carbene (see Chapter 7):

C(H)(X) $\xrightarrow{\alpha\ elimination}$ $-\ddot{C}-$ + HX
A carbene

In this section, we will examine the mechanisms of elimination reactions and also identify those factors that affect the balance between elimination and substitution and allow for synthetically viable reactions.

If ion pairing is ignored, elimination reactions can be described by three mechanistic classifications, E1 (elimination, unimolecular), E2 (elimination, bimolecular), and E1cB (elimination, unimolecular, conjugate base):

$$R_2\overset{H}{\underset{X}{C}}-CR_2 \underset{slow}{\rightleftharpoons} R_2\overset{H}{C}-\overset{+}{C}R_2+X^- \longrightarrow R_2C=CR_2+HX \qquad (E1)$$

$$\text{Base} \cdots \overset{H}{\underset{X}{|}}\ R_2C-CR_2 \longrightarrow R_2C=CR_2+\text{Base } H^+X^- \qquad (E2)$$

$$R_2\overset{H}{\underset{X}{C}}-CR_2+\text{Base} \underset{k_{-1}}{\overset{k_1}{\rightleftharpoons}} R_2\overset{}{\underset{X}{C^-}}-CR_2+\text{Base } H^+ \xrightarrow{k_2} R_2C=CR_2+X^- \quad (E1cB)$$

Note that the E1 and E2 mechanisms are closely analogous to the S$_N$1 and S$_N$2 mechanisms presented in the previous section; both E1 and S$_N$1 are two-step mechanisms in which the first, slow step is carbocation formation, and both E2 and S$_N$2 are one-step mechanisms involving base–substrate or nucleophile–substrate interaction.

Furthermore, there are two forms of the E1cB mechanism, one that is reversible ($k_{-1} \gg k_2$) and designated (E1cB)$_R$, and a second that is irreversible ($k_2 \gg k_{-1}$) and designated (E1cB)$_I$. Actually, reaction by either mechanism must be second-order kinetically, first-order in base, and first-order in substrate. The numeral one in the E1cB notation refers to the unimolecular reaction of the conjugate base, which may or may not be rate determining.

It can generally be assumed that simple primary and secondary derivatives react by an E2 mechanism, while simple tertiary derivatives react by an E1 mechanism:[56-58]

$$CH_3-CH-CH-CH_3 \longrightarrow CH_3-CH{=}CH-CH_3+H_2O+OT\bar{s}$$

$$(CH_3)_3C-Cl \xrightarrow[\substack{-Cl^- \\ slow}]{H_2O} (CH_3)_3\overset{+}{C} \longrightarrow CH_2{=}C(CH_3)_2+H_3O^+$$

The reason for this division is that tertiary derivatives undergo unimolecular ionization easily whereas primary and secondary derivatives do not; of course, borderline areas exist and these present difficulty.

A summary of some key kinetic criteria for classifying the four elimination mechanisms is given in Table 4.15. Note that E1 and E2 mechanisms

Table 4.15 *Kinetic Classification of Elimination Reactions*[59]

Mechanism	Kinetic Order	$\beta - \dfrac{k_H}{k_D}$	Leaving Group Effect
E1	First	~1.1	Large
$(E1cB)_R$	Second	~1.0	Large
$(E1cB)_I$	Second	2–8	Small
E2	Second	2–8	Large

appear to be readily distinguishable on the basis of β-deuterium isotope effects (β-ds) and kinetic order. In practice, this classification can become difficult for certain cases. For solvolysis reactions, for example, the solvent is the base and kinetic order cannot be determined. Furthermore, β-ds can sometimes prove difficult to interpret. The small β-d for the E1 mechanism is a secondary isotope effect; that is, an isotope effect due to some factor other than cleavage of the isotope bond. The isotope effect in this case apparently derives from less effective hyperconjugative stabilization of the incipient carbocation by a C–D bond. The large β-d for the E2 mechanism is a primary isotope effect resulting from slower cleavage of the stronger (relative to C–H) C–D bond. There is some debate as to the dividing line between magnitudes characteristic of secondary and primary β-ds; an effect of less than 1.1 is accepted as secondary and an effect of greater than 2.0 is accepted as primary but intermediate values are of questionable interpretability (V. J. Shiner, Jr., of Indiana University, a major contributor to the study of deuterium isotope effects, estimates the maximum secondary β-d on carbocation formation to be 1.46 for a β-CD_3 group or about 1.13 for a single β-deuterium).[60] This uncertainty regarding isotope effects coupled with the difficulty of determining reaction order in solvolysis reactions can make it difficult to distinguish between operation of E1 and E2 mechanisms.

Reaction by an E1 mechanism can be detected by a product independence of leaving group identity. For example, *tert*-butyl chloride, bromide, and iodide undergo hydrolysis to give 93% *tert*-butanol and 7% isobutylene:[1]

$$t\text{-BuX} \xrightarrow[-X^-]{} t\text{-Bu}^+ \xrightarrow{\text{sub}} t\text{-Bu-OH}$$

$$\downarrow \text{elim}$$

$$\text{isobutylene}$$

This criterion is confused by potential involvement of ion pairs that could result in product dependence on leaving group identity even for a first-order reaction (see section on "Ion Pairing"). Examination of product stereochemistry can also aid in distinguishing between operation of E1 and E2 mechanisms; this topic is discussed in a later section.

Reaction by the E1cB mechanisms is favored by the presence of strong base, by carbanion stabilizing groups in the β position, by poor leaving groups, and by substrates that form carbocations only with difficulty. For example, the following are interpreted to be E1cB reactions:[59]

The (E1cB)$_R$ mechanism can be detected readily using the criteria in Table 4.15 and the fact that reversal of the first step leads to deuterium incorporation by the substrate. The (E1cB)$_I$ mechanism, however, can be distinguished from the E2 mechanism only by close examination, since the gross features of the two mechanisms are identical. As noted in Table 4.15, reaction by an (E1cB)$_I$ mechanism should, however, show only a small rate variation upon leaving group variation (the bond to the leaving group is broken after the rate determining step). This technique has been used[59] to prove operation of the (E1cB)$_I$ mechanism for the cyclohexyl compounds above; 4-bromoacetoxy reacted only three times as fast as **4**-acetoxy, yet for an E2 mechanism this leaving group change would be expected to give a rate difference of 30 to 60. It is also noteworthy that *syn* elimination (H and OAc *cis*) from **4** occurs more rapidly that *anti* elimination (H and OAc *trans*) from **3**. Such a *syn–anti* relationship is inconsistent with an E2 mechanism since elimination in the *anti* fashion is substantially more facile for the concerted E2 process that follows. The fact that **4** reacts faster than **3** should not, however, be interpreted as evidence that *syn* eliminations are generally preferred for reaction by an E1cB mechanism. In this case, the equatorial proton of **4** is probably less sterically hindered.

The Spectrum of E2 Transition States

Reaction by an E1 mechanism is favored by substrates that undergo ready C–X cleavage to give carbocations. At the other extreme, the E1cB mechanism is favored by ready C–H cleavage to form a carbanion. Between these extremes lie those conditions conducive to reaction by an E2 mechanism. Thus, transition states for elimination form a graded series, depending on base strength and substrate, from E1 to E2 to E1cB:

The varying amounts of bond cleavage in the E2 transition states should be noted.

Evidence for such a spectrum of transition states comes from examination of β-deuterium isotope effects. Study of 2-phenylethyl derivatives shows a decrease in β-d with a decrease in leaving group ability (Table 4.16).[61] This trend is consistent with C–H cleavage becoming progressively more complete before the transition state is reached, that is, with the transition state becoming more E1cB-like.

Another illustration of the spectrum of E2 transition states is provided by observed variations in Hammett ρ values for elimination (also shown in

Table 4.16 β-Deuterium Isotope Effects and Hammett ρ Values for Ethoxide Catalyzed Elimination Reactions of $C_6H_5CD_2CH_2X$[61]

X	$\dfrac{k_H}{k_D}$	ρ^a
Br	7.1	2.14
OTs	5.7	2.27
SMe_2^+	5.1	2.75
NMe_3^+	3.0	3.77

[a] For substituted phenyls.

Table 4.16). As the leaving groups become poorer, the ρ values become more positive; that is, reaction is increasingly enhanced by electron-withdrawing substituents. The β-carbon of the transition state is indicated to be increasingly anionic and E1cB-like with the decrease in leaving group ability.

Hofmann and Saytzeff Elimination

Elimination frequently occurs to give isomeric products. For example, elimination from 2-bromobutane can proceed to give 2-butene or 1-butene (geometrical isomers will be discussed later):

$$CH_3—CH_2—CHBr—CH_3 \xrightarrow[\text{EtO}^-]{\text{EtOH}} CH_3—CH\!=\!CH—CH_3 + CH_3CH_2CH\!=\!CH_2$$

$$81\% \qquad\qquad\qquad 19\%$$

It is generally observed that β-eliminations give the more highly substituted and more stable alkene (the Saytzeff rule). Exceptions to this general statement are encountered in two specific instances. First, eliminations to give a bridgehead double bond are rarely observed (Bredt's rule). This is due to the large amount of angle strain associated with such double bonds:

(not formed)
Bridgehead

Second, elimination from alkylammonium salts or alkysulfonium salts usually gives the least substituted or least stable alkene (the Hofmann rule):[58]

$$CH_3CH_2\underset{\underset{\overset{|}{CH_3}\quad CH_3}{\overset{|}{S^+}}}{CHCH_3} \xrightarrow[\text{EtO}^-]{\text{EtOH}} CH_3CH\!=\!CHCH_3 + CH_3CH_2CH\!=\!CH_2$$

$$26\% \qquad\qquad 74\%$$

$$(CH_3CH_2CH_2)_2N^+(CH_2CH_3)_2 \xrightarrow[\text{OH}^-]{H_2O} CH_3CH\!=\!CH_2 + H_2C\!=\!CH_2$$

$$4\% \qquad\qquad 96\%$$

As might be expected from the foregoing discussion of leaving group effects, the balance between Hofmann and Saytzeff control is in large part a function of variation in transition state structure.[56–58] Since C–H and C–X bond cleavages are roughly equally advanced in the central E2 transition state, this transition state will have much double bond character. Reaction by a central-E2 mechanism then will lead to products by a transition state that reflects product stability and, therefore, favors formation of the most highly substituted alkene. The E1cB-like transition state has much carbanion character rather than double-bond character, and reaction by this transition state is largely controlled by carbanion stability. Carbanion stability, unlike carbocation stability (see Chapter 5), decreases with a higher degree of substitution; the E1cB-like transition state for 1-butene formation is favored over that for 2-butene formation:

Less stable carbanion (2°)   More stable carbanion (1°)

Thus, we see that the spectrum of E2 transition states, as reflected in the balance between carbanion stability and product stability, is a major factor in controlling the balance between product isomers: as the transition state becomes more E1cB-like, carbanion stability becomes more important than alkene stability and a shift is observed in favored products from most-highly-substituted to least-highly-substituted alkene.

Steric factors play an observable but less important role in determining product composition. Changes in alkyl group, leaving group, or base that lead to a more crowded transition state favor formation of terminal alkenes because of lessened steric effects for formation of this product. For example, reaction of 2-alkyl halides with *tert*-butoxide as base rather than ethoxide leads to increased formation of 1-alkene over 2-alkene.[56-58] Similarly, use of larger leaving groups frequently results in increased formation of terminal alkene. That steric factors are not dominant can be seen by the observation of increasing 2-hexene:1-hexene ratios in the following order: 2-hexyl-F < 2-hexyl-Cl < 2-hexyl-Br < 2-hexyl-I.[62] Despite an increase in leaving group size, an increasing amount of 2-alkene is formed; a decrease in E1cB character with an increase in leaving group ability is the controlling factor.

Stereochemistry

The E2 eliminations generally occur in an *anti* fashion:

anti—elimination   *syn*—elimination

The *anti* transition state permits electrons from the broken C–H bond to flow to the backside of the C–X bond and displace the leaving group. This electronic displacement from the backside is consistent with the ideas developed for our earlier treatment of nucleophilic displacement reactions and neighboring group participation. Observed examples of *anti*-elimination follow:[56-58]

Although the *anti*-eliminations are more commonly observed, *syn*-eliminations do occur:

It has been suggested that the *syn*-eliminations are the result of reaction by an E1cB mechanism.[56] However, it has been demonstrated that the *syn*-elimination from **9** and the *anti*-elimination from **8** are both concerted E2 processes.[63] This conclusion is based on the observation of similar ρ values for both reactions (phenyl substituents) and also on the observation of a large β-deuterium isotope effect (5.6) for the *syn*-elimination. Furthermore, not only can concerted *syn*-eliminations occur, but they can sometimes occur more rapidly than *anti*-eliminations. For example, the rate of *syn*-elimination from **10** is greater than the rate of *anti*-elimination from **11**. Thus, it appears that a dihedral angle of 0° between β C–H and C–X leads to more facile E2 elimination than does a dihedral angle such as that in **11**, which is about 120°. Studies of acyclic systems, where any dihedral angle is potentially available, show that a dihedral angle of 180° (so-called antiperiplanar) does lead to more facile elimination than any other angle; only *anti*-elimination is observed in these systems under E2 conditions. But even in the acyclic systems, the situation is confused because of the requirement that *syn*-eliminations must occur from a conformation in which there are two (excluding H and X) eclipsing interactions:

anti   syn

Steric factors rather than electronic factors may be the dominant force.

From these observations, we can conclude that the earlier analogy between E2 transition states and S_N2 transition states (a preference for electron flow to the backside of the C–X bond) is a poor one. Where the involved groups are properly oriented, concerted *syn*-eliminations occur readily. In this connection, it is also worthy of note that a particular substrate need not react by a concerted E2 mechanism simply because the groups to be eliminated are antiperiplanar. The most obvious example of a stepwise process being more facile than a concerted one is the E1 elimination of tertiary derivatives; certainly antiperiplanarity of hydrogen and chloride is readily available for *tert*-butyl chloride. More importantly, Bordwell[59] has shown that base-catalyzed, second-order eliminations may prefer a stepwise $(E1cB)_I$ mechanism even when eliminated groups are antiperiplanar. For example, elimination from **3** (p.182) is stepwise even though there is no obvious barrier to flow of electrons to displace the leaving group. It has frequently been proposed that there is something fundamental regarding the concerted nature of base catalyzed elimination; obviously, this point should be subjected to detailed investigation.

As was the case with the balance between Hofmann and Saytzeff elimination, the nature of the E2 transition state (E1-like, central, or E1cB-like) can also play a role in determining the stereochemistry of elimination. The central transition state, unlike the other two possibilities, has a large degree of double-bond character and reaction by such a transition state would be expected to reflect alkene stabilities. As an example, elimination from

2-hexyl halides can give *cis* or *trans* alkene:

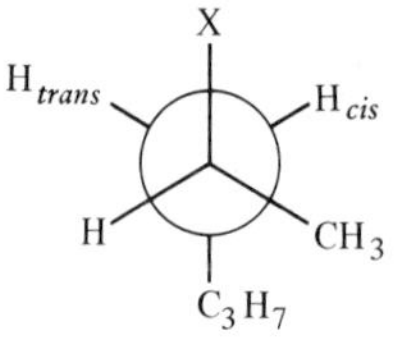

syn−elimination to	anti−elimination to	anti−elimination to
cis and *trans* alkene	*cis* alkene	*trans* alkene

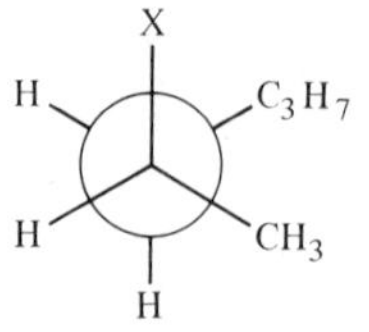

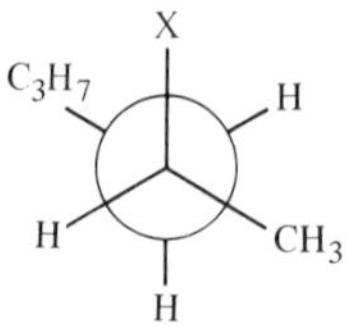

The *cis*-alkene is less stable than the *trans*-alkene because of steric effects. The *cis* : *trans* ratio for 2-hexyl halides is observed to increase (*trans* always major product) in the following order: $I < Br < Cl < F$.[62] This order can be rationalized on the basis that the poorer leaving groups lead to more E1cB-like transition states that more weakly reflect product alkene stability and more strongly reflect reactant stability; apparently the energy difference between transition states leading to *cis* and *trans* alkenes is lessened when the transition state has an increasing amount of reactant character.

As discussed above, eliminations from tetraalkylammonium salts occur via transition states with much E1cB character. Also for such substrates the conformation drawn above which leads to alkene by *syn*-elimination is the most stable because the large X group is bracketed by two hydrogens in this conformation. Since an E1cB-like transition state has much reactant character as opposed to product-alkene character, it is quite possible that *syn*-elimination from this most stable conformation is the most favorable reaction pathway for this system. That this may be the case can be deduced from the increasing amount of *cis* alkene formed from tetraalkylammonium salts with increasingly small bases:[56]

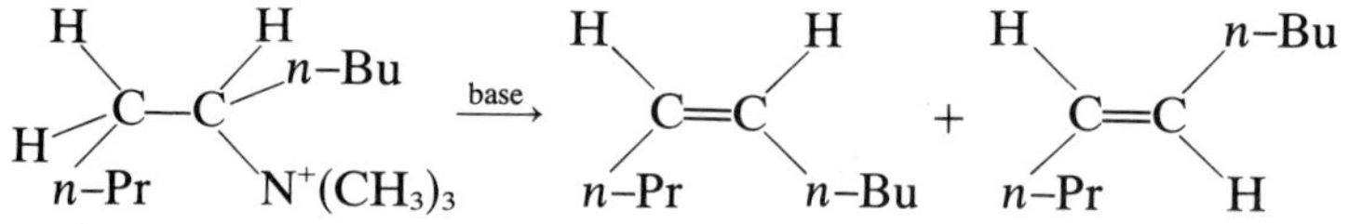

26%—*t*-BuO⁻ base—74%
74%—OEt⁻ base—26%
81%—OCH₃⁻ base—19%

A small base should be better able to remove the H_{cis} proton relative to the H_{trans} proton, but no such difference is evident for the *anti*-elimination conformations. This explanation does not, however, explain the formation of more *cis* than *trans* product with ethoxide and methoxide as base; the pathway to *cis* alkene would still appear to be less favorable than the pathway to *trans* alkene.

syn-Eliminations for synthetic purposes are probably best achieved by thermal decomposition of amine oxides (the Cope elimination):[56-58]

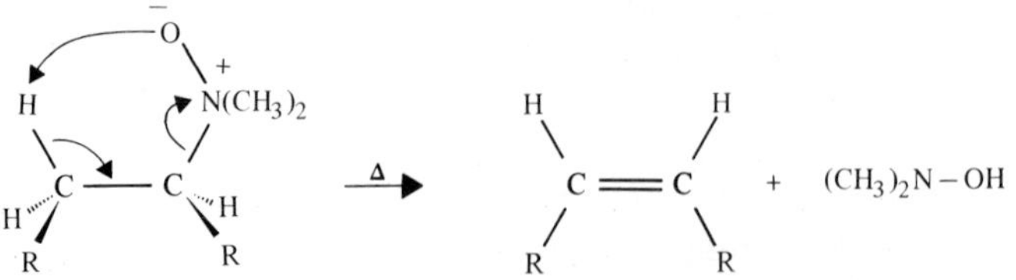

or carboxylate esters:

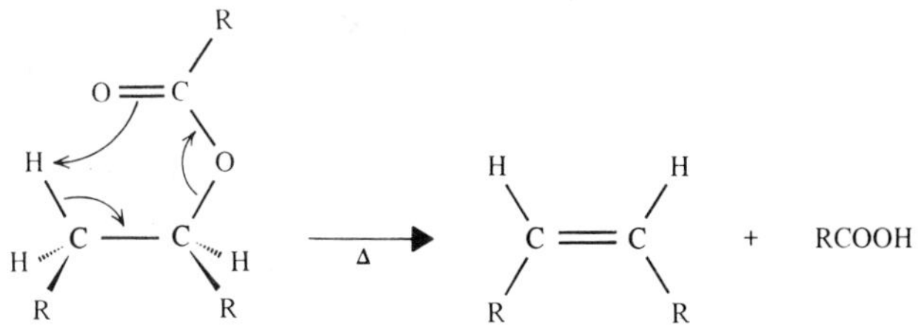

4.4 Fragmentation[64]

Certain γ-substituted alkyl derivatives can undergo a third process upon reaction with solvent, in addition to elimination and substitution, in which a carbon–carbon bond is broken and three fragments are formed:

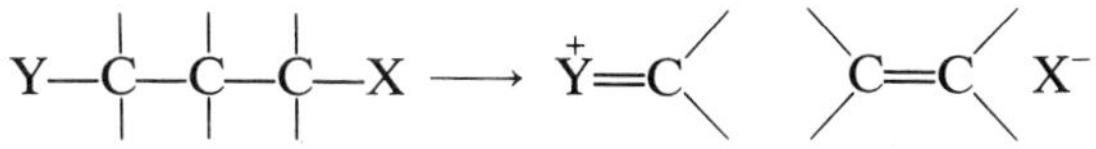

$$Y-\overset{|}{\underset{|}{C}}-\overset{|}{\underset{|}{C}}-\overset{|}{\underset{|}{C}}-X \longrightarrow \overset{+}{Y}=C \qquad C=C \quad X^- \tag{7}$$

For this reaction, Y contains a readily donatable lone pair of electrons (e.g., $-NR_2$, $-OH$) and X is the usual anionic leaving group (e.g., halogen or ester):

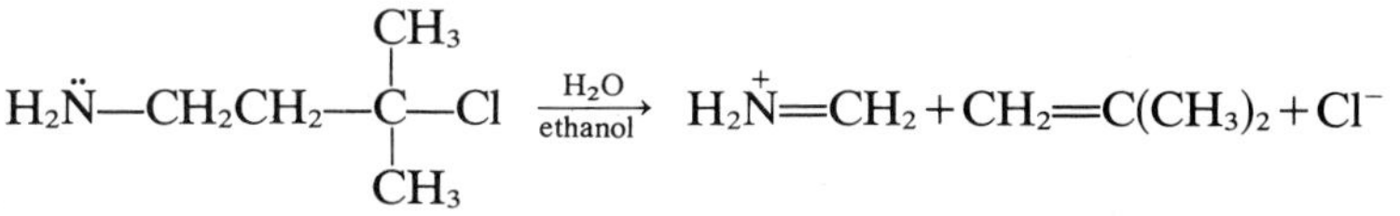

$$H_2\ddot{N}-CH_2CH_2-\overset{CH_3}{\underset{CH_3}{\overset{|}{\underset{|}{C}}}}-Cl \xrightarrow[\text{ethanol}]{H_2O} H_2\overset{+}{N}=CH_2 + CH_2=C(CH_3)_2 + Cl^-$$

These fragmentation processes occur by two mechanisms that are entirely analogous to the S_N1–S_N2 combination for substitution or the E1–E2 combination for elimination: that is, fragmentation may occur by a two-step, ionization–fragmentation mechanism or a one-step, internal-displacement mechanism:

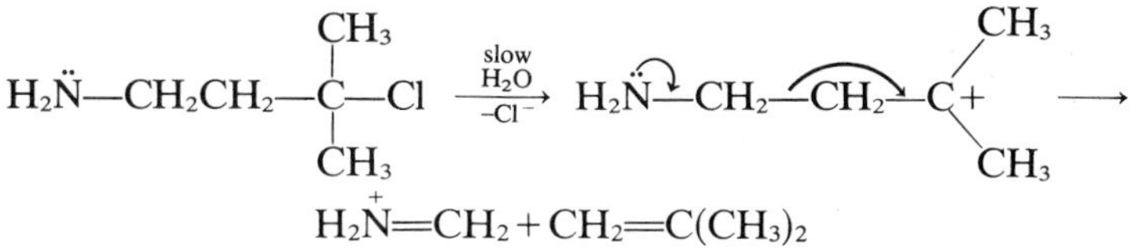

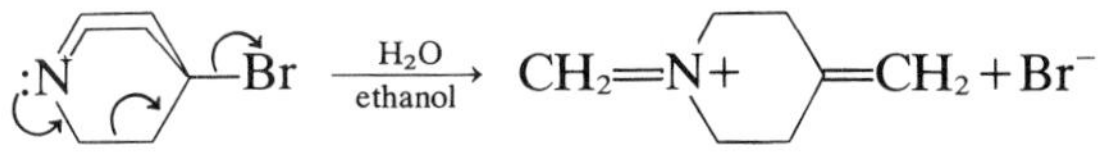

$$H_2\ddot{N}-CH_2CH_2-\overset{CH_3}{\underset{CH_3}{\overset{|}{\underset{|}{C}}}}-Cl \xrightarrow[-Cl^-]{\overset{\text{slow}}{H_2O}} H_2\ddot{N}-CH_2-CH_2-\overset{CH_3}{\underset{CH_3}{\overset{|}{\underset{|}{C}}}}+ \longrightarrow$$

$$H_2\overset{+}{N}=CH_2 + CH_2=C(CH_3)_2$$

12

The rate of reaction by the stepwise mechanism should vary only slightly with identity of the lone-pair substituent Y (eq. 7) and this has been observed for certain fragmentations:[64]

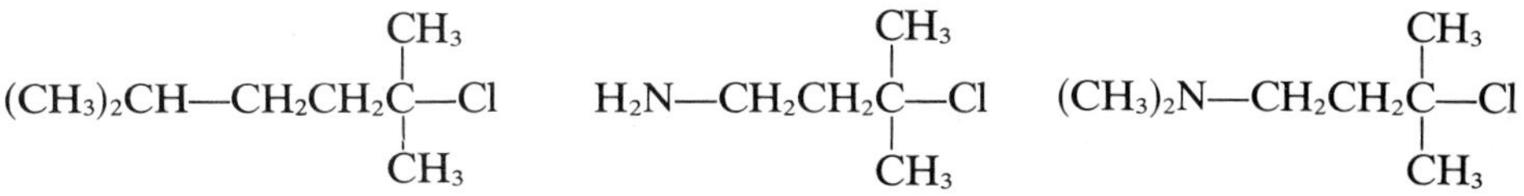

	$(CH_3)_2CH{-}CH_2CH_2\overset{CH_3}{\underset{CH_3}{\overset{|}{\underset{|}{C}}}}{-}Cl$	$H_2N{-}CH_2CH_2\overset{CH_3}{\underset{CH_3}{\overset{|}{\underset{|}{C}}}}{-}Cl$	$(CH_3)_2N{-}CH_2CH_2\overset{CH_3}{\underset{CH_3}{\overset{|}{\underset{|}{C}}}}{-}Cl$
k_{rel}	1.0	0.99	0.75
% Fragmentation	0	20	50

The concerted or one-step mechanism is actually a form of neighboring group displacement of the leaving group, and thus reaction by this mechanism should result in a rate acceleration relative to a nonassisted process. As an example, reaction of **12** occurs a factor of 10^4 faster than solvolysis of 1-bicyclo[2.2.2]octyl bromide. As would be expected on the basis of studies of neighboring group participation, there are stringent stereoelectronic requirements for the concerted fragmentation mechanism; the C–X and C_β–C_γ bonds must be antiperiplanar as must be the lone-pair of electrons on Y and the C_β–C_γ bond so that electrons can flow to the backside of the displaced bonds:

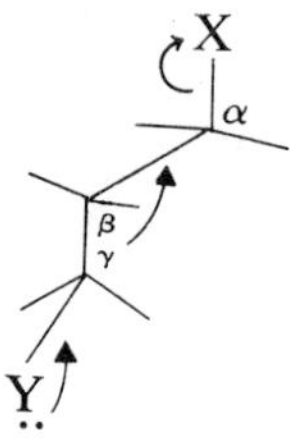

An illustration of this stereochemical requirement is provided by the reactions of 3α– and 3β–tropanyl chlorides, **13** and **14**.[64] The C_β–C_γ bond and the C–X bond are antiperiplanar for **14** but not for **13** (the leaving group for **14** is in a cyclohexyl equatorial position and that for **13** in an axial position; make a model to best illustrate the antiperiplanarity). As a consequence, **14** gives all fragmentation product while **13** gives only solvolytic substitution and elimination products:

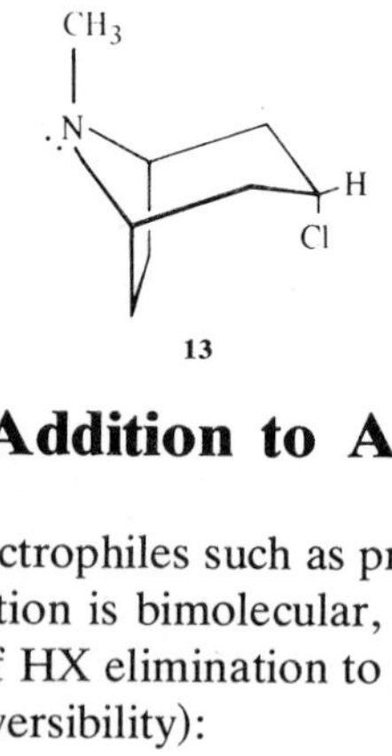

13 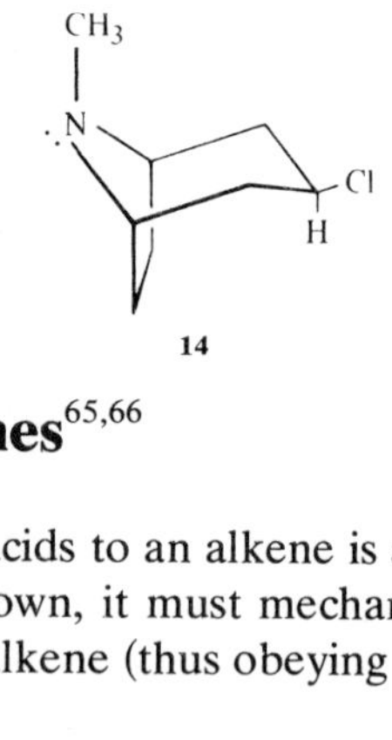   14

4.5 Electrophilic Addition to Alkenes[65,66]

Addition of electrophiles such as protic acids to an alkene is a facile process, and if the addition is bimolecular, as shown, it must mechanistically be the exact reverse of HX elimination to give alkene (thus obeying the principle of microscopic reversibility):

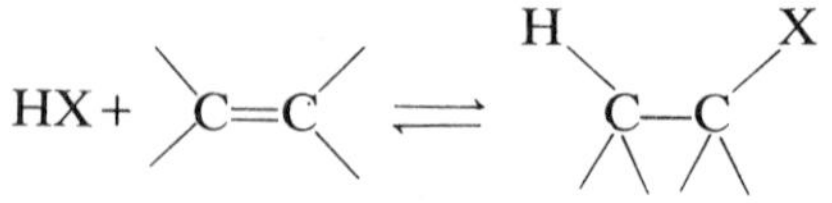

HX additions normally proceed via formation of an intermediate cationic species as shown by the observation of rearrangement products expected for cationic intermediates and by the observation of addition of the proton to

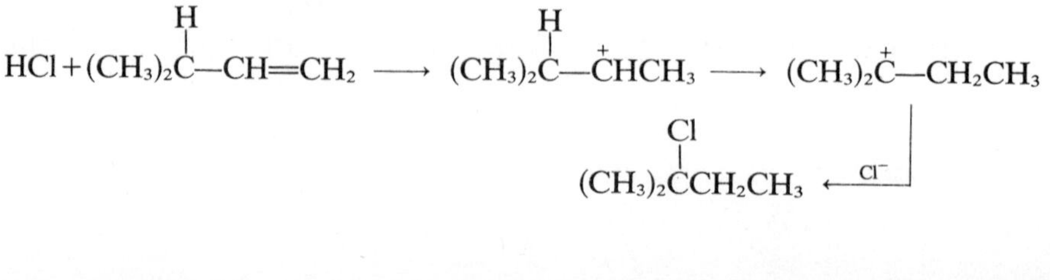

$$HCl + (CH_3)_2\overset{H}{C}-CH=CH_2 \longrightarrow (CH_3)_2\overset{H}{C}-\overset{+}{C}HCH_3 \longrightarrow (CH_3)_2\overset{+}{C}-CH_2CH_3$$

$$(CH_3)_2\overset{Cl}{C}CH_2CH_3 \xleftarrow{Cl^-}$$

the alkene carbon possessing the most hydrogens (so-called Markovnikov addition):

$$CH_3CH{=}CH_2 + HBr \longrightarrow CH_3{-}CH{-}CH_3$$
$$\underset{Br}{|}$$

$$\longrightarrow CH_3\overset{+}{C}HCH_3 \ Br^-$$

Markovnikov addition is most easily rationalized as being the result of formation of the most stable cationic intermediate followed by X^- attack on this intermediate. This mechanism is simply the reverse of the E1 elimination and is designated the Ad_E2 mechanism (for addition-electrophilic-bimolecular). Formation of the intermediate has been shown to be irreversible by the failure to observe proton exchange during DX addition:

$$\overset{H}{\underset{}{\bigg\rangle}}C{=}C\overset{/}{\underset{}{\bigg\langle}} + DX \longrightarrow D\overset{H}{\underset{|}{C}}{-}\overset{}{\underset{|}{C^+}} \quad X^- \not\longrightarrow \overset{D}{\underset{}{\bigg\rangle}}C{=}C\overset{/}{\underset{}{\bigg\langle}} + HX$$

Homolytic (free radical) mechanisms leading to anti-Markovnikov addition are sometimes observed if O_2 and trace peroxides are not rigorously excluded (see Chapter 6). Anti-Markovnikov addition for synthetic purposes can be best achieved by use of hydroboration and oxidation:

$$R{-}O{-}O{-}R \longrightarrow RO\cdot \xrightarrow{HBr} ROH + Br\cdot$$

$$CH_3CH{=}CH_2 \xrightarrow{Br\cdot} CH_3\dot{C}HCH_2Br \xrightarrow{HBr} CH_3CH_2CH_2Br + Br\cdot$$

$$CH_3CH{=}CH_2 \xrightarrow{B_2H_6} (CH_3CH_2CH_2)_3B \xrightarrow{H_2O_2} CH_3CH_2CH_2OH$$

In polar solvents, such as alcohols or water, acids are usually dissociated and protonated solvent molecules and solvent nucleophiles become involved:

$$HX + H_2O \rightleftharpoons H_3O^+X^- \overset{\underset{C=C}{\diagdown \diagup}}{\longrightarrow} \overset{X^-}{\underset{}{\diagdown \overset{+}{C}{-}C\diagup}}{-}H \longrightarrow \overset{OH}{\underset{}{\diagdown \overset{|}{C}{-}CH}}{+} \overset{X}{\underset{}{\diagdown \overset{|}{C}{-}CH}}$$

$$H_2O$$

To avoid this complicating factor, addition reactions are frequently conducted in nondissociating solvents such as pentane or methylene chloride. In these latter solvents, additional acid molecules are sometimes involved, as for example in the termolecular Ad_E3 mechanism:

$$2HX + \overset{}{\underset{}{\diagdown}}C{=}C\overset{/}{\underset{}{\diagup}} \longrightarrow \left[\overset{X{-}{-}H}{\underset{\underset{X{-}{-}{-}H}{|}}{-}\overset{}{\underset{}{C}}{-}{-}{-}{-}{-}{-}C\overset{}{\underset{}{\diagup}}} \right]^{\ddagger} \longrightarrow X^- + H\overset{|}{\underset{|}{C}}{-}\overset{|}{\underset{|}{C}}X + H^+$$

The Ad_E3 mechanism, proceeding either in a *syn* or *anti* fashion (discussed later), is similar to a reversed E2 mechanism. Other acid molecules may be involved to provide a more polar environment in the immediate vicinity of the polar transition state.

The addition of HX to alkenes can yield all stereochemical combinations from sterospecific *anti* addition to predominantly *syn* addition and including intermediate mixtures:

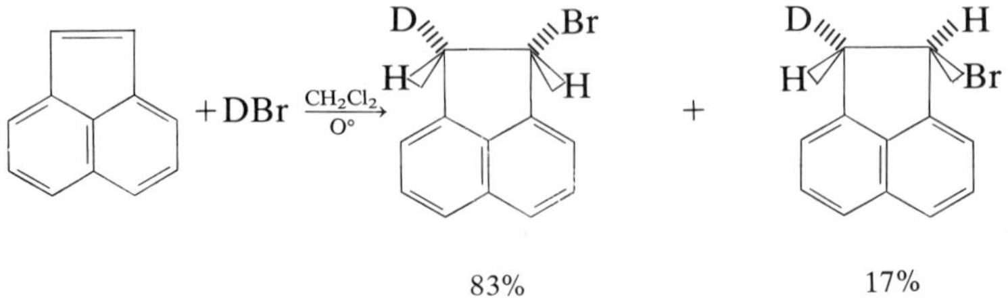

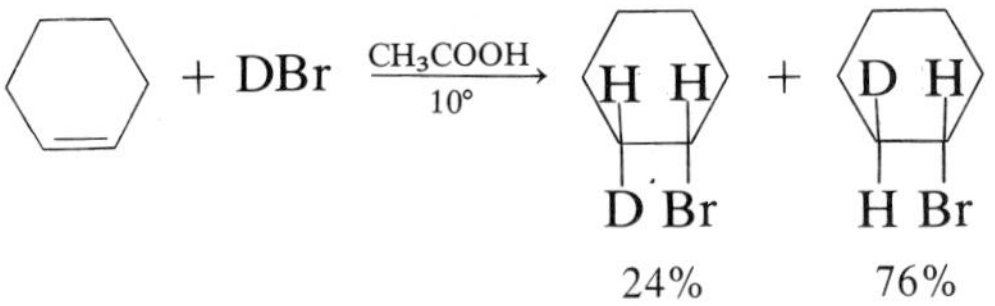

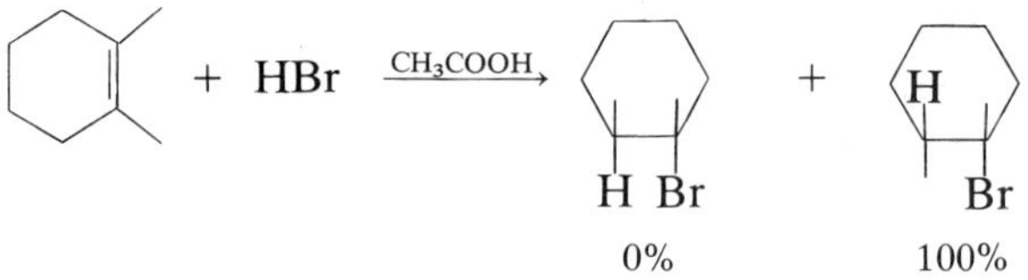

83%   17%

24%   76%

0%   100%

In order to best understand these stereochemical results, it is of benefit to consider the addition of molecular bromine and chlorine to alkenes.[67] These halogens have been shown to add heterolytically (light and O_2 must be excluded to avoid homolytic reaction) probably by an Ad_E2 mechanism. Consistent with this interpretation, rearranged products are observed and also mixed products from solvent attack on the intermediate:

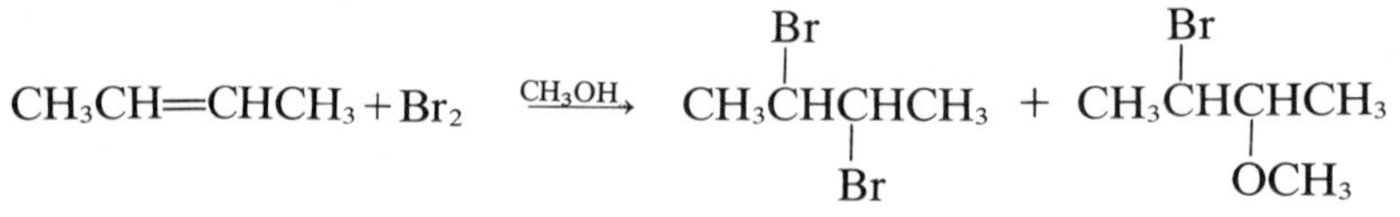

$$CH_3CH{=}CHCH_3 + Br_2 \xrightarrow{CH_3OH} CH_3\overset{Br}{\underset{Br}{C}}HCHCH_3 + CH_3\overset{Br}{\underset{OCH_3}{C}}HCHCH_3$$

Most important for the present consideration is the large amount of *anti* addition observed for the halogens:

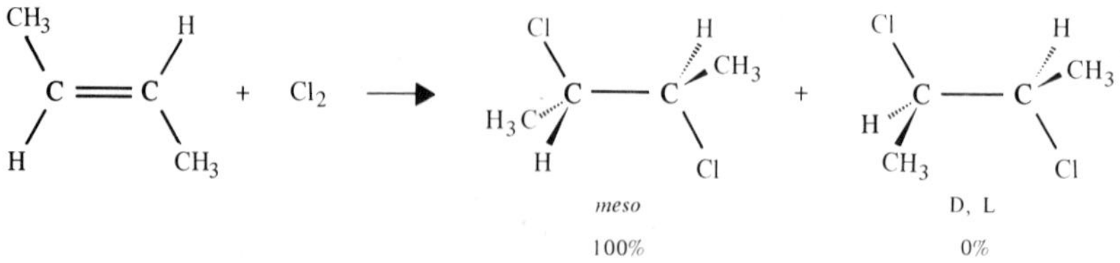

meso   D, L

100%   0%

The order of increasing importance of *anti* addition is HBr < Cl_2 < Br_2. For example, for addition to *trans*-1-phenylpropene, *anti* addition is 12% for

HBr, 33% for Cl_2, and 88% for Br_2.[67] The greater amount of *anti* addition of the halogens has been explained as due to increased favorability of formation of bromonium ion relative to chloronium ion and in turn relative to protonium ion:

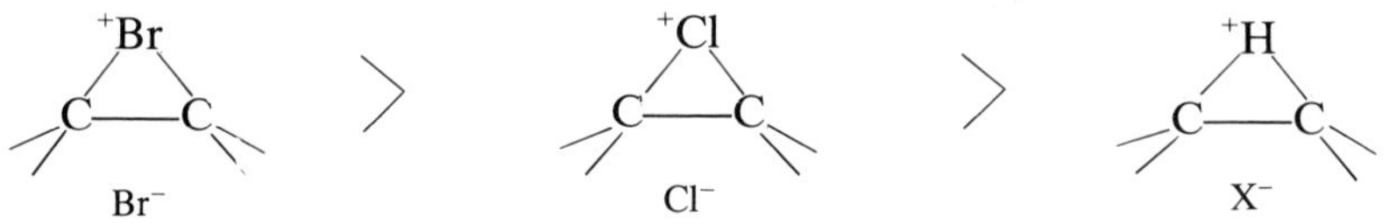

This stability order for bridged ions is consistent with observed abilities of bromine, chlorine, or hydrogen to act as neighboring groups in nucleophilic substitution reactions.

The increase in amount of *anti* addition with increased bridging tendency provides one key to understanding addition stereochemistry. A more complete theory results if it is further assumed that *syn* addition derives from collapse of an open (as opposed to bridged) ion pair, which is formed competitively with bridged ion, and whose formation is favored by conjugative and inductive effects which stabilize carbocations. Also, as the cationic fragment becomes increasingly stable, there should be increased dissociation to the free carbocation stage; at this point, the *syn* to *anti* ratio will be largely controlled by relative product stabilities. Thus, we have the following mechanism:[67]

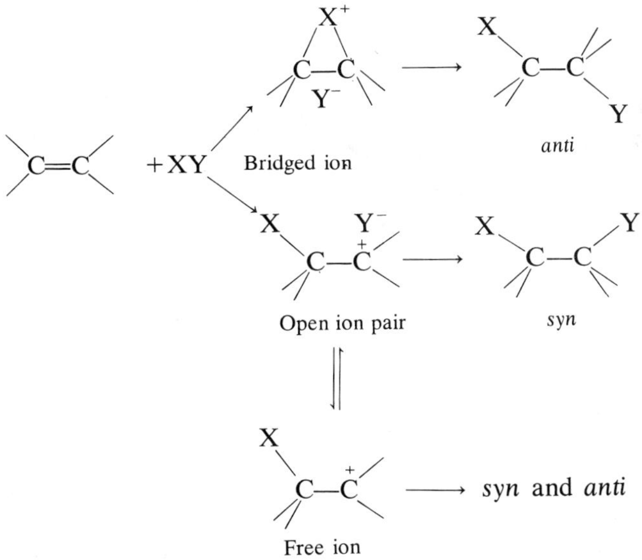

There is evidence that rapid, reversible formation of a π complex precedes formation of the bridged- or open-ion pairs; these complexes are discussed in the following section.

Referring to the examples given previously, we can see that the stereospecific *anti* addition of Cl_2 to *trans*-2-butene compared with only 33% *anti* addition to *trans*-1-phenylpropene can be rationalized as due to the greater stability of carbocation **15** relative to **16** and of **17** relative to **18.** In other words, the cation with the phenyl substituent has less need for bridging, and hence shows less *anti* addition.

$$
\underset{\textbf{15}}{\overset{\overset{\displaystyle Cl^- \quad Cl}{\overset{+}{|}\quad |}}{C_6H_5-CH-CH-CH_3}} \quad > \quad \underset{\textbf{16}}{\overset{\overset{\displaystyle Cl}{\diagup\ \overset{+}{\ }\ \diagdown}}{\underset{\displaystyle Cl^-}{C_6H_5-CH-CH-CH_3}}}
$$

$$
\underset{\textbf{17}}{\overset{\overset{\displaystyle Cl}{\diagup\ \overset{+}{\ }\ \diagdown}}{\underset{\displaystyle Cl^-}{CH_3-CH-CH-CH_3}}} \quad \gg \quad \underset{\textbf{18}}{\overset{\overset{\displaystyle Cl^- \quad Cl}{\overset{+}{|}\quad |}}{CH_3-CH-CH-CH_3}}
$$

Similarly, the larger amount of *syn* addition of DBr to acenaphthylene than to cyclohexene (p.192) can be explained as the result of the greater stability of the open carbocation for acenaphthylene.

As an open carbocation becomes increasingly stable it should become increasingly free of the counterion and a trend from near stereospecific *syn* addition to thermodynamic (product stability) control should be observed. Evidence for this reasoning comes from a study of chlorination of ring-substituted methyl cinnamates (Table 4.17);[68] as the electron donating ability of the ring substituent increases, the ratio of product isomers appears to become independent of starting substrate stereochemistry (*cis* or *trans*) as would be expected for chloride attack on free carbocation:

$$
\underset{H}{\overset{Ar}{\diagdown}}C=C\underset{H}{\overset{COOCH_3}{\diagup}} \xrightarrow{Cl_2}
$$

$$
\underset{Ar}{\overset{}{}}\overset{+}{C}H-\overset{\overset{\displaystyle Cl}{|}}{C}HCOOCH_3 \xrightarrow{Cl^-} \frac{threo}{erythro} = \text{constant}
$$

Free ion

$$
\underset{H}{\overset{Ar}{\diagdown}}C=C\underset{COOCH_3}{\overset{H}{\diagup}} \xrightarrow{Cl_2}
$$

Specifically, referring to Table 4.17, the *cis*- and *trans*-p-methoxy derivatives give product ratios more nearly identical than do the other substituents that are less able to stabilize an intermediate carbocation.

As is the case with any theory, a certain amount of caution must be used when applying this stereochemical theory. For example, the addition of HBr to cyclohexane and dimethylcyclohexane was shown previously to give more *anti* product for the latter alkene. Does this mean the bridged ion is more important for the dimethyl compound? Actually, one would expect bridging to be more important for the secondary carbocation formed from cyclohexene. The stereospecific *anti* addition to the dimethyl compound probably derives from a greater stability of the *anti* product.

There has been much debate regarding the involvement of protonium ions

Table 4.17. *Dichloride Products from Chlorination of Ring-substituted Methyl Cinnamates*[68]

Substituent	Cinnamate Isomer Used	% *Threo* Product	% *Erythro* Product
p-NO$_2$	trans	~100	~0
p-Cl	trans	72	28
	cis	17	83
H	trans	77	13
	cis	20	80
p-OCH$_3$	trans	23	77
	cis	6	94

(hydrogen-bridged cations) in addition reactions of acids and alkenes;[65,66] this debate, of course, parallels the controversy regarding the role of hydrogen-bridged species in solvolytic substitution reactions. There is evidence that product stereochemistry for HX addition is controlled by a competition between *syn* addition from collapse of an ion pair and *anti* addition, not from a bridged ion, but from a termolecular Ad$_E$3 process:[66]

To further complicate matters, the Ad$_E$2 mechanism can lead to *anti* addition if dissociation of the tight ion pair occurs, and the Ad$_E$3 mechanism can occur in a *syn* fashion. These complications serve to point out that the bridged-ion *versus* open-ion theory presented above is not by any means a complete theory. The theory can be used to explain many apparently disparate observations, but much remains to be learned about electrophilic addition reactions.

4.6 Electrophilic Aromatic Substitution[69-71]

Electrophiles react initially with a double bond of an aromatic molecule just as they react with the double bond of a simple alkene; in each case a carbocation is formed:

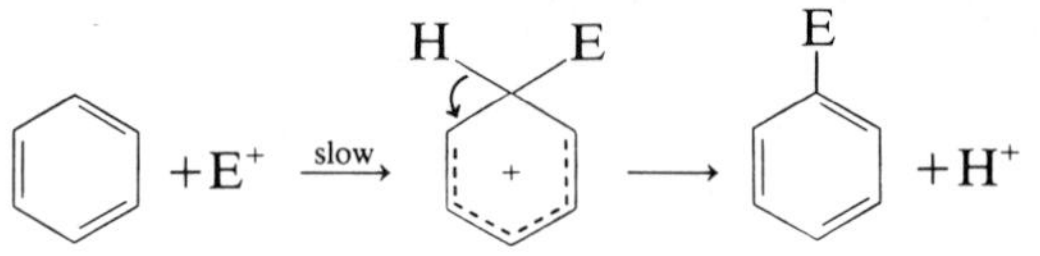

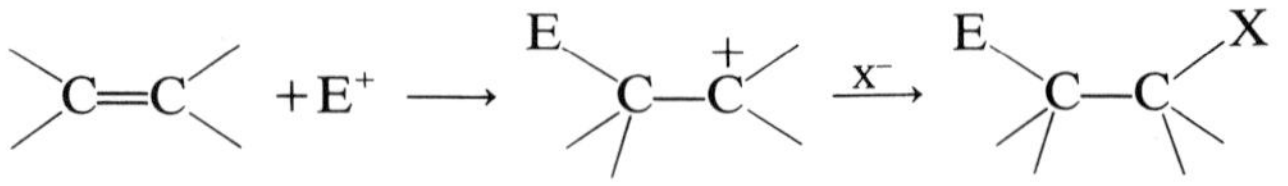

The reactions differ in the fates of the respective cations; the cyclohex-adienyl cation eliminates a proton to restore a double bond and regain an aromatic structure, but the alkyl cation is attacked by a nucleophile to give a saturated molecule. The overall result is substitution (E for H) for the aromatic molecule and addition for the alkene (recall this was one of the early definitions of aromaticity—see Chapter 2).

Many electrophiles react with benzenoid compounds to give electrophilic substitution. For example:

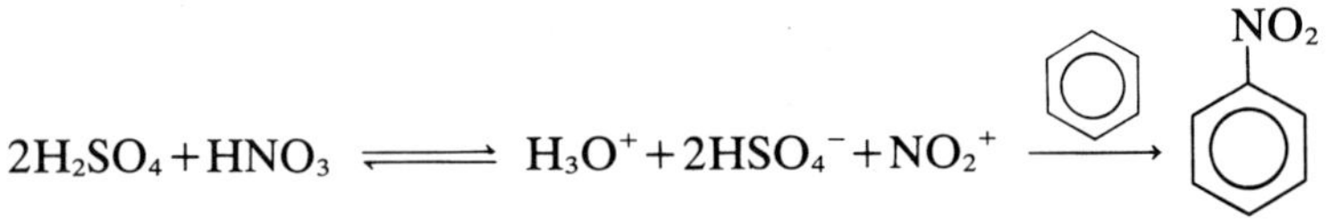

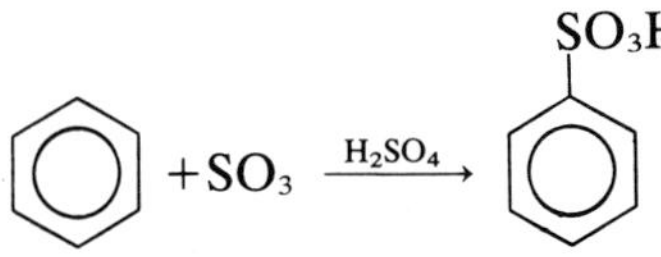

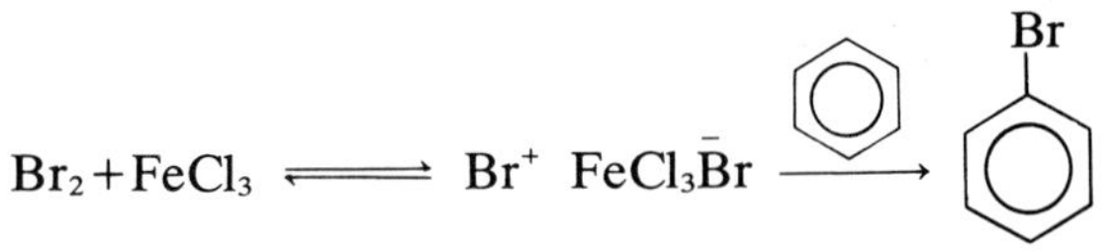

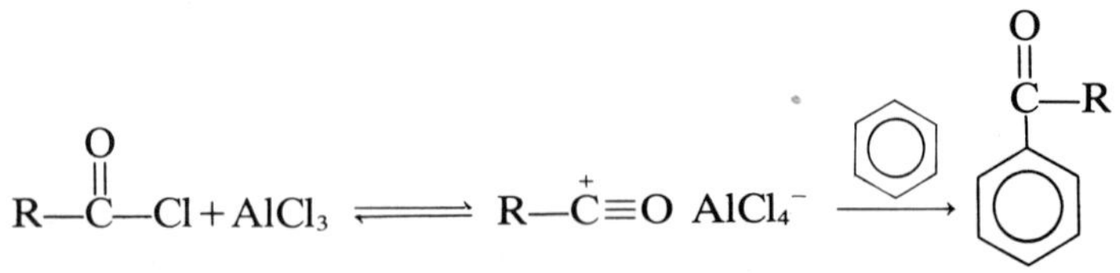

$$2H_2SO_4+HNO_3 \rightleftharpoons H_3O^+ +2HSO_4^- +NO_2^+$$

$$+SO_3 \xrightarrow{H_2SO_4}$$

$$Br_2+FeCl_3 \rightleftharpoons Br^+ \ FeCl_3\bar{Br}$$

$$R-\overset{O}{\overset{\|}{C}}-Cl+AlCl_3 \rightleftharpoons R-\overset{+}{C}\equiv O \ AlCl_4^-$$

The general reaction of an electrophile and benzene to give an inter-mediate cyclohexadienyl cation (usually referred to as a σ complex) fol-lowed by proton loss represents the mechanism of electrophilic aromatic substitution as it is usually presented. The intermediacy of the σ complex is supported by extensive evidence of which we will discuss four examples. First, the assumed involvement of a σ complex provides an explanation for the position of substitution on substituted benzenes. For example, nitration of methoxybenzene occurs predominantly in the *ortho* and *para* positions. If reaction does occur via endothermic formation of a σ complex, the transi-tion state for this step will resemble a σ complex. As can be seen from the partial list of resonance forms below, the methoxy group can stabilize the σ complex resulting from attack at *ortho* and *para* positions but not the one from *meta* attack.

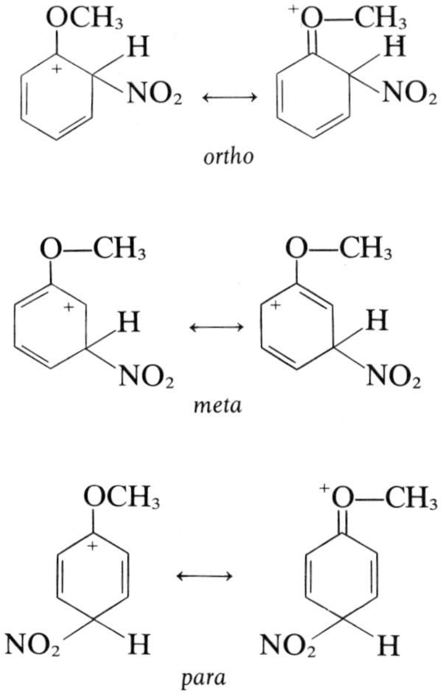

ortho

meta

para

Thus, the transition states for *ortho* and *para* attack will be favored. Unfavorable steric interactions for *ortho* attack generally result in *para* substitution being greater than *ortho* substitution. Similarly, the predominant *meta* substitution of benzenes containing deactivating substitutents can be rationalized as the result of the deactivating substituent exerting less of a destabilizing effect on the σ complex for *meta* attack since, in this case, positive charge is never adjacent to the substituent:

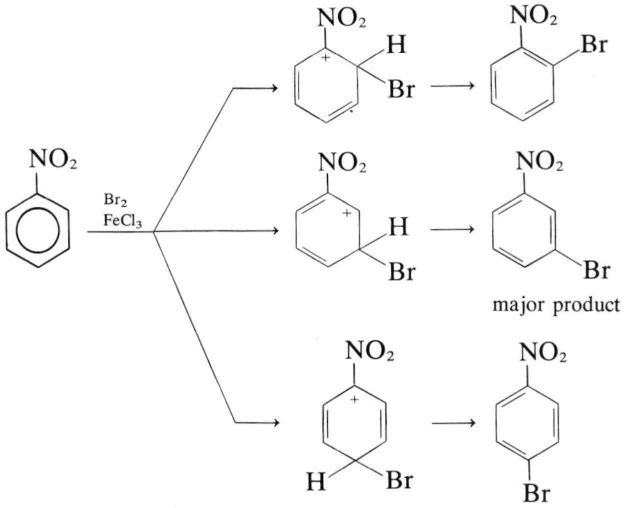

major product

Second, the involvement of σ complexes is indicated by the spectroscopic observation of stable σ complexes in solution and by the isolation of σ complexes that are stable at low temperature:

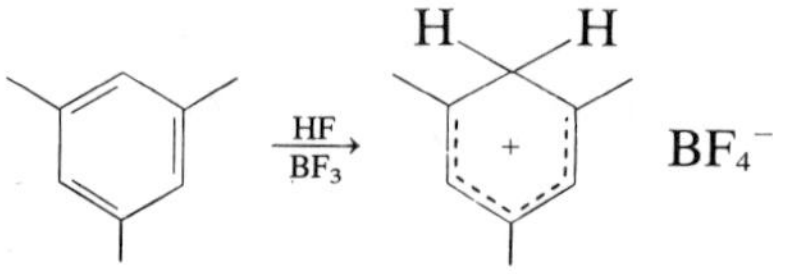

Isolable; decomposition point −15 °C

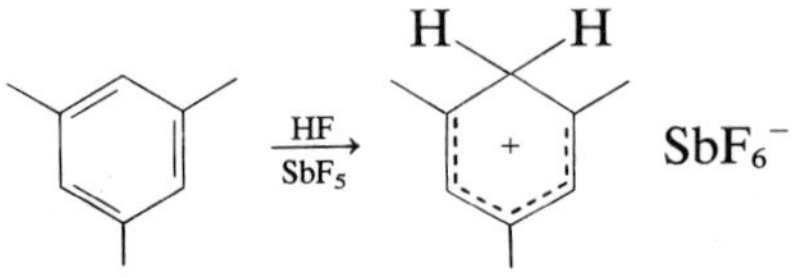

nmr spectrum obtained

Third, the relative rates of many substitution reactions are correlated by relative stabilities of observable σ complexes in strong acid solution (e.g., HF–BF$_3$). And finally, the intermediacy of σ complexes in these substitution reactions is strongly suggested by the results of isotope effect studies. If electrophilic substitution occurs by a two-step mechanism as shown, replace-

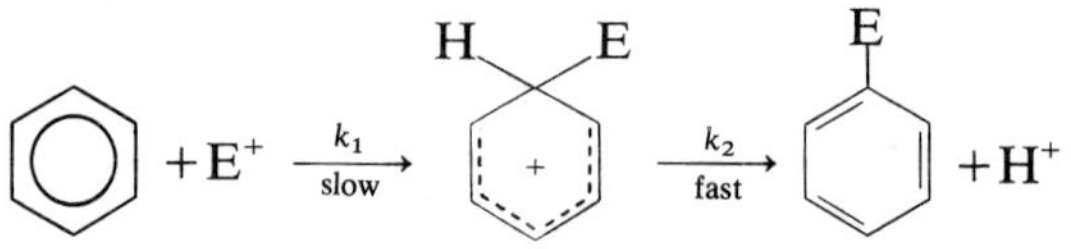

ment of hydrogen with deuterium should have little effect on the rate of reaction. In fact, kinetic deuterium isotope effects are normally not observed for these reactions. This observation clearly eliminates the possibility of a one-step mechanism. Also, it is implicit in this two-step mechanism ($k_1 < k_2$) that the reaction will not be reversible, and this is observed for most electrophilic substitutions. There are a few cases for which k_2 might be expected to be slower than k_1 and k_{-1}. For example, in the bromination of 1,3,5-tri-*tert*-butylbenzene, steric effects associated with proton loss (the second step) force the bromine and the *tert*-butyl group to become coplanar:

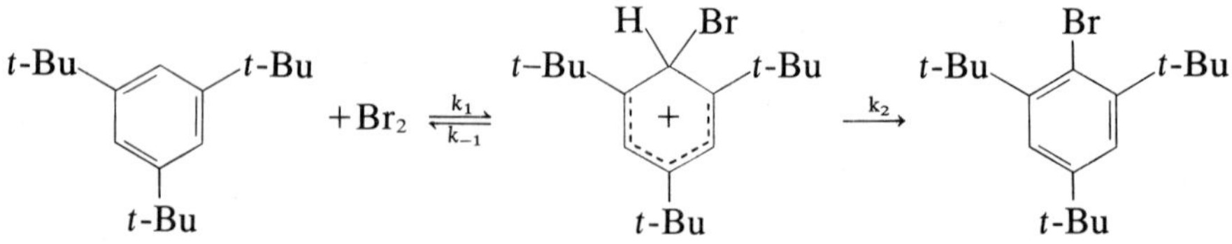

This slowing of k_2 leads to $k_{-1} > k_2$ and a kinetic deuterium isotope effect of about 3.6. Sulfonation of aromatic compounds is a reversible reaction (i.e., k_2 is the slow step) and, as expected, deuterium isotope effects are observed for this reaction.

π **Complexes and a Three-Step Mechanism**

The existence of weak complexes (called charge-transfer complexes) between electron donors and electron acceptors has been well documented.[66,71,72] When the complex is between a π donor and an electrophile, the term π complex is used. In view of the known importance of these weak complexes, it seems reasonable to assume that their formation is a preliminary step in electrophilic attack on an electron-rich π system:

π complex

π complex σ complex

For the usual electrophilic aromatic substitution, we can assume that formation of the π complex is rapid and reversible ($k_{-1} > k_2$), and that this step is followed by rate-determining σ complex formation and rapid product formation ($k_2 < k_3$). For such reactions, the involvement of π complexes cannot be determined by kinetic or spectroscopic techniques. In contrast, for reaction with highly reactive electrophiles leading to exothermic substitution, it is necessary to consider the rate-determining step to be π-complex formation ($k_{-1} < k_2 < k_3$). Evidence for this interpretation comes from the observation that certain highly exothermic substitutions show an unusual combination of low substrate selectivity (e.g., between toluene and benzene) and high positional selectivity (e.g., between *meta* and *para* substitution) (see Chapter 3 for a discussion of stability–selectivity relationships).[71] For the usual (not highly exothermic) substitutions, a linear relationship is found between substrate and positional selectivity; as the reactivity of the electrophile decreases, the electrophile becomes more selective, and both substrate and positional selectivity increase (Table 4.18).[73] H. C. Brown and his co-workers have shown that this stability–selectivity relationship is a quantitative one.[73]

$$\log p_f^X = b \log \left(\frac{p_f^X}{m_f^X} \right) \qquad (8)$$

Table 4.18 *Rates and Products for Substitution on Benzene and Toluene and ρ Values for Substitution on* C_6H_5X.[73,74]

Reaction	$\dfrac{k_t}{k_b}$	para/meta	ρ
Bromination	605	223	−12.1
Chlorination	350	79	−10.0
Benzoylation	110	60	−9.1[a]
Nitration	23	12	−6.0
Mercuration	7.9	7.3	−4.0
Isopropylation	1.8	1.8	−2.4

[a] Acetylation.

For this equation, stability or reactivity of the electrophile is given by a partial rate factor for attack at the *para* position of a X-substituted benzene (X = CH₃ for toluene); the partial rate factor is simply the ratio of the rate of attack at the *para* position of C_6H_5X and the rate of attack in benzene, statistically corrected for the six possible sites of attack on benzene and the one *para* position of C_6H_5X. For example, if bromination of an equimolar mixture of toluene and benzene yields

$$\frac{p\text{-Bromotoluene}}{\text{Bromobenzene}} = 25$$

then

$$\frac{k_p}{6k_b} = 25$$

and

$$p_f^{Me} = \frac{k_p}{k_b} = 150$$

Selectivity in eq. 8 is measured by the ratios of partial rate factors for *meta* and *para* attack.

Electrophilic substitution reactions on substituted benzenes can also be correlated by Brown's σ^+ substituent constants. It should be recalled that these constants are used for predicting the ability of a substituent to delocalize a positive charge when direct conjugation is possible (as in a σ complex):

$$\log \frac{k_z}{k_b} = \rho\sigma^+$$

$z = meta$ or *para*

The relationship discussed previously between stability and selectivity can also be seen in observed increases in ρ values (response to substituent effects) for more selective, less reactive electrophiles (Table 4.18). The transition states for the more endothermic reactions (less reactive electrophiles) will more resemble the σ-complex intermediates (the transition states will be "later") and conjugative effects will be more important.

Returning to the original question of the involvement of π complexes in electrophilic substitution, it was pointed out that certain highly reactive electrophiles gave the expected low substrate selectivity but also gave unexpectedly high positional selectivity. For example, nitration of toluene and benzene with nitric acid in nitromethane gives $k_t/k_b = 21$, and the isomer distribution for toluene of 59% *ortho*, 4% *meta*, and 37% *para*. These results fit the Brown selectivity relationship (eq. 8). In contrast, the rapid

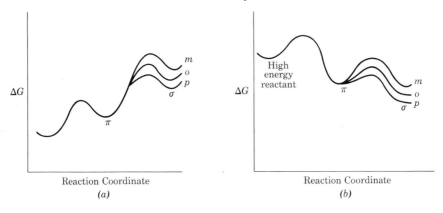

Fig. 4.6 *Potential energy diagrams for (a) endothermic and (b) exothermic electrophilic substitution.*

nitration of toluene and benzene with nitronium tetrafluoroborate ($NO_2^+BF_4^-$) in acetonitrile gives $k_t/k_b = 2.3$ and the isomer distribution of 69% *ortho*, 2% *meta*, and 29% *para*, much too selective to fit the selectivity relationship.[71] The σ^+ correlation fails for these exothermic reactions also. Could these deviations be the result of rate-determining π-complex formation? It would certainly be reasonable (if we can extend Hammond's postulate to a two-stage reaction) for the endothermic substitutions to have late, highest-energy transition states resembling σ complexes and for the exothermic reactions to have early, highest-energy transition states resembling π complexes (Figure 4.6).

For π complexes, little charge is developed and substituent effects are small, but for σ complexes a full positive charge is developed and the stabilizing or destabilizing effect of the substituent is fully felt (Table 4.19). Thus, reaction by rate-determining π-complex formation would be consistent with low substrate selectivity, but how can high positional selectivity result from this same pathway? Reference to Figure 4.6 shows that positional selectivity will always be determined by relative stabilities of transition states leading to σ complex regardless of which step (π-complex

Table 4.19 *Relative π- and σ-Complex Stabilities*[71]

Substituted Benzene	σ stability	π stability
H	1	1
Methyl	790	1.5
1,2-Dimethyl	7900	1.8
1,3-Dimethyl	10^6	2.0
1,4-Dimethyl	3200	1.6
1,2,3-Trimethyl	2×10^6	2.4
1,3,5-Trimethyl	6×10^8	2.6
1,2,4-Trimethyl	2×10^6	2.2

formation or σ-complex formation)·is rate determining; there is no positional orientation in the π complex but there is in the σ complex:

substrate selectivity $= k_b^\pi / k_t^\pi$

positional selectivity $= k_t^o / k_t^m / k_t^p$

Since σ-complex formation is subject to large substituent effects, positional selectivity will be larger than expected from substrate selectivity if the rate-determing step is π-complex formation. On the other hand, if σ-complex formation is rate determining and π-complex formation is rapid and reversible, both substrate and positional selectivity will be high (relative to the foregoing case) and determined by the same step:

Substrate and positional selectivity $= k_b^\sigma / k_t^o / k_t^m / k_t^p$

The failure of highly exothermic substitutions to correlate with σ^+ is also the result of rate-determining π-complex formation. The σ^+ substituent constants reflect the direct conjugative ability of a substituent. Since conjugative effects are unimportant for π complexes, reaction by rate-determining π-complex formation will not be correlated by σ^+ substituent constants.

The π-complex theory has been disputed on the basis that the highly reactive electrophiles that exhibit low substrate selectivity and high positional selectivity react at diffusion controlled rates such that substrate selectivity is controlled simply by the rate of electrophile-substrate encounters.[56,75] Olah has countered this argument by observing that selectivity values (positional and substrate) are independent of substrate concentration ratios and of mixing efficiency;[71] if substrate selectivity were controlled by a diffusion process, changing these two parameters should change the observed selectivity.

References

1. C. K. Ingold, "Structure and Mechanism in Organic Chemistry," 2nd. ed., Cornell Univ. Press, Ithaca, New York, 1969.
2. G. A. Olah, *J. Amer. Chem. Soc.*, **94,** 808 (1972).
3. D. G. Farnum, R. A. Mader, and G. Mehta, *J. Amer. Chem. Soc.*, **95,** 8692 (1973).
4. C. D. Hurd, *J. Chem. Educ.*, **48,** 490 (1971).
5. D. Bethell and V. Gold, "Carbonium Ions," Academic Press, London, 1967.
6. A. Streitwieser, Jr., "Solvolytic Displacement Reactions," McGraw-Hill, New York, 1962.
7. C. A. Bunton, "Nucleophilic Substitution at a Saturated Carbon Atom," Elsevier, Amsterdam, 1963.
8. E. R. Thorton, "Solvolysis Mechanisms," Ronald, New York, 1964.
9. S. R. Hartshorn, "Aliphatic Nucleophilic Substitution," Cambridge Univ. Press, London, 1973.
10. S. P. McManus and C. V. Pittman, Jr., in "Organic Reactive Intermediates," S. P. McManus, Ed., Academic Press, New York, 1973, Chapter 4.
11. M. H. Abraham, *Chem. Commun.*, 51 (1973).
12. L. Radom, J. A. Pople, and P. v. R. Schleyer, *J. Amer. Chem. Soc.*, **94,** 5935 (1972).
13. P. v. R. Schleyer and C. W. Woodworth, *J. Amer. Chem. Soc.*, **90,** 6528 (1968).
14. R. C. Bingham and P. v. R. Schleyer, *J. Amer. Chem. Soc.*, **93,** 3189 (1971).
15. H. C. Brown and M. Borkowski, *J. Amer. Chem. Soc.*, **74,** 1893 (1952).
16. R. C. Fort, Jr., in "Carbonium Ions," Vol. IV, G. A. Olah and P. v. R. Schleyer, Eds., Wiley-Interscience, New York, 1973, Chapter 32.
17. G. A. Olah and P. v. R. Schleyer, Eds., "Carbonium Ions," Vols. I–IV, Wiley-Interscience, New York, 1972.
18. P. E. Peterson, R. E. Kelley, Jr., R. Belloli, and K. A. Sipp, *J. Amer. Chem Soc.*, **87,** 5169 (1965).
19. T. W. Bentley, S. H. Liggero, M. A. Imhoff, and P. v. R. Schleyer, *J. Amer. Chem. Soc.*, **96,** 1970 (1974).
20. R. W. Taft, Jr., "Steric Effects in Organic Chemistry," M. S. Newman, Ed., Wiley, New York, 1956, p. 556 ff.
21. H. H. Freedman in "Carbonium Ions," Vol. IV, G. A. Olah and P. v. R. Schleyer, Eds., Wiley-Interscience, New York, 1973, Chapter 28.
22. H. C. Brown and Y. Okamoto, *J. Amer. Chem. Soc.*, **80,** 4979 (1958).
23. P. J. Stang, *Progr. Phys. Org. Chem.*, **10,** 205 (1973).
24. M. H. Abraham, *Progr. Phys. Org. Chem.*, **11,** 1 (1974).
25. S. Winstein and H. Marshall, *J. Amer. Chem. Soc.*, **74,** 1120 (1952); see also Ref. 34.
26. S. Winstein and A. H. Fainberg, *J. Amer. Chem. Soc.*, **79,** 5937 (1957).
27. D. N. Kevill, K. G. Kolwyck, D. M. Shold, and C-B. Kim, *J. Amer. Chem. Soc.*, **95,** 6022 (1973); see also Ref. 36.
28. J. M. Harris, *Progr. Phys. Org. Chem.*, **11,** 90 (1974).
29. H. C. Brown, "Boranes in Organic Chemistry," p. 99, Cornell Univ. Press, Ithaca, N.Y., 1972.
30. D. J. Raber, unpublished results.
31. H. C. Brown and A. Cahn, *J. Amer. Chem. Soc.*, **77,** 1715 (1955).
32. A. J. Parker, *Chem. Rev.*, **69,** 1 (1969).
33. P. R. Wells, "Linear Free Energy Relationships," Academic Press, New York, 1968.

34. T. W. Bentley, F. L. Schadt, and P. v. R. Schleyer, *J. Amer. Chem. Soc.*, **94,** 993 (1972).

35. J. Stulsky, R. C. Bingham, P. v. R. Schleyer, W. C. Dickason, and H. C. Brown, *J. Amer. Chem. Soc.*, **96,** 1969 (1974).

36. H. M. R. Hoffmann, *J. Chem. Soc.*, 6762 (1965).

37. D. J. Raber, J. M. Harris, and P. v. R. Schleyer, "Ions and Ion Pairs in Organic Reactions," Vol. 2, M. Szwarc, Ed., Wiley-Interscience, New York, 1974, Chapter 3.

38. S. Winstein, B. Appel, R. Baker, and A. Diaz, *Chem. Soc. (London) Spec. Pub.,* **19,** 109 (1965).

39. L. C. Bateman, E. D. Hughes, and C. K. Ingold, *J. Chem. Soc.*, 974 (1940).

40. J. M. Harris, A. Becker, J. F. Fagan, and F. A. Walden, *J. Amer. Chem. Soc.*, **96,** 4484 (1974).

41. C. J. Lancelot, D. J. Cram, and P. v. R. Schleyer, "Carbonium Ions," Vol. III, G. A. Olah and P. v. R. Schleyer, Eds., Wiley-Interscience, New York, 1972, Chapter 27.

42. B. Capon, *Quart. Rev. (London)*, **18,** 45 (1964).

43. P. D. Bartlett, "Nonclassical Ions," Benjamin, New York (1965).

44. P. D. Bartlett and T. T. Tidwell, *J. Amer. Chem. Soc.*, **90,** 4421 (1968).

45. H. Goering and K. Humski, *J. Amer. Chem. Soc.*, **90,** 6213 (1968).

46. J. P. Dirlam and S. Winstein, *J. Amer. Chem. Soc.*, **91,** 5905 (1969).

47. C. J. Kim and H. C. Brown, *J. Amer. Chem. Soc.*, **91,** 4289 (1969).

48. J. E. Nordlander and T. J. McCrary, Jr., *J. Amer. Chem. Soc.*, **94,** 5133 (1972).

49. P. D. Bartlett, W. D. Closson, and T. J. Cogdell, *J. Amer. Chem. Soc.*, **87,** 1308 (1965).

50. S. Winstein and M. Shatavsky, *J. Amer. Chem. Soc.*, **78,** 592 (1956).

51. G. D. Sargent, "Carbonium Ions," Vol. III, Wiley-Interscience, New York, 1972, Chapter 24.

52. (a) H. C. Brown, *Chem. Eng. News,* **45,** (7), 86 (1967); (b) H. C. Brown and K-T. Liu, *J. Amer. Chem. Soc.*, **97,** 600 (1975).

53. S. Winstein and D. Trifan, *J. Amer. Chem. Soc.*, **74,** 1127, 1154 (1952).

54. G. A. Olah, G. Liang, G. D. Mateescu, and J. L. Riemenschneider, *J. Amer. Chem. Soc.*, **95,** 8698 (1973).

55. J. M. Harris and S. P. McManus, *J. Amer. Chem. Soc.*, **96,** 4693 (1974).

56. R. W. Alder, R. Baker, and J. M. Brown, "Mechanism in Organic Chemistry," Wiley, New York, 1971.

57. D. V. Banthorpe, "Elimination Reactions." Elsevier, Amsterdam, 1963.

58. W. H. Sanders, Jr., and A. F. Cockerill, "Mechanisms of Elimination Reactions," Wiley-Interscience, New York, 1973.

59. F. G. Bordwell, *Acc. Chem. Res.,* **5,** 374 (1972).

60. V. J. Shiner, Jr. in "Isotope Effects in Chemical Reactions." C. J. Collins and N. S. Bowman, Eds., Von Nostrand Reinhold, New York, 1970.

61. D. B. Banthorpe, in "Studies on Chemical Structure and Reactivity," J. H. Ridd, Ed., Methuen, London, 1966.

62. R. A. Bartsch and J. F. Bunnett, *J. Amer. Chem. Soc.*, **90,** 408 (1968).

63. C. H. Depuy, G. F. Morris, J. S. Smith, and R. J. Smat, *J. Amer. Chem. Soc.*, **87,** 2421 (1965).

64. C. A. Grob, *Angew. Chem. Int. Ed. Engl.,* **8,** 535 (1969).

65. P. B. de la Mare and R. Bolton, "Electrophilic Additions to Unsaturated Systems," Elsevier, Amsterdam, 1966.

66. R. C. Fahey, *Top. Stereochem.,* **3,** 237 (1968).

67. W. R. Dolbier, *J. Chem. Educ.,* **46,** 324 (1969).

68. M. D. Johnson and E. N. Trachtenberg, *J. Chem. Soc. (B)*, 1018 (1068).

69. R. O. C. Norman and R. Taylor, "Electrophilic Substitution of Benzenoid Compounds," Elsevier, Amsterdam, 1965.

70. G. A. Olah, Ed., "Friedel-Crafts and Related Reactions," Wiley, New York, 1963.
71. G. A. Olah, *Acc. Chem. Res.*, **4**, 240 (1971).
72. M. J. S. Dewar, "The Molecular Orbital Theory of Organic Chemistry," McGraw-Hill, New York, 1969.
73. L. M. Stock and H. C. Brown, *Adv. Phys. Org. Chem.*, **1**, 35 (1963).
74. R. D. Gilliom, "Introduction to Physical Organic Chemistry," Addison-Wesley, New York, 1970.
75. R. G. Coombes, R. B. Moochie, and K. Schofield, *J. Chem. Soc. (B)*, 800 (1969).

Bibliography

Nucleophilic Displacement Reactions

C. K. Ingold, "Structure and Mechanism in Organic Chemistry," 2nd ed., Cornell Univ. Press, Ithaca, N.Y., 1969.
Contains a thorough treatment by one of the pioneer workers in this area.

D. Bethell and V. Gold, "Carbonium Ions," Academic Press, London, 1967.
A recent work containing much discussion of nucleophilic displacement.

A. Streitwieser, Jr., "Solvolytic Displacement Reactions," McGraw-Hill, New York, 1962. C. A. Bunton, "Nucleophilic Substitution at a Saturated Carbon Atom," Elsevier, Amsterdam, 1963. E. R. Thorton, "Solvolysis Mechanisms," Ronald, New York, 1964.
These three books present general reviews of the subject.

S. P. McManus and C. V. Pittman, Jr. in "Organic Reactive Intermediates," S. P. McManus, Ed., Academic Press, New York, 1973, Chapter 4.
A recent and concise review.

S. R. Hartshorn, "Aliphatic Nucleophilic Substitution," Cambridge Univ. Press, London, 1973.
An excellent introductory treatment.

J. M. Harris, *Progr. Phys. Org. Chem.*, **11**, 89 (1974).
A recent review of ion pairing and neighboring group participation in simple alkyl systems.

G. A. Olah and P. v. R. Schleyer, Eds., "Carbonium Ions," Wiley-Interscience, New York, 1968–1973.
A four volume collection of reviews on almost every aspect of carbocation chemistry.

R. A. Sneen and J. W. Larsen, *J. Amer. Chem. Soc.*, **91**, 6031 (1969). V. J. Shiner, Jr., R. D. Fisher, and W. Dowd, *J. Amer. Chem. Soc.*, **91**, 7748 (1969). W. M. Schubert and W. L. Henson, *J. Amer. Chem. Soc.*, **93**, 6299 (1971).
These papers present a novel theory for anchimeric assistance where the neighboring group acts by removing ion pair return.

β- Elimination Reactions

R. W. Alder, R. Baker, and J. M. Brown, "Mechanism in Organic Chemistry," Wiley-Interscience, New York, 1971.
Contains the most recent introductory treatment.

D. V. Banthorpe, "Elimination Reactions," Elsevier, Amsterdam, 1963.
A general treatment.

W. H. Saunders, Jr., and A. F. Cockerill, "Mechanisms of Elimination Reactions," Wiley-Interscience, New York, 1973.

F. G. Bordwell, *Acc. Chem. Res.*, **5,** 374 (1972). W. T. Ford, *Acc. Chem. Res.*, **6,** 410 (1973).
These two reviews provide a probing examination of the generally accepted elimination mechanisms.

Fragmentation

C. A. Grob, *Angew. Chem. Int. Ed. Engl.*, **8,** 535 (1969).
A short review by the pioneer worker in this area.

Electrophilic Addition to Alkenes

P. B. de la Mare and R. Bolton, "Electrophilic Additions to Unsaturated Systems," Elsevier, Amsterdam, 1966.
A general review.

R. C. Fahey, *Top. Stereochem.*, **3,** 237 (1968).
Concentrates on stereochemical aspects.

W. R. Dolbier, *J. Chem. Educ.*, **46,** 324 (1969).
Presents the mechanism of competition between formation of bridged and open carbocations.

Electrophilic Aromatic Substitution

R. O. C. Norman and R. Taylor, "Electrophilic Substitution of Benzenoid Compounds," Elsevier, Amsterdam, 1965.
A general review.

G. A. Olah, Ed., "Friedel-Crafts and Related Reactions, Wiley, New York, 1963.
A six-volume treatment of alkylation and acylation reactions.

G. A. Olah, *Acc. Chem. Res.*, **4,** 240 (1971).
A detailed treatment of the role of π complexes.

Carbocation Nomenclature

G. A. Olah, *J. Amer. Chem. Soc.*, **94,** 808 (1972).
D. G. Farnum, R. A. Mader, and G. Mehta, *J. Amer. Chem. Soc.*, **95,** 8692 (1973).
C. D. Hurd, *J. Chem. Educ.*, **48,** 490 (1971).

Problems

1. Complete the following equations:

(*a*) $(CH_3)_3C$—$OH + A \rightarrow (CH_3)_3C$—$Cl + B$

(*b*) CH_3CH_2—$OTs + A \rightarrow CH_3CH_2OCOCH_3 + B$

(*c*) $(CH_3)_2CH$—$Cl + A \rightarrow (CH_3)_2CH$—$N_3 + B$

(*d*) $+ OH^- \xrightarrow{\text{H}_2\text{O}}$

2. Explain the fact that optically active 2-iodooctane undergoes substitution by radioactive iodide only half as fast as optical activity is lost.

3. The following reaction is a second-order reaction:

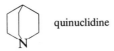

—OPNB = *p*—nitrobenzoate

Explain the rearrangement and stereochemistry of substitution. (For reaction of the above ester, which was optically active and contained carbonyl ^{18}O, the following order of rate constants was observed: $k_\alpha = k_{equil} > k_t$. Also, if ester is recovered after partial reaction and resolved, the separated enantiomers have different amounts of ^{18}O scrambling. Explain these results.

4. Given the solvolysis rates of *exo*-2-norbornyl tosylate in 60 and 80% aqueous ethanol at 25°C as 1.60×10^{-3} and 2.31×10^{-4} sec^{-1}, respectively, calculate the solvolysis rate in acetic acid.

5. Rationalize the observation that triethylamine reacts with methyl and isopropyl iodides to give $k_{CH_3}/k_{isopropyl} = 2.9 \times 10^4$, and quinuclidine reacts to give $k_{CH_3}/k_{isopropyl} = 2.4 \times 10^3$

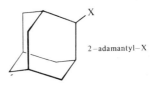

quinuclidine

6. Assuming that both cyclopentyl and cyclohexyl tosylates react in acetic acid by an S_N1 mechanism, which tosylate will undergo acetolysis most rapidly? (See ref. 6, p. 95.) Will this order hold for an S_N2 mechanism?

7. The relative rates of solvolysis in 80% ethanol of the compounds C$_6$H$_5$–CHCl–R for different R groups are methyl, 540; ethyl, 125; isopropyl, 27; *tert*-butyl, 1. Explain.

8. Nucleophiles with adjacent lone pairs (such as hydrazine) appear to show enhanced nucleophilicity. Explain this effect on the basis of orbital interaction within the nucleophile (see PMO theory for interaction of degenerate orbitals).

9. Given the rates of ethanolysis of 2-adamantyl *p*-methoxy- and *p*-methyl-benzenesulfonates as 4.3×10^{-5} and 5.7×10^{-5} sec^{-1}, respectively, calculate the ethanolysis rate of the *p*-nitro derivative.

2—adamantyl—X

10. Arrange the following compounds in order of solvolytic reactivity:

11. Does failure to observe a common ion rate depression for a particular reaction rule out involvement of a free carbocation in the reaction? Explain.

12. Give a mechanism for the following:

$$\triangleright\!\!-CH_2-OTs \xrightarrow{\text{AcOH}} \triangleright\!\!-CH_2-OAc + \square\!\!\!\overset{\overset{\text{H}}{|}}{-OAc}$$

$$+ \quad \diagup\!\!\!\!\diagdown\!\!\!\!\diagdown\!\!\!_{OAc}$$

13. Explain the order of m values for methyl, isopropyl, and 2-adamantyl tosylates in terms of charge development in the respective transition states (Table 4.6).

14. Given that

$$\frac{k_N[N_3^-]}{k_W[H_2O]} = \frac{RN_3}{ROH} \text{ for R—X} \begin{cases} \xrightarrow{N_3^-} RN_3 \\ \xrightarrow{H_2O} ROH \end{cases}$$

and

$$\text{rate acceleration} = \text{r.a.} = \frac{k_N[N_3^-] + k_W[H_2O]}{k_W[H_2O]}$$

derive

$$1 - \frac{1}{\text{r.a.}} = \frac{\% \, RN_3}{100}$$

Using this equation, treat the following data and determine which of the reactions is second-order and which is first-order.

	2-Propyl-OTs			2-Adamantyl-OTs	
r.a.	$[N_3^-]$	% RN$_3$	r.a.	$[N_3^-]$	% RN$_3$
0	0.00	0	0	0.00	0.0
1.44	0.02	31	1.08	0.02	0.1
2.18	0.04	54	1.12	0.04	0.4
2.94	0.06	65	1.16	0.06	0.7

15. Predict the products of solvolysis of the following compounds. Both follow k_Δ processes.

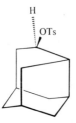

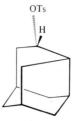

16. Predict the final position of ^{14}C in the acetolysis of **I.** It rearranges.

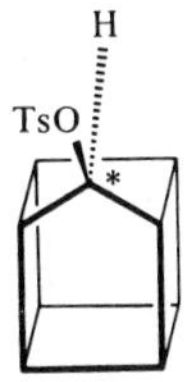

I

17. Suggest mechanisms for the following reactions:

$$CH_3-\underset{CH_3}{\overset{OCH_3}{C}}-CH_2-OBs \xrightarrow{H_2O} \begin{matrix} CH_3 \\ CH_3 \end{matrix}\!\!>CH-CHO$$

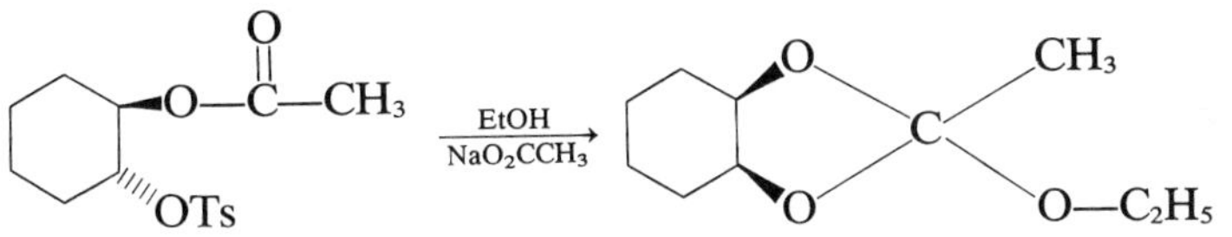

18. Predict the products and stereochemistry for the following reaction:

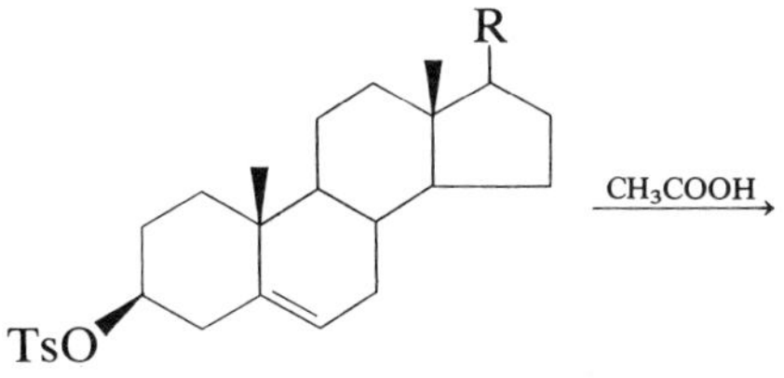

Hint: Draw the first two rings in the chair form.

19. The proton nmr spectrum of the 2-norbornyl cation in strong acid solution at 0°C shows only one absorbance. Explain (refer to p. 177).

20. Protonation of cyclooctatetraene gives a carbocation displaying absorptions in the nmr spectrum at τ 1.4, 3.4, 4.8, and 10.6 (areas 5:2:1:1). The protons giving the latter two absorptions are attached to the same carbon; explain the extraordinary difference in chemical shift for hydrogens attached to the same carbon.

21. Propose mechanisms for the following reactions:

(*a*) $$C_6H_5\underset{CH_3}{\overset{*}{C}}HOH + SOCl_2 \xrightarrow{\Delta} C_6H_5\underset{CH_3}{\overset{*}{C}}HCl$$

retention of configuration
increased solvent polarity facilitates reaction

(*b*) $$(CH_3)_2\underset{OH}{C}-\underset{Cl}{C}(CH_3)_2 \xrightarrow[\text{acetone}]{H_2O} CH_3\overset{O}{\overset{\|}{C}}C(CH_3)_3$$

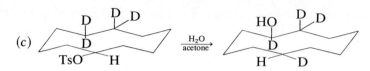

(c)

$$\xrightarrow[\text{acetone}]{\text{H}_2\text{O}}$$

(d) $\underset{\overset{|}{\text{C}_2\text{H}_5}}{\text{C}_4\text{H}_9\text{CH}}{-}\underset{}{\overset{\overset{\displaystyle\text{N-OH}}{\|}}{\text{C}}}{-}\text{CH}_3$ $\xrightarrow[\text{ether}]{\text{H}_2\text{SO}_4}$ $\underset{\overset{|}{\text{C}_2\text{H}_5}}{\text{C}_4\text{H}_9\text{CH}}{-}\text{NH}\overset{\overset{\displaystyle\text{O}}{\|}}{\text{C}}{-}\text{CH}_3$

a Beckmann rearrangement
retention of configuration

22. The double bond in the following compound acts as a neighboring group. Predict the products of solvolysis.

23. Use the following plot for 3-aryl-2-butyl tosylate acetolysis to predict the product stereochemistry for Compound **II**.

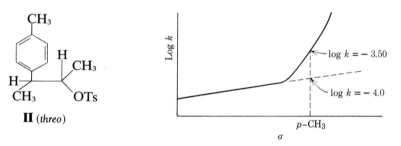

II (threo)

$\log k = -3.50$

$\log k = -4.0$

24. Predict elimination products from reaction of the following compounds with base:

(a) $\underset{\quad\quad\quad\overset{|}{+\text{N}(\text{CH}_3)_3}}{\text{CH}_3\text{CH}_2\text{CH}_2\text{C}(\text{CH}_3)_2}$

(b) $\underset{\quad\quad\quad\quad\overset{|}{\text{Cl}}}{\text{CH}_3\text{CH}_2\text{CH}_2\text{CHCH}_3}$

(c)

(d)

25. Rationalize the observation that the reactions of 2-Bu-SMe$_2^+$ with ethoxide and 2-BuBr with *tert*-butoxide both give ratios of 1-butene to 2-butene of three.

26. Assuming operation of an E2 mechanism for elimination, predict the order of elimination:substitution ratios for ethyl, 2-propyl, and *tert*-butyl chlorides.

27. For the second-order reaction of 2-propyl bromide with hydroxide, predict the effect on elimination:substitution ratio of a solvent change from water to ethanol (refer to discussion of solvent effects on substitution reaction, p. 151).

28. Explain:

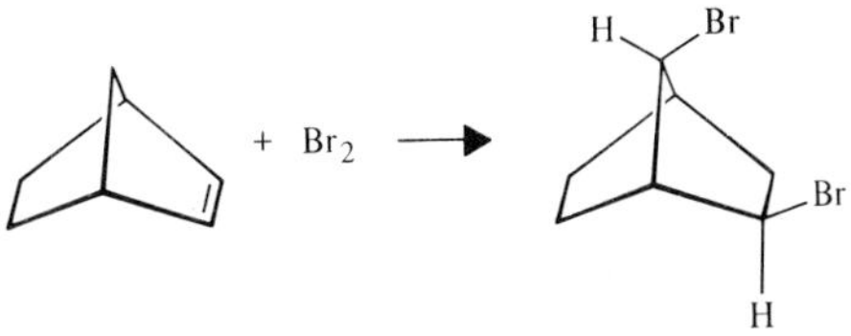

29. The addition of hydrogen halides to α,β-unsaturated ketones is acid catalyzed and retarded by alkyl substitution on the olefinic carbons. These properties are not those expected for the electrophilic addition mechanism; propose a reasonable mechanism.

30. Explain the following:

(a)

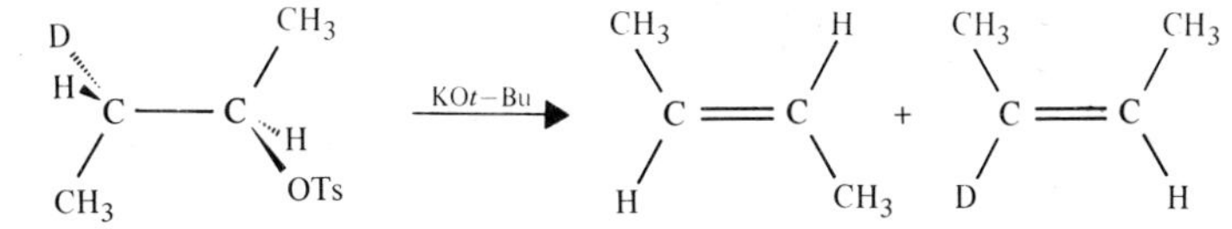

(b)

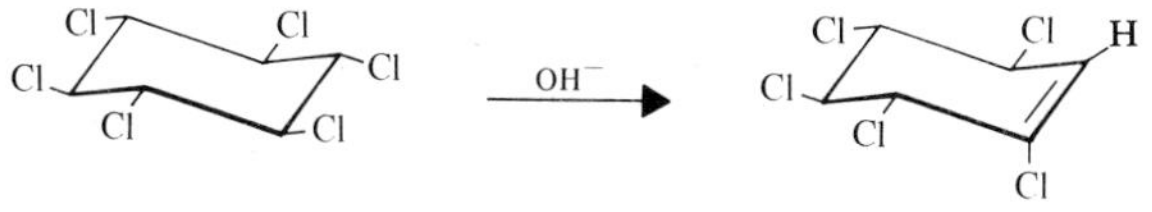

10^4 slower than any other isomer

(c) $Cl(CH_2)_3CN$ $\xrightarrow[\text{NH}_3]{\text{NaNH}_2}$

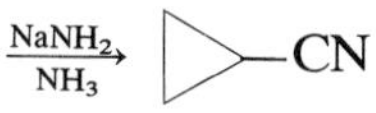

(d) $CH_2{=}CHCH_2NH\overset{\displaystyle O}{\overset{\displaystyle \|}{C}}C_6H_5$ $\xrightarrow[\text{AcOH}]{\text{Br}_2}$

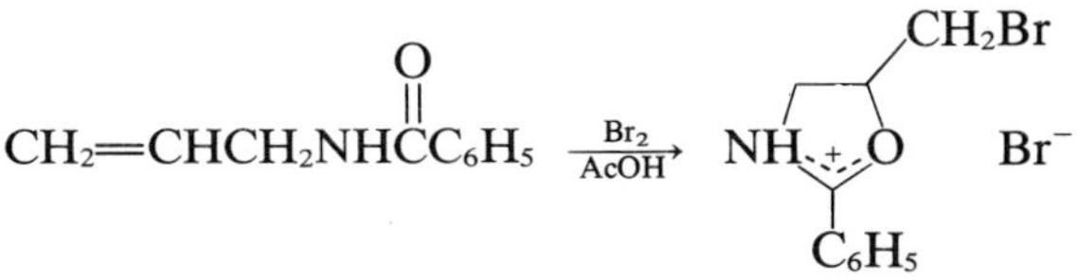

(stable—why?)

31. Rationalize the observations that hydroboration-oxidation of norbornene gives exo-OH whereas reaction with 7,7-dimethylnorbornene gives predominantly *endo*-OH. Addition of HCl gives *exo*-Cl for both norbornene and 7,7-dimethylnorbornene.

32. Assuming the *para* : *meta* ratio for nitration of toluene is 11.3, calculate ρ for this reaction and *para* : *meta* for nitration of chlorobenzene.

33. Propose mechanisms for the following rearrangements:

(a)

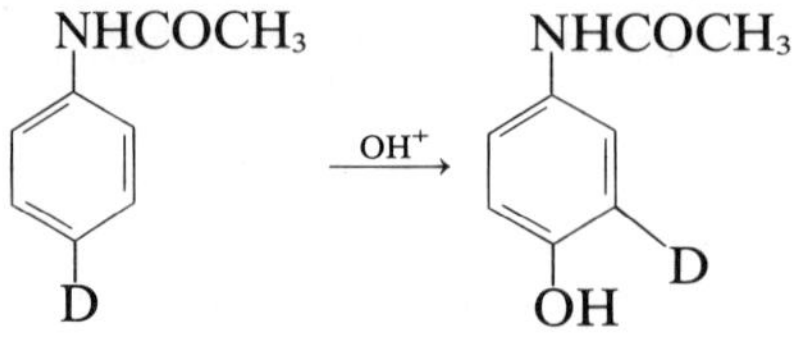

$$\overset{NHCOCH_3}{\underset{D}{\bigcirc}} \xrightarrow{\ OH^+\ } \overset{NHCOCH_3}{\underset{OH}{\bigcirc}} {}^D$$

(b)

$$\overset{NHNO_2}{\bigcirc} \xrightarrow[H^{15}NO_3]{H^+} \overset{NH_2}{\underset{}{\bigcirc}} NO_2 \quad + \quad \overset{NH_2}{\underset{NO_2}{\bigcirc}}$$

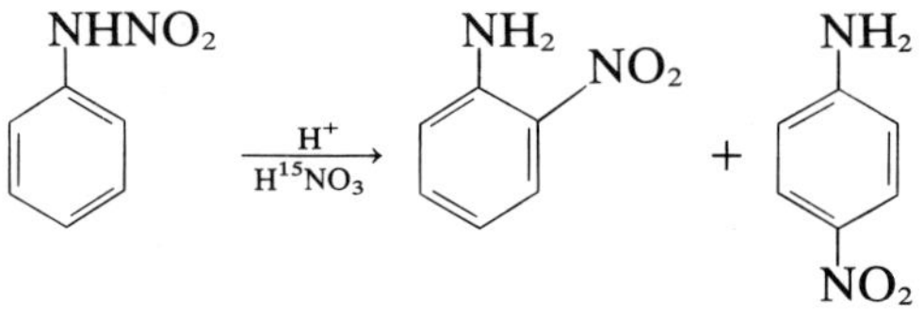

No ^{15}N incorporation
Hint: A nitrite is involved.

(c)

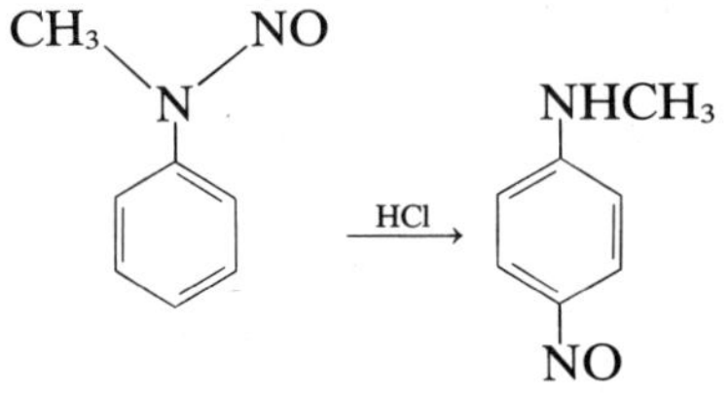

$$\overset{CH_3 \diagdown \underset{N}{} \diagup NO}{\bigcirc} \xrightarrow{\ HCl\ } \overset{NHCH_3}{\underset{NO}{\bigcirc}} \qquad \text{(A Fisher–Hepp rearrangement)}$$

Hint: If $C_6H_5NMe_2$ is added, it is nitrosated.

(d)

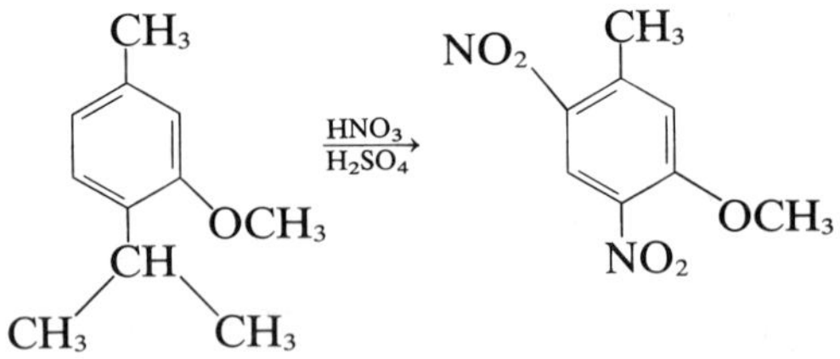

$$\underset{\underset{CH_3}{\overset{CH_3}{\diagup}}\underset{\diagdown}{\overset{CH}{|}} CH_3}{\overset{CH_3}{\bigcirc}{OCH_3}} \xrightarrow[H_2SO_4]{HNO_3} \overset{NO_2 \diagdown \; CH_3}{\underset{NO_2}{\bigcirc} OCH_3}$$

(e)

$$\underset{CH_3O}{\overset{OCH_3}{\bigcirc}} \overset{CHCH_2CH_2CH_2OBs}{\underset{CH_3}{|}} \xrightarrow{\ HCOOH\ } \underset{CH_3O}{\overset{OCH_3}{\bigcirc}}\underset{CH_3}{}$$

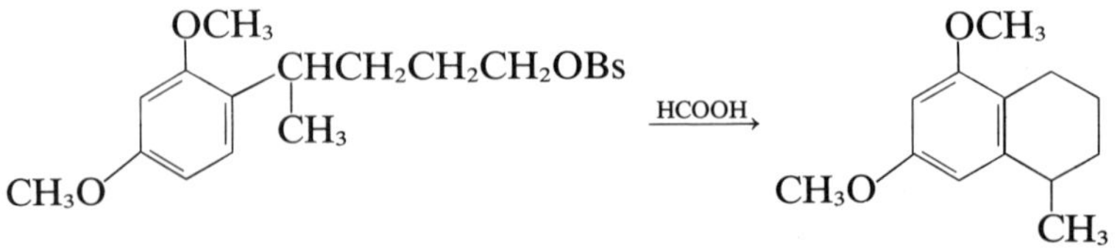

5

Carbanions

Carbanions are intermediates in which carbon is negatively charged. The elimination discussed in Chapter 4 which proceeded by the E1cB mechanism provided one example of carbanion intermediates. In the present chapter, we will discuss the possible involvement of carbanion intermediates in nucleophilic aromatic substitution

$$\qquad (1)$$

and in electrophilic aliphatic substitution reactions (such as protonation of organometallic compounds)

$$CH_3\!-\!MgBr + H_2O \rightarrow CH_4 + HOMgBr \qquad (2)$$

and in the aldol condensation

$$\qquad (3)$$

In this chapter, we also introduce methods for the study of acid–base catalysis.

5.1 Structure and Stability

Structure

There are two reasonable structures for alkyl carbanions, a planar sp^2 hybridized configuration and a pyramidal sp^3 hybridized configuration:

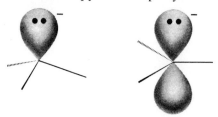

Increasing the amount of s character in an orbital (e.g., $sp^2 = 33\frac{1}{3}\%$ s and $sp^3 = 25\%$ s) results in the orbital being closer to the nucleus and lower in energy. Since the lone-pair electrons of a carbanion would be more stable if they were closer to the nucleus, it seems probable that the pyramidal sp^3 hybridized configuration will be preferred for carbanions. Also, there is a repulsive interaction between the lone-pair electrons and the three pairs of bonding electrons that is minimized in the pyramidal form. Evidence for this interpretation comes from the study of optically active 2-octyllithium prepared by metal-halogen exchange:

$$\underset{\text{Optically active}}{\overset{\overset{\displaystyle I}{|}}{C_6H_{13}-CH-CH_3}} \xrightarrow[-2-BuI]{2-BuLi} \overset{\overset{\displaystyle Li}{|}}{C_6H_{13}-CH-CH_3} \xrightarrow[(2)H_3O^+]{(1)CO_2} \underset{\substack{-70°;\ 20\%\ \text{retention} \\ 0°;\ \text{racemic}}}{\overset{\overset{\displaystyle CO_2H}{|}}{C_6H_{13}-CH-CH_3}}$$

Reaction of this compound with CO_2 at $-70°C$ gave the carboxylic acid with overall 20% retention (i.e., 60% retention, 40% inversion) of configuration, but reaction at 0°C gave a racemic product.[1] These results are consistent with dissociation of the covalent[2] lithium compound to give a rapidly equilibrating (by inversion) pair of pyramidal carbanions that attack CO_2; at $-70°C$ attack occurs before equilibration is complete to give partial retention of configuration, but at 0°C complete equilibration and racemization result:

Enantiomers

The barrier to carbanion inversion can be increased by incorporating the carbanion in a three-membered ring. There is substantial ring strain in a cyclopropane; the bond angles are restricted to 60°, although tetrahedral angles of 109° would be preferred. Upon inversion, the carbanion must pass through a planar state in which the bond angles remain restricted to 60° although angles of 120° would be preferred:

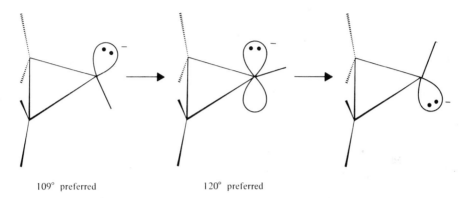

109° preferred   120° preferred

Inversion must, therefore, be accompanied by a major strain increase. Not surprisingly then, carbonation of *cis*-2-methylcyclopropyl lithium gives stereospecific formation of the *cis* acid:[3]

Similar effects are observed for base-catalyzed racemization and hydrogen isotope exchange of optically active alkanes. Treatment of optically active **1** with $CH_3OD/CH_3O^-K^+$ gives an exchange to racemization ratio of unity,[4] but the same conditions for **2** gives a ratio of 8080;[5] inversion and racemization must be much more rapid than return and deuteration for **1** but not for **2**. Alternatively, **3** may be planar.

$$C_6H_5CH_2-\underset{\underset{H}{|}}{\overset{\overset{CH_3}{|}}{C}}-CN \xrightarrow{CH_3O^-} C_6H_5CH_2-\underset{\underset{}{|}}{\overset{\overset{CH_3}{|}}{\underset{-}{C}}}-CN \xrightarrow{CH_3OD} C_6H_5CH_2-\underset{\underset{D}{|}}{\overset{\overset{CH_3}{|}}{C}}-CN$$

Optically active	Rapidly inverting or planar	Racemic
1	**3**	

Optically active	Slow inversion	Largely retained
2		

Conjugative overlap is most effective for a planar, sp^2-hybridized center and overlap between the cyano group and negative carbon quite possibly leads to dominance of the planar form for **3**. Other conjugatively-stabilized carbanions react similarly with loss of chirality:

$$\underset{C_6H_5CH_2}{\overset{CH_3}{\diagdown}}C{=}C{=}N^-$$

3

The evidence for a preferred pyramidal electronic configuration for carbanions is further strengthened by the observation that bridgehead organolithium compounds form readily and react normally with electrophiles;[6] if the preferred conformation were planar, as for carbocations, this result would be surprising. Also, carbanions are isoelectronic with amines, and amines are known to exist in rapidly inverting pyramidal forms.

Stability—Methods of Determination[7-11]

Relative carbanion stabilities are most readily evaluated by measurement of hydrocarbon acidities. Although we do not generally think of hydrocarbons as acids, they may be, when in the presence of strong bases:

$$\underset{}{\diagup}\overset{}{\diagdown}C{-}H + B{:}^- \underset{k_{-1}}{\overset{k_1}{\rightleftarrows}} \underset{}{\diagup}\overset{}{\diagdown}C{:}^- + BH$$

$$K_a = \frac{k_1}{k_{-1}}$$

Since carbon acids are so weak, it is necessary to use techniques more complicated than a simple titration to determine their acidities. A commonly used method is based on the equilibrium established between two carbon acids and their metal (e.g., potassium) salts.[7] This competitive method gives the difference in pK_as of the hydrocarbons, and if the pK_a of

$$R{-}H + R_1^-K^+ \rightleftharpoons R^-K^+ + R_1{-}H$$

$$pK_a - pK_a^1 = \log\frac{[RH][R_1^-K^+]}{[R^-K^+][R_1H]}$$

one acid is known, the other can be determined by difference. This competitive method has also been applied to the metal–halogen exchange reaction:[12]

$$R{-}Li + R_1{-}I \rightleftharpoons R{-}I + R_1{-}Li$$

One difficulty with the competitive method derives from the involvement of ion pairs in these reactions and from solvent effects on this ion pairing. For example, potassium salts of carbon acids may exist as free ions in one solvent, and as tight ion pairs in another solvent. This effect can lead to a dependence of acidity on solvent as is described later. Also, many of the organolithium and organomagnesium compounds used in competitive studies are known to be covalent, rather than ionic, and to exist in solution as aggregates.[2] It would be best for the present purposes if equilibria between free, monomeric carbanions could be studied; such conditions are achieved for potassium salts in dimethyl sulfoxide.[2d]

The pK_a is defined in terms of activities (a_x) and is, therefore, independent of medium if properly determined:

$$AH \rightleftharpoons A^- + H^+$$

$$pK_a = -\log K_a = -\log\frac{a_{H^+}a_{A^-}}{a_{HA}} \tag{4}$$

$$a_x = \gamma_x[X] \tag{5}$$

In dilute aqueous solution, activities can be equated to concentrations (i.e., activity coefficient $\gamma = 1$), and pK_a can be directly determined. However, to obtain measurable equilibria with weakly acidic carbon acids it is necessary to use strongly basic media (frequently nonaqueous or mixed solvents) where activities and concentration cannot be equated, and true thermodynamic or solvent-independent pK_as cannot be directly determined. A general and frequently used approach to this problem was developed by Hammett[13] for determination of acidities of concentrated acid solutions.[14–17] This approach results in a so-called acidity function as a measure of acidity (and basicity) of a medium. The acidity function approach is based on the observation that the relative extents of proton donation are essentially constant for a given pair of protonated aromatic amines in different mixtures of water and sulfuric acid; that is, for $Ar{-}NH_3^+$ and $Ar_1{-}NH_3^+$ in all acid mixtures:

$$\log\frac{[Ar{-}NH_3^+]}{[Ar{-}NH_2]} - \log\frac{[Ar_1{-}NH_3^+]}{[Ar_1{-}NH_2]} = C \tag{6}$$

To understand the importance of this observation, it is helpful to consider the equation of pK_a of an acid of the type $A\text{–}H^+$:

$$pK_a = -\log\frac{[H^+][A]}{[AH^+]} - \log\frac{\gamma_{H^+}\gamma_A}{\gamma_{HA^+}} \tag{7}$$

The difference in pK_as of two acids AH^+ and BH^+ is then

$$pK_a - pK_a' = \log\frac{[AH^+]}{[A]} - \log\frac{[BH^+]}{[B]} + \log\frac{\gamma_{AH^+}\gamma_B}{\gamma_A\gamma_{BH^+}} \tag{8}$$

According to eq. 6, the first two terms on the right-hand side of eq. 8 must be independent of solvent variation. Thus, the activity-coefficient term must also be independent of solvent, and, since this term is zero in water, it must be zero in all solvents. Therefore,

$$\log\frac{\gamma_{AH^+}\gamma_B}{\gamma_A\gamma_{BH^+}} = 0 \tag{9}$$

and

$$\log\frac{\gamma_{AH^+}}{\gamma_A} = \log\frac{\gamma_{BH^+}}{\gamma_B} \tag{10}$$

In other words, the constant difference in pK_as of similar $Ar\text{–}NH_3^+$ in various water–sulfuric acid mixtures appears to be the result of similar variations in the activity coefficients for the two acids. A large series of aromatic amines studied by Hammett gave this behavior.

Adherence to eq. 6 for a series of acids in various solvents leads to two important results. First, it permits calculation of pK_as of acids in media other than dilute aqueous solution. Consider two acids, $Ar\text{–}NH_3^+$ and $Ar'\text{–}NH_3^+$, and assume that $Ar\text{–}NH_2$ is sufficiently basic to provide experimentally (e.g., by ultraviolet spectroscopy) determinable dissociation in water, but $Ar'\text{–}NH_2$ is not quite basic enough to provide a measurable equilibrium. Also, assume that both acids are measurably dissociated in 5% H_2SO_4 such that eq. 6 can be determined. The known pK_a of $Ar\text{–}NH_3^+$ in water permits solution of eq. 8 to give the pK_a of $Ar'\text{–}NH_3^+$, which was not measurable in water. If $Ar\text{–}NH_2$ is such a strong base that equilibrium cannot be measured in 10% H_2SO_4, it cannot be used to determine pK_as in this solvent. But, assume that $Ar'\text{–}NH_2$ being a weaker base can be used to determine pK_as of still stronger acids (or weaker bases) in the more strongly acidic medium. By using this overlapping technique of weaker and weaker bases, it is possible to measure pK_as of bases that are only protonated significantly in 100% H_2SO_4.

The second important result from eq. 6 is the quantitative definition of acidity of the various water–sulfuric acid mixtures. The pK_a of a particular acid is a constant irrespective of solvent and is given by

$$pK_a = \log\frac{[AH^+]}{[A]} - \log\frac{a_{H^+}\gamma_A}{\gamma_{AH^+}} \tag{11}$$

and eq. 7. If the pK_a is a constant, the extent of protonation given by $\log[AH^+]/[A]$ is determined by the activity term, $\log a_{H^+}\gamma_A/\gamma_{AH^+}$. This last term, therefore, determines the ability of the solvent to accept protons from AH^+ and to donate protons to A. In other words, $\log a_{H^+}\gamma_A/\gamma_{AH^+}$ is a measure of acidity of the solvent system (in this case water–sulfuric acid mixtures). Hammett has designated this term the acidity function H_0, where the subscript indicates a neutral base (Ar–NH$_2$) is used as A in eq. 12. The H$_0$ values are determined for a particular solvent mixture by measuring the [AH$^+$]/[A] ratio in the solvent for an acid of known pK_a (eq. 13).

$$H_0 = -\log \frac{a_{H^+}\gamma_A}{\gamma_{AH^+}} \tag{12}$$

$$pK_a = \log \frac{[AH^+]}{[A]} + H_0 \tag{13}$$

To summarize Hammett's work with aromatic amines in water–sulfuric acid, the observation of offsetting activity-coefficient ratios (eq. 10) and constant acidity differences (eq. 6) for similar protonated aromatic amines in different water–sulfuric acid mixtures permitted calculation of pK_as of weak bases, which were not measurable in water, by using an "anchor" pK_a of a base measurable in water and then overlapping pK_as of decreasingly basic amines in increasingly acidic solvent mixtures. Second, the known pK_a values of the amines (referred to as Hammett indicators) could be used to determine acidities, in terms of H$_0$, for the water–sulfuric acid mixtures. The Hammett indicators have also been used to determine H$_0$ values for other solvent systems (e.g., HClO$_4$, FSO$_3$H, and CF$_3$SO$_3$H), and the H$_0$ values have been used to determine basicities of weak bases other than aromatic amines.

Other indicators and acidity functions have also been used. In line with the present subject of acidities of weak carbon acids, an H$_-$ acidity function has been derived for the measurement of acidities of weak acids in strongly basic media.[18] This method utilizes nitroanilines as weak acids in strongly basic solutions such as dimethylsulfoxide-methanol-sodium methoxide:

$$pK_a = \log \frac{[Ar–NH_2]}{[Ar–NH^-]} + H_- \tag{14}$$

Once the pK_as of the anilines are determined by the overlap method, the H$_-$ values of various concentrated base solutions can be determined. These concentrated base solutions of known H$_-$ can then be used to determine the pK_as of carbon acids.

It should be recalled that the acidity function approach is an approximate approach that is dependent on the extent to which eq. 6 is obeyed. The method has been shown to fail in certain solvent mixtures and to depend on the choice of indicators.[16,17] Also, as is discussed later, variations in extent of ion pairing in different solvents can lead to different species being studied in different solvents.

Hydrocarbon acidities have frequently been measured by kinetic rather than thermodynamic techniques.[7-11] The usual method is to measure the rate of base-catalyzed hydrogen isotope exchange:

$$R—H + B^{\ominus} \underset{-BH}{\longrightarrow} R^{\ominus} \xrightarrow{A—T} R—T + A^{\ominus}$$

A number of attempts have been made to correlate kinetic and thermodynamic acidities, and despite difficulties, progress has been made.[7]

In Table 5.1 we have given some carbon-acid pK_as determined by the several methods discussed. The data in this table should not be regarded as final, since new, more accurate results are constantly being published.

Table 5.1 *Acidities of Carbon Acids*[7,8,10,19]

Compound	pK_a	Compound	pK_a
$CH_2(CHO)_2$	5	Triphenylmethane	30
$HC(CN)_3$	5.1	Diphenylmethane	32
Acetylacetone	9	Toluene (α position)	35
$CH_2(CN)_2$	11.2	Cycloheptatriene	36
Ethyl malonate	13	Ethylene	36.5
Nitromethane	15	Benzene	37
Cyclopentadiene	15	Cyclopropane	39
Fluorene	21	Methane	40
Acetone	24	Ethane	42
Acetylene	25	Cyclobutane	43
$CH_3SO_2CH_3$	28.5	Cyclopentane	44
Acetonitrile	29	Cyclohexane	45

Factors Affecting Stability of Carbanions

S-CHARACTER EFFECTS

As discussed earlier, the greater stability of a lone pair of electrons in an orbital of greater s character (a sp^3 orbital as compared to a p orbital) plays a role in favoring the pyramidal configuration for simple carbanions. Consistent with this interpretation, a good correlation is observed between per cent s character of a C–H bond and the acidity of this bond.[7] For example, there is a steady increase in acidity from ethane to ethylene to acetylene as per cent s in the respective C–H bonds goes from 25 to 33 to 50 (Table 5.1).

The per cent s character of a C–H bond is also a function of angle strain; as one bond angle of a tetrahedral carbon is reduced by incorporation in a small ring, angle strain is introduced and per cent s character of the exocyclic C–H bond is increased:

Angle variations
change per cent s character

This fact has been verified by molecular orbital calculations and by observations of trends in C_{13}–H nmr coupling constants (which are directly proportional to s character of a C–H bond).[20] As would then be expected, there is a steady increase in acidity along the series cyclohexane, cyclopentane, cyclobutane, and cyclopropane. Streitwieser has shown the rates of proton abstraction for a series of cycloalkanes from cyclopropane through cyclotetradecane to be linearly correlated with C_{13}–H coupling constants.[21] It is important to note that these proton abstractions are correlated by a ground state property (C_{13}–H coupling) and that consideration need not be given to product or transition state properties; this is certainly not the case for carbocation formation from cyclic precursors, where changes in torsional and angle strain upon ionization are most important. The key difference is that both carbanion and alkane have similar configurations. This lack of change results in essentially constant strain effects for ground and transition state and permits correlation of reactivity by a ground state property.

A rather extreme example of the angle strain effect is seen in the ready exchange of tertiary hydrogens of tricyclo [3.1.0.0^{2,6}] hexane:[22]

CONJUGATIVE EFFECTS[7]

As shown by the relatively high acidities of acetone, acetonitrile, nitromethane, and toluene (Table 5.1), the negative charge on carbon of a carbanion can be effectively delocalized and stabilized by conjugation:

The high acidity of cyclopentadiene (pK_a = 15) is also the result of a conjugative effect. In this case, the lone pair of electrons on the cyclopentadienyl anion are incorporated into an aromatic system (six π electrons) and delocalized throughout the ring. In dramatic contrast, cycloheptatriene

is found to be 21 pK_a units less acidic than cyclopentadiene. This difference is due to formation of the antiaromatic eight π electron carbanion upon removal of a proton from cycloheptatriene:

For conjugation to be effective, it is necessary for the interacting groups to become coplanar. Thus, **4** is observed to be much more acidic than **5**, which cannot form a planar carbanion.

4

5

Although we usually think of conjugative overlap between p orbitals, overlap between p and d orbitals can also be important. Consider, for example, the observation that the tetramethylphosphonium ion undergoes deuterium exchange in D_2O/OD^- 10^6 times as fast as does the tetramethylammonium ion.[23] This difference has been explained as due to the stabilizing effect of $p–d$ overlap for the phosphorus that is impossible for

$$^-CH_2{-}\overset{+}{P}{\Big\langle} \longleftrightarrow CH_2{=}P{\Big\langle}$$

$$^-CH_2{-}\overset{+}{N}{\Big\langle} \overset{\times}{\longleftrightarrow} CH_2{=}N{\Big\langle}$$

nitrogen; second-row elements have no low-energy empty d orbitals to accept the lone-pair electrons. These intermediates containing both positive and negative charges are referred to as ylids. Many other examples of carbanion stabilization by $p–d$ conjugation, usually involving sulfur or phosphorus, are known.[7]

To this point, we have considered only $\pi–\pi$ conjugation. Conjugation between σ and π systems, or hyperconjugation, is also potentially important in stabilization of carbanions. For example, fluorines substituted β to a negative center have a major carbanion stabilizing effect (Table 5.2) that is frequently attributed to hyperconjugation:[7,24]

$$F{-}\overset{\overset{\displaystyle F}{|}}{\underset{\underset{\displaystyle F}{|}}{C}}{-}\bar{C}{\Big\langle} \longleftrightarrow \overset{\displaystyle F}{\underset{\displaystyle F}{\Big\rangle}}C{=}C{\Big\langle}$$
$$F^-$$

INDUCTIVE EFFECTS[7-11]

Obviously, electron-withdrawing substituents will stabilize an electron-rich carbon. For example, attachment of a quaternary ammonium group to an alkane results in its being more acidic than an alkene. Consistent with the proposed electron-donating effect of alkyl groups, the observed acidity order for alkanes is primary > secondary > tertiary.

Table 5.2 *Rates of Deuterium Exchange in* $CH_3OD–CH_3O^-$ *for a Series of Monohydrofluorocarbons*[24]

Compound	Relative Rate
CF_3H	1
$CF_3(CF_2)_5CF_2H$	6
$(CF_3)_2CFH$	10^5
$(CF_3)_3CH$	10^9

SOLVENT EFFECTS

As with all reactions involving ions, solvent plays a highly significant role in stabilizing the participant ions. In the absence of solvent, ionic reactions in the gas phase are highly energetic; free radical reactions are much more common. For example, dissociation of HCl to radicals requires 103 kcal/mole, while dissociation to ions requires 333 kcal/mole. Since HCl readily dissociates to ions in polar solvents, the profound effect of solvent in stabilizing ions is evident.

The relatively new technique of ion cyclotron resonance spectroscopy has been utilized to monitor gas-phase ionic reactions. A major application has been the determination of relative gas-phase acidities, in some cases with surprising results. For example, the order of acidity of alcohols in the gas phase is found to be t-BuOH $> i$-PrOH $>$ EtOH $>$ MeOH $>$ H$_2$O,[25] exactly the reverse order of what is normally observed in solution. Although alkyl groups are normally considered to be inductively electron-donating, it is proposed that the greater polarizability of alkyl groups (relative to H) allows them to be effective in delocalizing negative charge (again relative to H). The reversal of the order in solution is then considered to reflect steric solvation effects. That is, the bulkier alkyl groups prevent the alkoxide anion from being well solvated. By destabilizing the anion, larger alkyl groups lead to decreased acidity in solution.

Hydrocarbon acidities and carbanion stabilities in the gas phase also show some surprises. For example, the acidity of toluene in the gas phase is greater than that of methanol, and propylene is more acidic than water.[25] In the absence of solvent, the dependence upon substituents for resonance or inductive stabilization seems to be greatly enhanced. Substituents that are particularly effective for carbanion stabilization, such as aryl, acetyl, benzoyl, and cyano, have been shown to exert their effects in roughly the same order in the gas phase as in solution.[26]

Given the critical effect of solvent upon ionic stabilities, in many instances solvents may be selected for specific purposes. For example, polar protic solvents, such as water and other hydroxylic solvents, are capable of effectively solvating both cations and anions; cations are solvated by dipolar interactions with lone pairs, and anions are solvated by hydrogen bonding. Polar aprotic solvents, however, are capable of solvating cations but do not effectively solvate anions. Thus, anions tend to be more reactive in polar aprotic solvents.[7]

$$\underset{\text{Dimethylsulfoxide (DMSO)}}{\overset{\overset{\displaystyle O}{\|}}{CH_3SCH_3}} \qquad \underset{\text{Dimethylformamide (DMF)}}{\overset{\overset{\displaystyle O}{\|}}{HCN(CH_3)_2}} \qquad \underset{\text{Hexamethylphosphoramide (HMPA)}}{[(CH_3)_2N]_3P{=}O}$$

Some common polar aprotic solvents

A good example of such a solvent effect is the observation of the greatly enhanced basicity of methoxide in DMSO, where it is poorly solvated, as compared to methanol, where it is heavily hydrogen-bonded. In terms of the H$_-$ acidity function (see p. 219), for 0.035 M sodium methoxide in methanol, H$_-$ is measured to be 12.2, while an equal concentration in 95% DMSO–5% methanol solvent has an H$_-$ value of 19.4.[27] In kinetic terms, optically active

2-methyl-3-phenylpropionitrile undergoes methoxide-catalyzed racemization and deuterium exchange about 10^8 times faster in 98.5% DMSO–1.5% methanol than in pure methanol:

$$\phi\text{—CH}_2\text{—}\overset{\displaystyle \text{CH}_3}{\underset{\displaystyle \text{CN}}{\overset{\displaystyle |}{\underset{\displaystyle |}{\text{C}^*}}}}\text{—H}$$

A comparable technique that leads to the same effect is the solubilization of the cation by a complexing or chelating agent, most commonly by macrocyclic polyethers called crown ethers.[28] The anion is thereby brought into solution as the necessary counterion, even though it may not be effectively solvated. For example, potassium permanganate may be dissolved in benzene (commercially available as "purple benzene") by complexing the potassium ions with a crown ether. There are many synthetic advantages of having a highly reactive oxidizing agent (poorly solvated MnO_4^-) in a nonpolar organic solvent.

A crown ether complex of potassium ion (soluble in organic solvents)

Crown ethers have now been used to solubilize salts of all kinds in nonpolar solvents. Not only basicity but nucleophilicity can be enhanced under these conditions. Direct nucleophilic substitution of methoxide on o-dichlorobenzene was accomplished when the potassium methoxide was solubilized with a crown ether.[29] Normally, such an aromatic nucleophilic substitution would take place only on an aryl halide that was activated by strong electron-withdrawing groups (see p. 241):

$$\text{K}^+(\text{crown ether})\ ^-\text{OCH}_3 +$$

(40 to 50%)

Solvent variation can also have an effect on the reactivity of carbanionic species by affecting ion pair equilibria and ion aggregation. Ion pairing was discussed in detail in Chapter 4, but we have not discussed ion aggregation. Ion aggregation is quite important for organometallic species and is simply the association in solution of several molecules, $(M\text{–}R)_n$.[2b]

The effect of solvation on ion pairing and ion aggregation, and in turn on the properties of the carbanions, is readily seen in solvent effects on pK_a values. For example, phenylacetylene is found to have a pK_a of 16 in ether, 21 in cyclohexylamine, and 27 in dimethylsulfoxide.[19] This enormous acidity dependence on solvent has been attributed to ion-pairing effects.[19] The three foregoing pK_as were all determined by equilibrium measurement:

$$\text{K}^+\text{In}^- + \text{R—H} \rightleftharpoons \text{H—In} + \text{K}^+\text{R}^-$$

where H–In is some indicator of known acidity, and R is $C_6H_5C{\equiv}C{-}$. In dimethylsulfoxide the ions are fully dissociated. However, ion pairing and ion aggregation are known to be important in the other two solvents, being most important in ether. It is generally observed, as in this instance, that the equilibrium will shift to favor the ion pair of a weak acid as ion pairing becomes more important. Thus, measured acidities of weak acids are much greater in weakly dissociating solvents.[19] Apparently, the more unstable carbanions stand more to gain as ion pairing or ion aggregation become more important. Also, it is probable that different types of ion pairs or ion aggregates are involved in the two solvents, for example, possibly a tight ion pair, K^+R^-, in cyclohexylamine and an ion aggregate, $(K^+R^-)_n$, in ether. If this is the case, different species and equilibria are being studied in the different solvents, and the lack of agreement between pK_as is not surprising.

5.2 Electrophilic Aliphatic Substitution

Many substitution reactions on alkyl groups occur by electrophilic attack on an intermediate carbanion:

$$R{-}X \underset{\text{slow}}{\rightleftharpoons} R^- + X^+$$
$$\phantom{R{-}X}\;\;\xrightarrow{\;E^+\;} R{-}E \qquad\qquad (S_E1)$$

This mechanism is referred to as the S_E1 mechanism—substitution, electrophilic, unimolecular. The most common examples of the S_E1 mechanism are probably the base-catalyzed reactions of various carbon acids, especially carbonyl containing ones:

$$\underset{CH_3\overset{\displaystyle\;O\;\parallel}{C}CH_3 + OH^-}{} \xrightarrow{\text{slow}} CH_3\overset{\displaystyle\;O\;\parallel}{C}CH_2^- \xrightarrow{Br_2} CH_3\overset{\displaystyle\;O\;\parallel}{C}CH_2Br + Br^-$$

These reactions are actually second-order kinetically, first-order in base, and first-order in substrate. The S_E1 designation is used in the broader sense of a stepwise reaction independent of electrophile in the rate-determining step; the S_N1 designation was used similarly. Certain electrophilic substitutions of organometallic compounds may also occur by an S_E1 mechanism:

$$CH_3CH_2Li \xrightarrow{\text{pentane}} CH_3CH_2^- + Li^+$$
$$\;\;\xrightarrow{\;CO_2\;} CH_3CH_2CO_2^-Li^+$$

Electrophilic aliphatic substitutions can also occur by bimolecular concerted mechanisms of which two have been suggested, the S_E2 mechanism and the four-center S_Ei mechanism.

$$R{-}X + E^+ \longrightarrow \left[\overset{\delta}{R}\cdots\overset{+\delta}{\underset{\overset{+\delta}{E}}{X}}\right]^{\ddagger} \longrightarrow R{-}E + X^+ \qquad (S_E2)$$

$$R{-}X + E{-}Y \longrightarrow \left[R\overset{\cdots E\cdots}{\underset{\cdots X\cdots}{}}Y\right]^{\ddagger} \longrightarrow R{-}E + XY \qquad (S_Ei)$$

Reaction by the $S_{E}i$ mechanism must take place with retention of configuration. There is no such stereochemical restriction on the $S_{E}2$ mechanism, but the great majority of second-order electrophilic substitutions have been observed to occur with retention of configuration, and it must be supposed that the $S_{E}2$ mechanism is best drawn as giving retention.[30] Actually, the involvement of $S_{E}2$ mechanisms has been questioned.[31] Electrophilic substitution by a concerted mechanism is thought to be important for many organometallic compounds including organomercury and organomagnesium compounds:

$$2\text{-BuHgBr} + 2\text{-OctylLi} \xrightarrow{\text{pentane}} 2\text{-Bu-Hg-2-Octyl} + \text{LiBr}$$

$$2\text{-PrMgBr} + \text{HCl} \longrightarrow 2\text{-Pr-H} + \text{MgClBr}$$

In the next two sections, we discuss two typical and well-studied examples of electrophilic aliphatic substitution; these are the base-catalyzed condensation reactions of carbonyl compounds and electrophilic substitutions on organomercurials.

The Aldol Condensation

Base-catalyzed substitutions on carbonyl compounds must be classified as among the most important synthetic reactions of organic chemistry; some examples of these reactions follow:

$$2\text{CH}_3\text{CHO} \xrightarrow{\text{OH}^-} \overset{\overset{\displaystyle \text{OH}}{\displaystyle |}}{\text{CH}_3\text{CH}}\text{—CH}_2\text{CHO}$$

Aldol Condensation

$$2\text{CH}_3\text{CO}_2\text{Et} \xrightarrow{\text{OEt}^-} \overset{\overset{\displaystyle \text{O}}{\displaystyle ||}}{\text{CH}_3\text{C}}\text{CH}_2\text{CO}_2\text{Et}$$

Claisen Condensation

$$\overset{\overset{\displaystyle \text{O}}{\displaystyle ||}}{\text{CH}_3\text{C}}\text{CH}_3 + \text{Br}_2 \xrightarrow{\text{OH}^-} \overset{\overset{\displaystyle \text{O}}{\displaystyle ||}}{\text{CH}_3\text{C}}\text{CH}_2\text{Br}$$

Halogenation

$$\text{CH}_2(\text{CO}_2\text{Et})_2 + \text{CH}_3\text{Br} \xrightarrow{\text{OEt}^-} \text{CH}_3\text{CH}(\text{CO}_2\text{Et})_2$$

$$\xrightarrow[\text{H}_2\text{O}]{\text{H}^+} \text{CH}_3\text{CH}(\text{CO}_2\text{H})_2 \xrightarrow{\Delta} \text{CH}_3\text{CH}_2\text{CO}_2\text{H}$$

Malonic ester synthesis

The mechanisms of these various substitutions are quite similar, and as representative examples, we have chosen to examine the aldol condensations of acetaldehyde and acetone. This examination is also important in that some general concepts of the study of acid- or base-catalyzed reactions are introduced.

CATALYSIS BY ACIDS OR BASES

The base-catalyzed condensations of aldehydes and ketones are thought to proceed by formation of an enolate that then attacks the carbonyl group of another molecule of aldehyde or ketone:[32-34]

The relative rates of the different steps vary, depending on the nature of the carbonyl compound and the concentrations of the reactants. Also, as shown, the reaction is reversible; the equilibrium favors products for acetaldehyde and reactants for acetone. Thus, it is more convenient to study the acetone reaction from the diacetone alcohol side; according to the principle of microscopic reversibility, the mechanism for the reversal process must simply be the reverse of that for the forward process.

The reaction sequence for the aldol condensation is a typical one for base-catalyzed reactions and certain labels have evolved to classify these mechanisms.[33] If the first step, proton transfer, is rate-determining, the reaction is said to be general base catalyzed because any base present in the solution will have a catalytic effect. If the second step, reaction of enolate, is slow, the reaction is said to be specific base catalyzed because in this instance, as we will show later, catalysis is a function only of the concentration of a single base, OH$^-$; in other words, the degree of specific base catalysis is pH dependent. As noted previously, for general base catalysis, the rate is a function of the concentration of all bases in solution. This difference permits identification of the rate-determining step. Assume that the aldol condensation is catalyzed by phenolate anion as a base. The pH equation for this solution then is given by

$$pH = pK + \log \frac{[C_6H_5O^-]}{[C_6H_5OH]}$$

As long as the ratio of phenolate to phenol remains a constant, the concentration of phenol and basic phenolate in solution can be varied without changing the pH. If the rate of our hypothetical aldol condensation is unchanged at a constant pH despite changes in $[C_6H_5O^-]$, the reaction must be specific base catalyzed, and k_2 must be the rate-determining step. On the other hand, if the rate does vary with $[C_6H_5O^-]$ despite the constant pH the reaction must be general base catalyzed and proton transfer must be rate-determining. This criterion is a general one for all base-catalyzed reactions.

The concepts of general and specific base catalysis can be best understood in kinetic terms. The preceding mechanistic scheme can be written more simply as

$$HA + B \underset{k_{-1}}{\overset{k_1}{\rightleftharpoons}} A^- + BH^+ \qquad A^- + HA \overset{k_2}{\longrightarrow} product$$

where HA is the carbonyl compound, B is the base, and A^- is the enolate ion. The final protonation step in the detailed mechanistic scheme is always faster than the other steps and can be ignored from a kinetic standpoint.

The rate of product formation for this scheme is given by

$$\text{rate} = \frac{d[\text{product}]}{dt} = k_2[A^-][HA]$$

Applying the steady-state approximation to the intermediate A^- gives

$$\frac{d[A^-]}{dt} = 0 = k_1[HA][B] - k_2[A^-][HA] - k_{-1}[A^-][BH^+]$$

$$[A^-] = \frac{k_1[HA][B]}{k_2[HA] + k_{-1}[BH^+]}$$

Therefore,

$$\frac{d[\text{product}]}{dt} = \frac{k_1 k_2[HA]^2[B]}{k_2[HA] + k_{-1}[BH^+]}$$

We have considered two limiting situations, one in which $k_2 \gg k_{-1}$ (general base catalysis) and another in which $k_2 \ll k_{-1}$ (specific base catalysis). For general base catalysis, the kinetic equation reduces to

$$\frac{d[\text{product}]}{dt} = \frac{k_1 k_2[HA]^2[B]}{k_2[HA]} = k_1[HA][B]$$

and for specific base catalysis to

$$\frac{d[\text{product}]}{dt} = \frac{k_1 k_2[HA]^2[B]}{k_{-1}[BH^+]}$$

This equation may be simplified by considering that, for any base in aqueous solution, the equilibrium constant is given by

$$K_b = \frac{[BH^+][OH^-]}{[B]}$$

The equation for specific base catalysis becomes

$$\frac{d[\text{product}]}{dt} = \frac{k_1 k_2[HA]^2[OH^-]}{k_{-1} K_b}$$

The important thing to note about the preceding equation is that hydroxide, not B, is the only base represented. As stated previously, if a reaction is subject to specific base catalysis (reaction of the conjugate base of the substrate—enolate in this case—is rate-determining), this catalysis will be a function of pH only and not of other base concentrations. This does not mean that bases other than hydroxide do not remove protons from the substrate; however, this participation reduces mathematically to a simple dependence on hydroxide concentration. Thus, for basic action by B before the rate-determining step, we have

$$B + HA \rightleftharpoons BH^+ + A^-$$

But we must also consider

$$BH^+ + OH^- \rightleftharpoons B + H_2O$$

And combination of these two equations gives

$$OH^- + HA \rightleftharpoons H_2O + A^-$$

As noted, it is necessary only to consider hydroxide as a base. Or stated another way, the substrate (carbonyl compound) is just another acid, and the amount of anion formed and ready to undergo the slow step from it will be a function only of its pK and the pH:

$$pH = pK + \log \frac{\left[R-\overset{\overset{\displaystyle O}{\|}}{C}-\bar{C}\diagup \right]}{\left[R-\overset{\overset{\displaystyle O}{\|}}{C}-\overset{|}{\underset{|}{C}}-H \right]}$$

Returning to the consideration of the condensations of acetaldehyde and acetone, it has been shown that varying the concentrations of bases and their conjugate acids while holding the ratios $[B]/[BH^+]$ constant (a buffered solution), has no effect on the rate of the acetone condensation; the rate is pH dependent, however. It must then be concluded that the reaction is specific base catalyzed. This interpretation is supported by the observation of rapid deuterium uptake by acetone in D_2O/OD^-; that is, k_{-1} is greater than k_2.[33]

For reaction of acetaldehyde, reaction rates are also observed to be a function of pH but not of buffer concentration, and rapid proton transfer is again indicated. However, the product from this reaction in D_2O/OD^- is deuterium free, and $k_2 > k_{-1}$ is indicated.[33,34] This example is an important one, since it illustrates a situation in which the buffer-ratio technique fails to detect general base catalysis. To understand the failure of the buffer-ratio technique it is necessary to consider the frequently observed linear free energy relationship (the Bronsted relationship) between the strength of a base (K_b) and its catalytic effect (k_{cat}):

$$\log k_{cat} = \beta \log K_b + \text{constant}$$

The constant β measures the dependence of a reaction on base strength. For reactions in which β approaches unity, strong bases are much more effective catalysts than weak bases; this is the case for the condensation of acetaldehyde where β is approximately 0.9. In dilute aqueous solution, the strongest possible base is hydroxide. The high value of β means that catalysis by bases other than hydroxide is simply difficult to detect. In these cases, it is necessary to use techniques such as deuterium exchange studies to detect rate-determining proton transfer.

The aldol condensation is also catalyzed by acids. A possible mechanism for this reaction is given by

$$R-\overset{O}{\underset{\parallel}{C}}-CH_3+H^+ \underset{k_{-1}}{\overset{k_1}{\rightleftharpoons}} R-\overset{\overset{+}{O}-H}{\underset{\parallel}{C}}-CH_3 \underset{k_{-2}}{\overset{k_2}{\rightleftharpoons}} R-\overset{OH}{\underset{\parallel}{C}}=CH_2+H^+$$

enol

$$R-\overset{OH}{\underset{\parallel}{C}}=CH_2 \xrightarrow[k_3]{\overset{+O-H}{\underset{R-C-CH_3}{}}} R-\overset{\overset{+}{O}-H}{\underset{\parallel}{C}}-CH_2-\overset{CH_3}{\underset{\underset{R}{|}}{\overset{|}{C}}}-OH \xrightarrow{k_4} R-\overset{O}{\underset{\parallel}{C}}-CH_2-\overset{CH_3}{\underset{\underset{R}{|}}{\overset{|}{C}}}-OH$$

In this case, the enol form of the ketone acts as the nucleophile and attacks the carbonyl carbon of another ketone molecule. Catalysis derives from facilitation of enol formation and from enhanced electrophilicity of the protonated ketone. Also, as before, protonation or reaction of the protonated species may be rate-determining, and the reaction may be general or specific acid catalyzed:

$$HX+S \underset{k_{-1}}{\overset{k_1}{\rightleftharpoons}} S-H^+ +X^-$$

$$S-H^+ \xrightarrow{k_2} product$$

$$k_{-1} \ll k_2 \qquad \text{general acid catalysis}$$

$$k_{-1} \gg k_2 \qquad \text{specific acid catalysis}$$

The same criteria hold as for base catalysis; a rate depending on buffer concentration and pH indicates general acid catalysis, and a rate dependency only on pH indicates specific acid catalysis. Also a Bronsted relationship between $\log k_{cat}$ and $\log K_a$ is observed for general acid catalysis. Both types of catalysis are observed for these reactions.

The other condensation reactions mentioned previously occur by mechanisms similar to that of the aldol condensations. In addition to these condensations, there are two related reactions of carbonyl compounds that are of much synthetic importance—these are the Wittig reaction and the reaction of enamines. The Wittig reaction involves attack of a phosphorous ylid on an aldehyde or ketone and provides an excellent method for the preparation of alkenes:[35]

$$\phi_3P=CH_2 \longleftrightarrow \phi_3\overset{+}{P}-CH_2^- \xrightarrow{\overset{O}{\overset{\parallel}{R-C-R}}} \begin{matrix} \phi_3\overset{+}{P}-CH_2 \\ \overset{|}{\underset{\underset{R}{|}}{\overset{-}{O}-C-R}} \end{matrix}$$

An ylid   A betaine

$$\overset{R}{\underset{R}{>}}C=CH_2+\phi_3PO \longleftarrow$$

Enamines are readily prepared from a secondary amine and a ketone, and because of their nucleophilicity (analogous to enols), they provide an excellent route for α alkylation of ketones:[35]

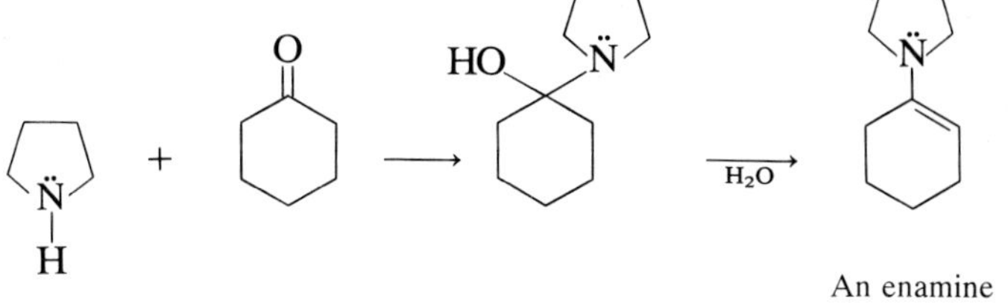

An enamine

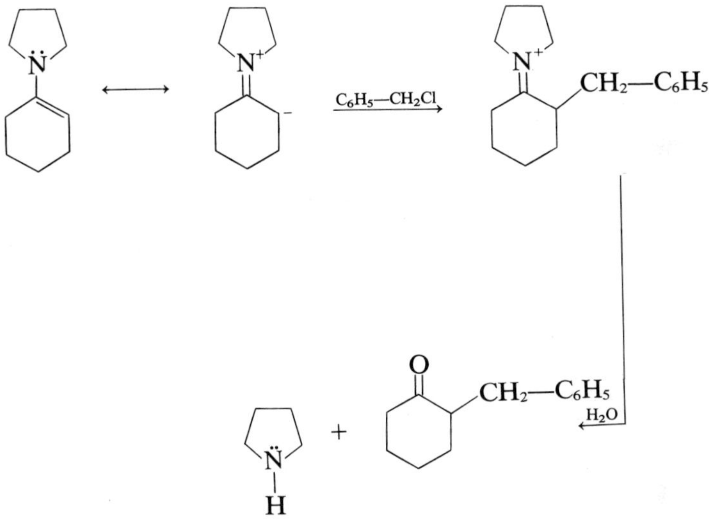

Direct base-catalyzed alkylation of ketones is a poor synthetic reaction because polysubstitution frequently occurs:[35]

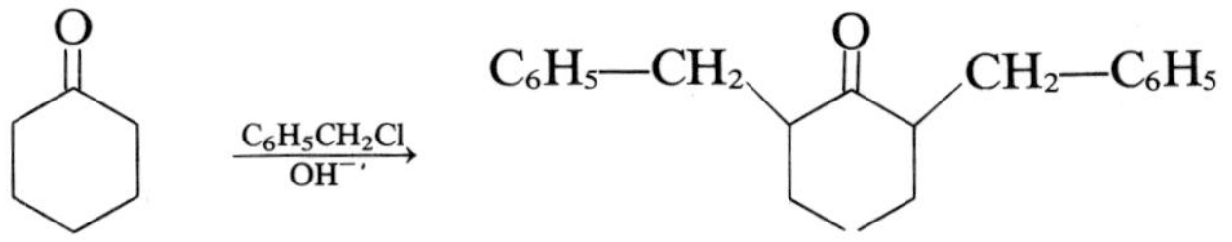

THE CARBONYL GROUP AS AN ELECTROPHILE—ESTER HYDROLYSIS[36-38]

Condensation reactions, and several other reactions, involve nucleophilic attack on a carbonyl group. The great susceptibility of the carbonyl group to nucleophilic attack is due to the ability of the carbonyl oxygen to accept negative charge and stabilize the resulting transition state:

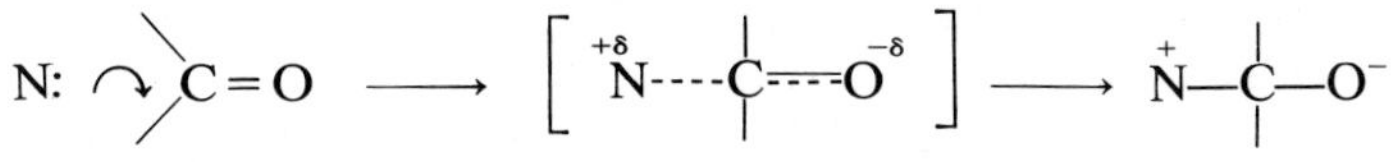

The tetrahedral intermediate can subsequently be protonated (resulting in overall addition of H and N), or the carbonyl group may be reformed by elimination of N or one of the other attached groups (resulting in overall substitution).

$$N^+\!-\!\overset{|}{\underset{|}{C}}\!-\!O^- \xrightarrow{H_2O} N^+\!-\!\overset{|}{\underset{|}{C}}\!-\!OH \qquad \text{(addition)}$$

$$N^+\!-\!\overset{|}{\underset{|}{C}}\!-\!O^- \longrightarrow N^+\!-\!\overset{O}{\overset{\|}{C}}\!-\!+X^- \qquad \text{(substitution)}$$

$$\longrightarrow -\overset{O}{\overset{\|}{C}}\!-\!X+N \qquad \text{(reversal)}$$

The addition pathway is observed for condensation reactions

$$CH_3\!-\!\overset{O}{\overset{\|}{C}}\!-\!CH_2^- + CH_3\!-\!\overset{O}{\overset{\|}{C}}\!-\!H \xrightarrow[H_2O]{OH^-} CH_3\!-\!\overset{O}{\overset{\|}{C}}\!-\!CH_2\!-\!\overset{O^-}{\underset{H}{\overset{|}{C}}}\!-\!CH_3$$

$$\xrightarrow{H_2O} CH_3\!-\!\overset{O}{\overset{\|}{C}}\!-\!CH_2\!-\!\overset{OH}{\overset{|}{C}H}\!-\!CH_3$$

and the substitution pathway for ester hydrolysis and other reactions in which the ejected group X^- is relatively stable

$$C_6H_5\overset{O}{\overset{\|}{C}}OCH_3 + OH^- \longrightarrow C_6H_5\!-\!\overset{O^-}{\underset{HO}{\overset{|}{C}}}\!-\!OCH_3 \longrightarrow C_6H_5\overset{O}{\overset{\|}{C}}OH + CH_3O^-$$

Ester hydrolyses are among the most important reactions of organic chemistry and especially of biological chemistry. In view of this importance, a closer examination of this reaction type is warranted. Ester hydrolyses are of two general types: those in which acyl-oxygen cleavage occurs and those in which alkyl-oxygen cleavage occurs:

$$R\!-\!\overset{O}{\overset{\|}{C}}\!-\!O\!-\!R \qquad\qquad R\!-\!\overset{O}{\overset{\|}{C}}\!-\!O\!-\!R$$

Acyl-oxygen cleavage   Alkyl-oxygen cleavage

Furthermore, reaction may be acid- or base-catalyzed, and kinetically unimolecular or bimolecular. Combinations of these possibilities yields eight possible mechanisms. Ingold[38] has provided mechanistic labels as follows: A or B for acid or base catalysis, AL or AC for alkyl-oxygen or acyl-oxygen cleavage, and 1 or 2 for unimolecular or bimolecular. Thus, the eight possibilities are: $B_{AC}1$, $B_{AC}2$, $B_{AL}1$, $B_{AL}2$, $A_{AC}1$, $A_{AC}2$, $A_{AL}1$, and $A_{AL}2$. Reaction in neutral solution is similar to the base catalyzed processes in that attack occurs on neutral ester, and reaction in neutral solution is, therefore, given the B label. Of the eight possibilities, only four are commonly observed: $A_{AL}1$, $B_{AL}1$, $B_{AC}2$, and $A_{AC}2$.

Reaction by the $A_{AL}1$ and $B_{AL}1$ mechanisms is important for esters of tertiary alcohols and benzhydrols and is simply equivalent to reaction by the S_N1 mechanism described in Chapter 4:

$$R-\overset{\overset{\displaystyle O}{\|}}{C}-O-CH(C_6H_5)_2 \xrightarrow[\text{acetone}]{H_2O} RCO_2^- + {}^+CH(C_6H_5)_2$$

$$\downarrow H_2O$$

$$B_{AL}1 \equiv S_N1 \qquad\qquad HOCH(C_6H_5)_2$$

$$R-\overset{\overset{\displaystyle O}{\|}}{\underset{+}{C}}-\overset{\overset{\displaystyle H}{|}}{O}-C(CH_3)_3 \xrightarrow[\text{acetone}]{H_2O} RCO_2H + {}^+C(CH_3)_3$$

$$\downarrow H_2O$$

$$A_{AL}1 \equiv S_N1 \qquad\qquad HOC(CH_3)_3$$

This process is a facile one in these instances because of the relatively great stability of the resulting carbocations. Alkyl-oxygen cleavage in these reactions can be proved by conducting the reactions in $H_2{}^{18}O$; for alkyl-oxygen cleavage, the resulting alcohol should then contain ^{18}O:[36]

$$CH_3CO_2-C(CH_3)_3 \xrightarrow{H_2^{18}O} CH_3CO_2^- + {}^+C(CH_3)_3$$

$$\downarrow H_2^{18}O$$

$$H^{18}O-C(CH_3)_3$$

Additionally, carbocation formation should result in racemized alcohol:

$$C_6H_5CO_2-{}^*CH\overset{\displaystyle C_6H_5}{\underset{\displaystyle \text{tolyl}}{\big\langle}} \xrightarrow[\text{acetone}]{H_2O} HO-CH\overset{\displaystyle C_6H_5}{\underset{\displaystyle \text{tolyl}}{\big\langle}}$$

Optically active   Largely racemic

The similar $B_{AL}2$ and $A_{AL}2$ processes, which correspond to the S_N2 mechanism, are not important for carboxylate esters. However, for sulfonate esters, the S_N2 displacement mechanism is commonly observed:

$$R-SO_2-O\frown CH_2 \quad \overset{\frown}{O}H_2 \xrightarrow[B_{AL}2]{S_N2} RSO_3^- + CH_3CH_2\overset{+}{O}H_2$$
$$\underset{\displaystyle CH_3}{|}$$

Presumably, this is because of the greater stability of RSO_3^-, relative to RCO_2^-, as the leaving group.

The base- or acid-catalyzed hydrolyses of carboxylate esters of alcohols other than tertiaries or stabilized secondaries generally proceed by $B_{AC}2$ or $A_{AC}2$ mechanisms, respectively. Evidence for operation of the $B_{AC}2$ mechanism for most base-catalyzed hydrolyses is as follows: The reactions are second-order; in $H_2{}^{18}O$, no $R^{18}OH$ is formed; and hydrolyses of RCO_2R'

for which R^+ would rearrange (e.g., neopentyl) show no rearranged alcohol. This evidence suggests acyl-oxygen cleavage and the following mechanism:[36-38]

$$RC(=O)-O-R' + OH^- \xrightarrow{\text{slow}} R-\underset{\underset{OH}{|}}{\overset{\overset{O^-}{|}}{C}}-OR' \longrightarrow R-C(=O)-OH + R'O^-$$

That the tetrahedral intermediate is an intermediate, and not a transition state, is shown by the observation that basic hydrolysis of carbonyl-labeled ethyl benzoate yields ester in which some ^{18}O has been lost:[36]

$$C_6H_5-\overset{^{18}O}{\overset{||}{C}}-OEt \xrightarrow[H_2O]{OH^-} C_6H_5-\overset{O}{\overset{||}{C}}-OEt$$

This could only result by reversal of the first step, an impossibility if the intermediate were actually a transition state:

$$C_6H_5-\overset{^{18}O}{\overset{||}{C}}-OEt \underset{}{\overset{OH^-}{\rightleftharpoons}} C_6H_5-\underset{\underset{OH}{|}}{\overset{\overset{^{18}O^-}{|}}{C}}-OEt \longrightarrow C_6H_5\overset{^{18}O}{\overset{||}{C}}-OH + EtO^-$$

$$\updownarrow \quad \begin{matrix}\text{rapid} \\ \sim H\end{matrix}$$

$$C_6H_5-\overset{O}{\overset{||}{C}}-OEt \underset{}{\overset{^{18}OH^-}{\rightleftharpoons}} C_6H_5-\underset{\underset{O^-}{|}}{\overset{\overset{^{18}OH}{|}}{C}}-OEt \longrightarrow C_6H_5\overset{O}{\overset{||}{C}}-{^{18}OH} + EtO^-$$

Evidence for the $A_{AC}2$ mechanism is similar to that for the $B_{AC}2$ mechanism. Again, lack of $R^{18}OH$ formation is observed in $H_2{^{18}O}$, and loss of ester ^{18}O is observed for acyl-labeled compound:[36-38]

$$HOOCCH_2CH_2\overset{O}{\overset{||}{C}}-OEt + H_2{^{18}O} \xrightarrow{H^+} HOOCCH_2CH_2-\overset{O}{\overset{||}{C}}-{^{18}OH} + EtOH$$

$$C_6H_5-\overset{^{18}O}{\overset{||}{C}}-OEt \xrightarrow[H_2O]{H^+} C_6H_5-\overset{O}{\overset{||}{C}}-OEt + EtOH + C_6H_5-CO_2H$$
$$^{18}O \text{ lost}$$

Combined with a rate dependence on acidity, these results suggest the following detailed $A_{AC}2$ mechanism:

$$R-\overset{O}{\overset{||}{C}}-OR' \underset{}{\overset{H^+}{\rightleftharpoons}} R-\overset{^+O-H}{\overset{||}{C}}-OR' \underset{}{\overset{H_2O}{\rightleftharpoons}} R-\underset{\underset{^+OH_2}{|}}{\overset{\overset{OH}{|}}{C}}-OR' \rightleftharpoons R-\underset{\underset{OH}{|}}{\overset{\overset{HO \quad H}{|}}{C}}-\overset{+}{O}R'$$

$$\updownarrow$$

$$H^+ + R-\overset{O}{\overset{||}{C}}-OH \rightleftharpoons R-\overset{^+OH}{\overset{||}{C}}-OH + R'OH$$

Note that this reaction is reversible. Thus, acid-catalyzed esterification will occur and, according to the principle of microscopic reversibility, it will occur by the same mechanism. Base-catalyzed hydrolysis is not reversible because carboxylic acids exist in base as carboxylate ions that are not susceptible to nucleophilic attack.

ELECTROPHILIC SUBSTITUTIONS OF ORGANOMETALLIC COMPOUNDS

Electrophilic substitution reactions of organic derivatives of lithium, sodium, potassium, magnesium, mercury, and other metals are among the most important organic reactions for interconversion of various functional groups:[7,11,16,30,31]

$$R-M+E^+ \rightarrow R-E+M^+$$

where $M = Li$, Na, or K;

$$R-M-X+E^+ \rightarrow R-E+MX^+$$

where $M = Mg$ or Hg and $X =$ halide;

$$R_2M+E^+ \rightarrow R-E+RM^+$$

where $M = Hg$, Cd, or Mg.

Mechanistic studies of these reactions are difficult because of the stereochemical instability of the metal compounds. Metal–carbon bonds are highly ionic, and ionization-inversion is an almost constant problem. In fact, the first optically active Grignard reagent was prepared only recently,[39] and

this particular compound was a cyclopropyl derivative that, as pointed out previously, will invert only with difficulty as a carbanion. Also, loss of stereochemistry can occur during preparation of organometallic compounds; presumably, this is the result of formation of radical intermediates. For

example, formation of cyclopropyllithiums occurs with retention of configuration upon bromine–lithium exchange[3] and with extensive racemization upon direct treatment with lithium;[40] a radical mechanism is indicated for the racemization:

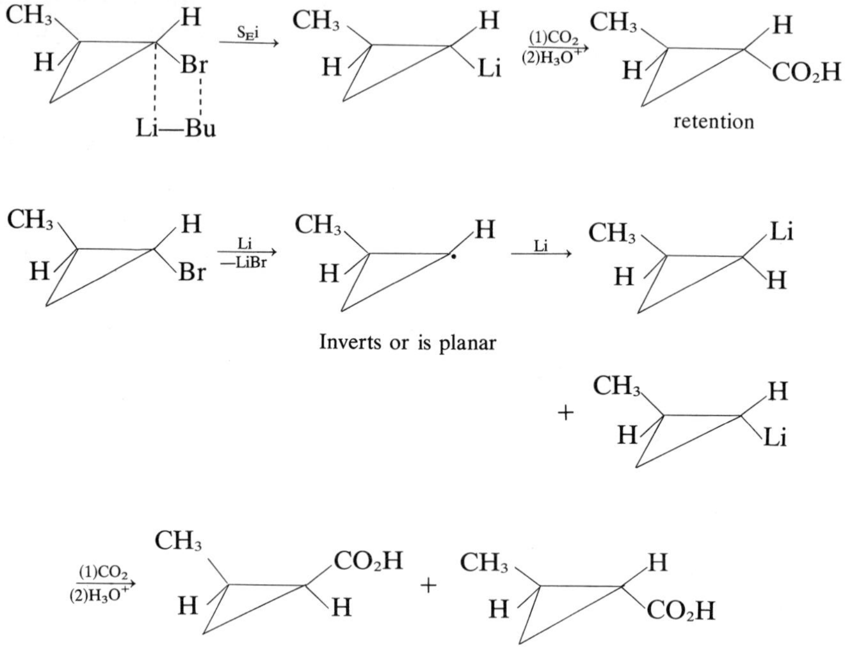

retention

Inverts or is planar

A similar mechanism can account for racemization of the Grignard reagent.

Organomercury compounds are ideally suited for the study of electrophilic substitution reactions because the mercurials are resolvable (e.g., by recrystallization of mandelates) and stereochemically stable:[30]

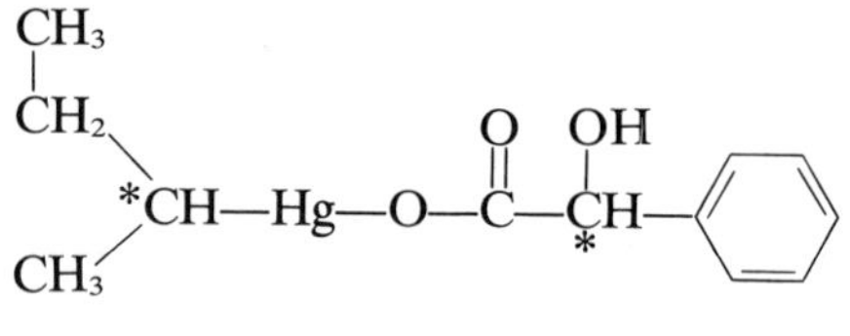

An optically active mercury mandelate

Also, organomercury compounds are monomeric in solution, whereas organolithiums and organomagnesiums frequently exist as aggregates in solution.[30] Although they are less ionic than the commonly used lithium and magnesium compounds (the order of decreasing percentage of ionic character of C—M is K>Li>Mg>Hg), the reactions of these compounds are similar in many ways. A great deal of mechanistic work has been performed on the organomercury compounds that provides much insight into the concerted S_Ei mechanism of electrophilic aliphatic substitution.[30,31]

Dialkylmercury compounds and alkylmercuric halides undergo ready cleavage with acids:

$$RHgR + HX \xrightarrow{\text{fast}} RH + RHgX$$

$$RHgX + HX \xrightarrow{\text{slow}} RH + HgX_2$$

The reactions are second-order and take place with overall retention of configuration, accompanied by extensive racemization;[30,31] experimental difficulties make determination of exact stereochemistries difficult. For example, the reaction of optically active di-*sec*-butylmercury with DCl in dioxane gave slightly retained 2-deuterobutane, but examination of the starting dialkylmercury compound showed it to be about 70% racemized during reaction, thus making calculation of cleavage stereochemistry impossible;[30,42] racemization was thought to possibly involve a free radical process:

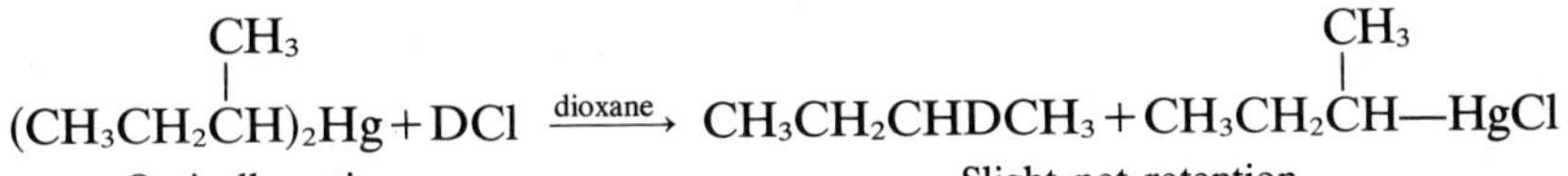

$$\underset{\text{Optically active}}{(CH_3CH_2\overset{\overset{\displaystyle CH_3}{|}}{C}H)_2Hg + DCl} \xrightarrow{\text{dioxane}} \underset{\text{Slight net retention}}{CH_3CH_2CHDCH_3 + CH_3CH_2\overset{\overset{\displaystyle CH_3}{|}}{C}H-HgCl}$$

In acetic acid, containing perchloric acid, the same starting material was almost completely racemized. *Trans*-di-4-methylcyclohexylmercury was found to be stereochemically stable to reaction conditions in dioxane-DCl, and reaction gave 4-methylcyclohexane-1-*d* that was 80% *trans*:[43]

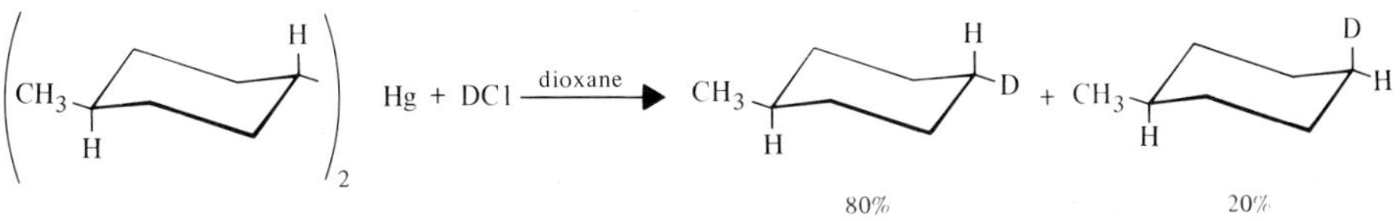

The observed second-order kinetics and retention of configuration are suggestive of operation of an S_E2 or S_Ei mechanism:

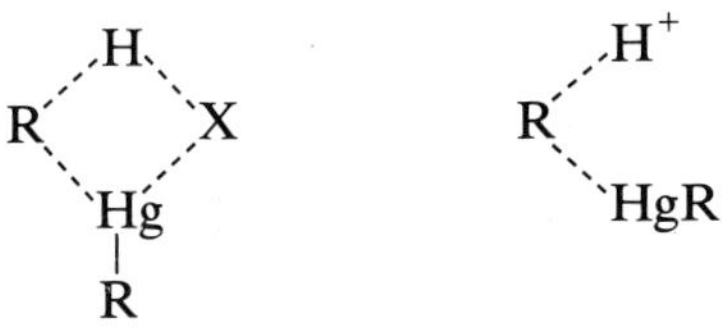

Since the rates of protic acid cleavage are greater for the acids that are least ionized (e.g., $HCl > H_2SO_4$), it must be concluded that the anion is important and the S_Ei mechanism most likely.[30]

The nature of polarization in the S_Ei transition state is not known. By drawing dotted lines between the electrophile and carbon in the S_Ei transition state, it is implied that the electrophile attacks the carbon nucleus. This picture is misleading since the electrophile actually attacks the bond between carbon and mercury (i.e., the bonding electrons, not the nucleus). Retention of configuration is readily understood on this basis. Also, since the electrophile is withdrawing electrons from the C–Hg bond, the carbon might be expected to be partially positive in the S_Ei transition state:

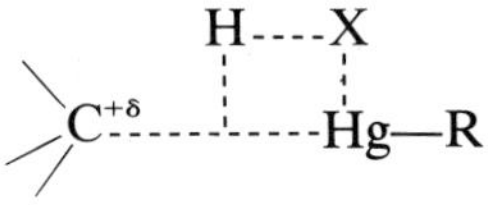

Electron flow from X should lessen the electron deficiency on mercury. Experiment has shown the mercury to be electron deficient; for example, dialkyl derivatives react more rapidly than alkylmercury halides. Determination of the charge on carbon has not been achieved with certainty (conflicting results have been reported), but the following reactivity order for R–Hg cleavage indicates that carbon may be partially negative in the S_Ei transition state for protic acid cleavage:

$$Me > Et > n\text{-}Pr > benzyl > cyclohexyl$$

Possibly, in this case, nucleophilic attack by X is more advanced than electrophilic attack by H, but this rationale does not agree with the apparent positive charge on mercury. Obviously further work, including study of steric effects, needs to be done.[30]

Aryl-mercury bonds are always cleaved more rapidly than alkyl mercury bonds. Aryl-mercury cleavages are also accelerated by electron-donating ring substituents, and reaction by a σ complex is suggested:[44]

Another important electrophilic reaction of organomercurials is the reaction with halogens. These reactions are also found to be second-order and to occur with retention of configuration:[45,46]

Radical reactions were found to be important, but they could be suppressed by addition of a radical inhibitor such as molecular oxygen or by using a complex of the halogen such as pyridine–X_2 or X_2–ZnX_2.[47]

As with polar acid cleavage, the effects of substituents are not easily explained. For the cleavage of alkylmercuric iodide with iodine the following rate order is obtained: ϕ_3CCH_2— $< n$-butyl $< sec$-butyl $\ll$ phenyl.[47] Also, for iodination of benzylmercuric chlorides, both electron-withdrawing and electron-donating substituents give a rate enhancement.[48] Obviously, a simple model of the transition state having either a partially positive or a partially negative carbon is not sufficient for explanation of these substituent effects.

The observed second-order kinetics and retention of configuration again suggest an S_Ei mechanism:

As a final example of the reactions of organomercurials, we consider the cleavage of organomercurials by mercury salts:

$$R_2Hg + HgX_2 \rightarrow 2RHgX$$

$$(CH_3CH_2)_2Hg + HgI_2 \xrightarrow{\text{dioxane}} 2CH_3CH_2HgI$$

The reactions are second-order and take place with retention of configuration:[49]

100%

Substituent effects are small, and as with the previous two organomercurial reactions, the substituent effect studies that have been done are of little use in determining the degree or nature of polarization in the transition state; different trends are observed depending on solvent, salt, and temperature.[30]

It has been suggested that the reactions of dialkylmercurials with mercury salts are S_E2 reactions. For example, the more ionic mercury salts react more rapidly:[50]

$$Hg(NO_3)_2 > Hg(OAc)_2 > HgBr_2$$

Also, the more nucleophilic $HgBr_3^-$ reacts more slowly than $HgBr_2$. These results indicate a greater importance of electrophilicity relative to nucleophilicity of the salt. For a four-center S_Ei transition state, both nucleophilicity and electrophilicity are important, and in fact, for protic acid cleavages, the less-ionized acids react most rapidly; it is important, in this case, that the anionic portion of the acid be present in the transition state:

$$R_2Hg + E-N \xrightarrow{S_Ei} \left[\begin{array}{c} \overset{+\delta}{E} \\ R \diamond \overset{-\delta}{N} \\ \underset{|}{Hg} \\ R \end{array} \right]^{\ddagger} \longrightarrow RE + NHgR$$

$$R_2Hg + E^+ \xrightarrow{S_E2} \left[\begin{array}{c} \overset{+\delta}{E} \\ R \cdots HgR \end{array} \right]^{\ddagger} \longrightarrow RE + RHg^+ \xrightarrow{N^-} RHgN$$

This difference between mercuric salt cleavage and protic acid cleavage, a supposed S_Ei reaction, has led to the proposal that mercuric salt cleavages proceed by an S_E2 mechanism:[30]

$$R_2Hg + {}^+HgX \longrightarrow \left[\begin{array}{c} \overset{+\delta}{HgX} \\ R \cdots HgR \end{array} \right]^{\ddagger} \longrightarrow RHgX + RHg^+ \xrightarrow{X^-} RHgX$$

However, the results could be explained equally as well by assuming that S_Ei transition states can be unsymmetrical; in some cases electrophilicity is important and in others nucleophilicity is important:

$\overset{+\delta}{E}\text{----}\overset{-\delta}{N}$

R⸍ ⸌Hg

|

R

$\overset{+\delta}{E}$

R⸍ ⸌N $^{-\delta}$

Hg

|

R

Electrophilicity   Nucleophilicity
important   important

It should be noted that the reactions of organomercurials, while extensively investigated, are not as well understood as the other reactions we have discussed to this point. The reactions are quite difficult to study and conflicting results have frequently been reported from different laboratories. One of the major reasons for these conflicts is the ready occurrence of radical reactions that, incidently, have also received study; radical reactions are discussed in Chapter 6.

5.3 Nucleophilic Aromatic Substitution[51-54]

In the previous chapter, it was pointed out that nucleophilic substitution reactions occur readily by S_N1 and S_N2 mechanisms for aliphatic systems but not for aromatic systems:

Cl

⬡ $\xrightarrow[\text{acetone--water}]{150°C}$ no reaction

Nucleophilic substitutions on aromatic derivatives are unfavorable, in part because the carbon–chlorine bond involving an sp^2 hybridized carbon is stronger and more difficultly broken than that with an sp^3 hybridized carbon. Also, reaction by an S_N1 mechanism to give a phenyl cation is disfavored because the positive charge in this species is on a relatively electronegative sp^2 hybridized carbon; it is important to note that the vacant sp^2 orbital in this cation is orthogonal to the π system, and thus the positive charge cannot be delocalized by resonance:

Nucleophilic substitution on aromatic systems does occur, but mechanistically these reactions are more varied than nucleophilic substitution on aliphatic systems. The S_N1 and S_N2 mechanisms do operate for aromatic substitution but only under rather limited conditions. In addition, nucleophilic aromatic substitution frequently takes place by two carbanionic mechanisms, the addition–elimination mechanism:

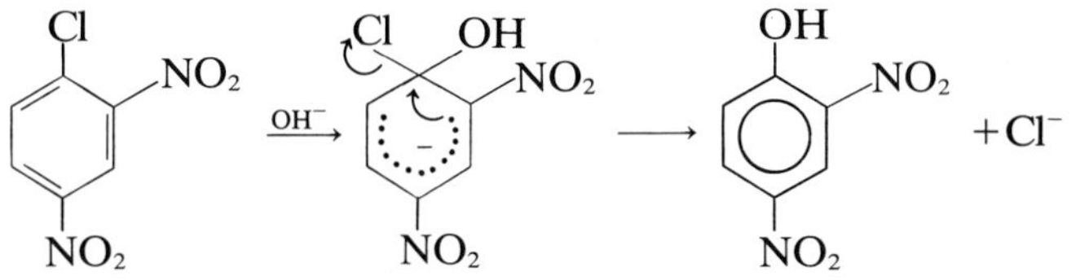

and the elimination–addition or benzyne mechanism:

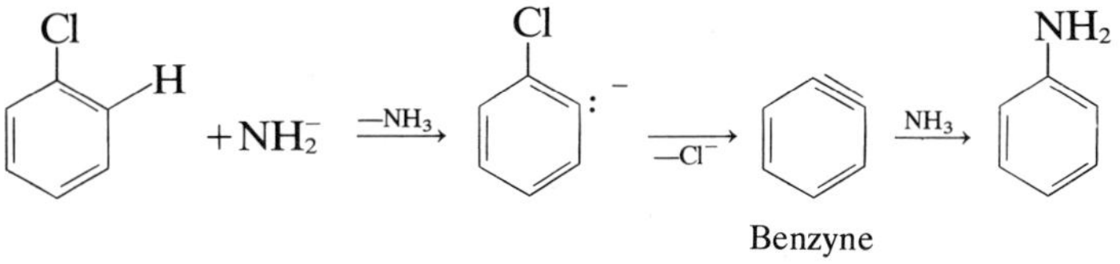

Benzyne

In the remainder of this section we will consider evidence for these four mechanistic types.

The Addition–Elimination Mechanism

The first step in the addition–elimination mechanism is formation of a cyclohexadienyl anion, analogous to the cyclohexadienyl cation formed in electrophilic aromatic substitution:

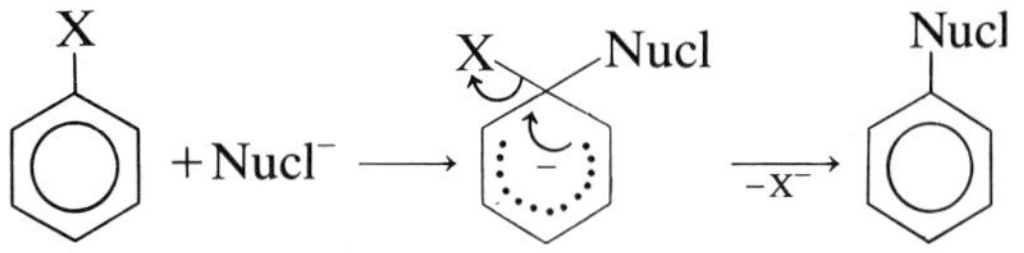

Such a carbanion, of course, would be effectively stabilized by electron withdrawing substituents, and the effect of properly oriented (*ortho* and *para* rather than *meta*) electron withdrawing substituents on nucleophilic aromatic substitution forms some of the strongest evidence for involvement of the addition–elimination mechanism. For example, contrast the conditions necessary for hydroxide substitution on chlorobenzene and trinitro-chlorobenzene:[51,55]

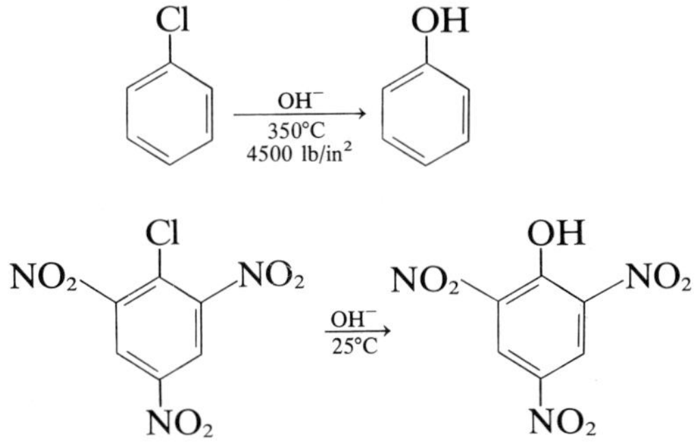

This reactivity order is certainly consistent with reaction of the trinitro compound by an addition–elimination mechanism. Actually, substrates without electron withdrawing substituents (such as chlorobenzene) probably do not react by this mechanism; these reactions are considered later. It has also been observed that a *para* nitro group is more effective than a *meta* nitro group; a *para* electron withdrawing group can stabilize a cyclohexadienyl carbanion by a resonance effect:[51,55]

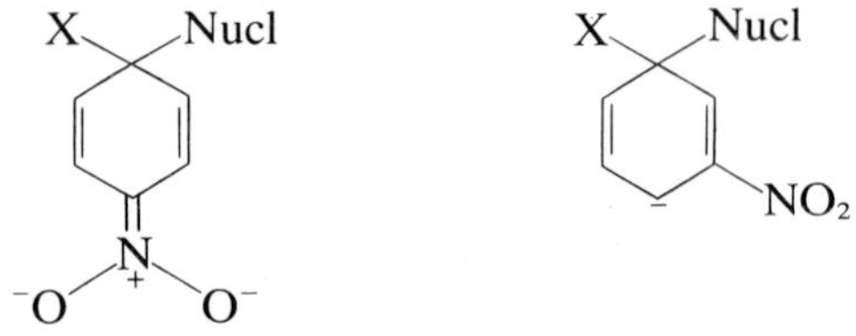

Further evidence for operation of the addition–elimination mechanism for nitro-substituted aromatics comes from the observed rate order for piperidine displacement on 1-halo-2,4-dinitrobenzenes:[56]

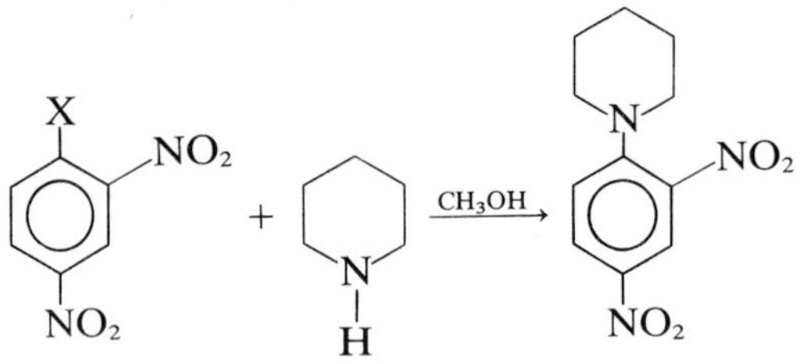

$$X = F > Cl > Br > I$$

If this substitution were a simple, one-step displacement, the opposite reactivity order would be expected (p. 145) on the basis of leaving group abilities. The present result is consistent with formation of a cyclohexadienyl anion in a slow step in which the carbon–halogen bond is not broken; the observed reactivity order then can be explained as due to the more electronegative halogen better stabilizing the carbanion:

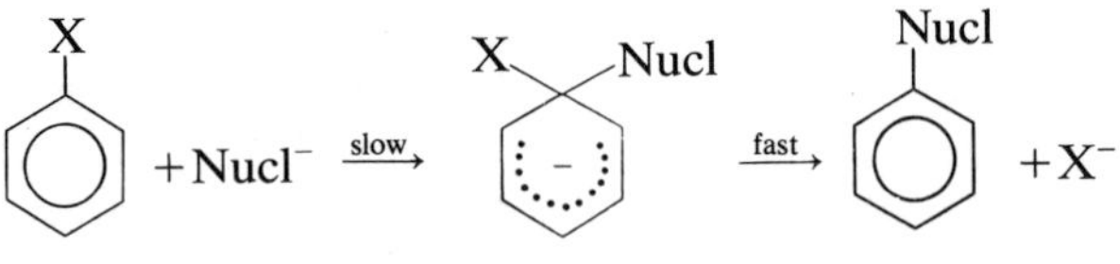

As a final piece of evidence for the addition–elimination mechanism, cyclohexadienyl anions have actually been isolated in cases in which the leaving group is a poor one:[52]

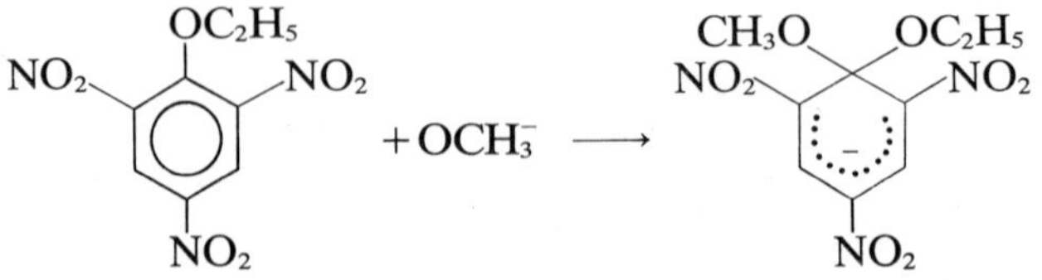

Isolable
(a Meisenheimer complex)

The Benzyne Mechanism[52]

Nucleophilic aromatic substitutions also can occur when the substrate contains no activating groups, if the nucleophile is also a strong base such as

NH_2^-. These reactions form a distinct mechanistic class from those considered in the previous section:

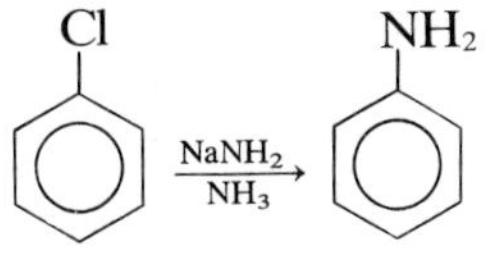

Two properties which clearly rule out operation of the addition–elimination mechanism in this case are the requirement that *ortho* hydrogens be present, and the observation of substitution at positions different from the location of the leaving group (so-called *cine* substitution):[54,57]

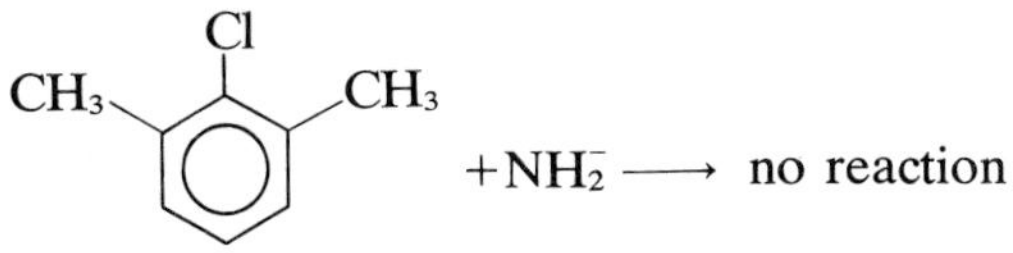

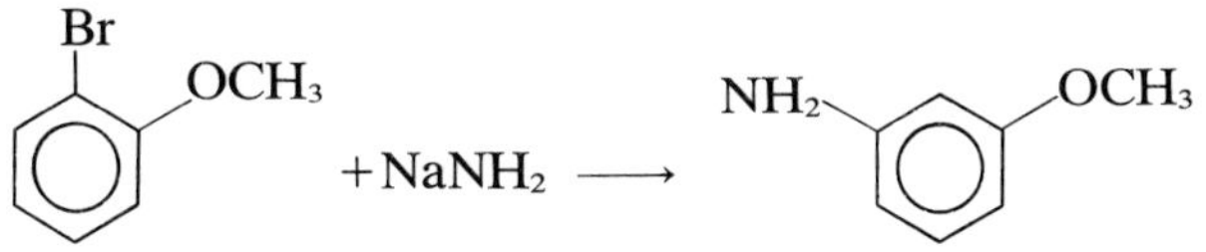

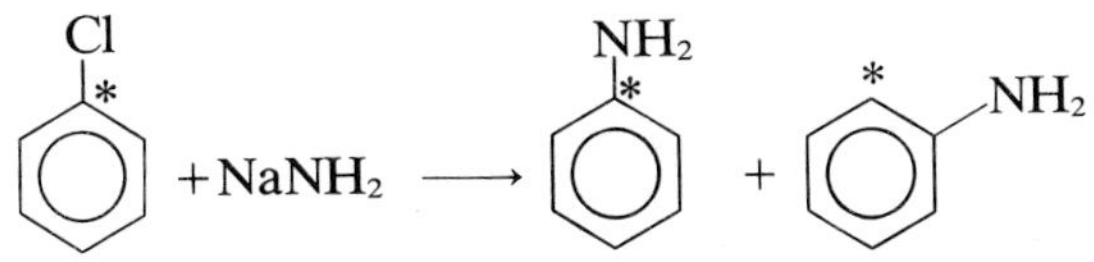

$$CH_3 \quad \overset{Cl}{\underset{}{\bigcirc}} \quad CH_3 \quad + NH_2^- \longrightarrow \text{ no reaction}$$

$+ NaNH_2 \longrightarrow$

$+ NaNH_2 \longrightarrow \qquad + $

$* = {}^{14}C$ label

These two results suggest an elimination–addition mechanism involving a benzyne intermediate:

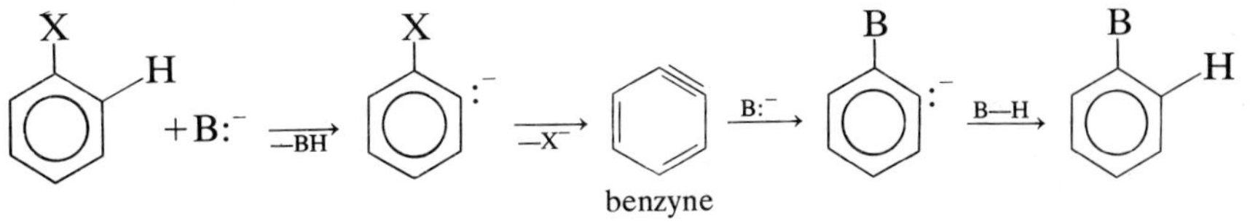

benzyne

This mechanism was hotly contested at first, primarily because it was believed that benzyne could not exist; incorporating a triple bond into such a small ring certainly does introduce ring strain (cyclooctyne is the smallest

cycloalkyne yet prepared). Of course, it is necessary for benzyne (or a substituted benzyne) to have only a transitory existence.

As further evidence for the intermediacy of benzyne, several different methods that logically could give benzyne give a common intermediate with the chemical properties of benzyne.[58,59]

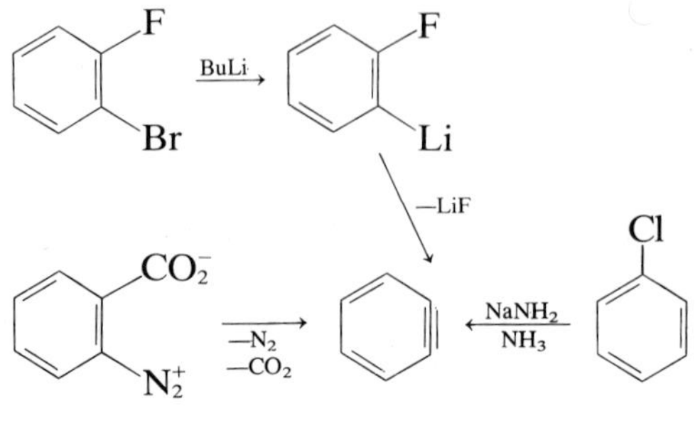

Finally, if a diene is added to a benzyne reaction, products are formed which must result from Diels–Alder trapping of the benzyne intermediate:[60]

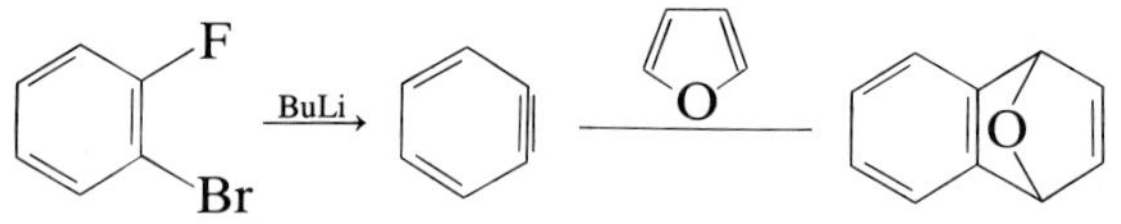

The elimination of HX upon reaction of an aryl halide with strong base can potentially be concerted or stepwise:

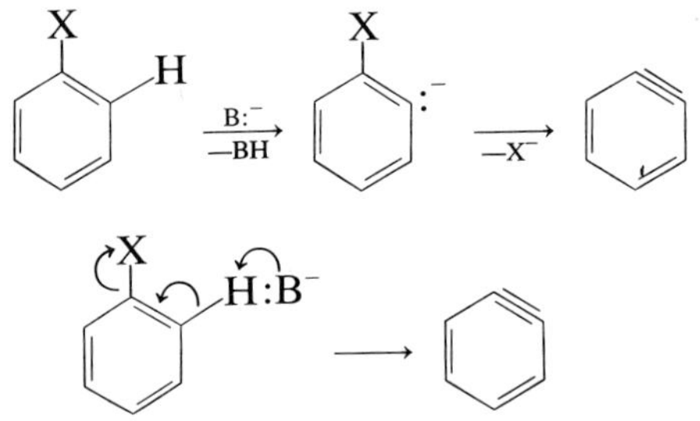

stepwise (ElcB)

concerted (E2)

Also, either the first or the second step of the stepwise mechanism could be the slow step. A deuterium isotope effect has been observed for reaction of bromobenzene, so fast proton removal by a stepwise mechanism can be ruled out for this reaction:[54,61]

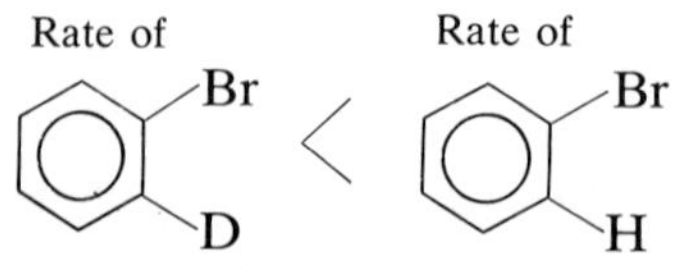

Changing the leaving group yields the rate order $Br > Cl \gg F$. Reaction by the stepwise mechanism, with proton removal rate determining, should show little rate variation with leaving group changes. Thus, the isotope and leaving group effects point to a concerted mechanism. However, the reaction of deuterated fluorobenzene with sodium amide to give aniline is accompanied by rapid formation of nondeuterated fluorobenzene, probably by rapid and reversible carbanion formation; the concerted mechanism apparently occurs only with the better leaving groups.

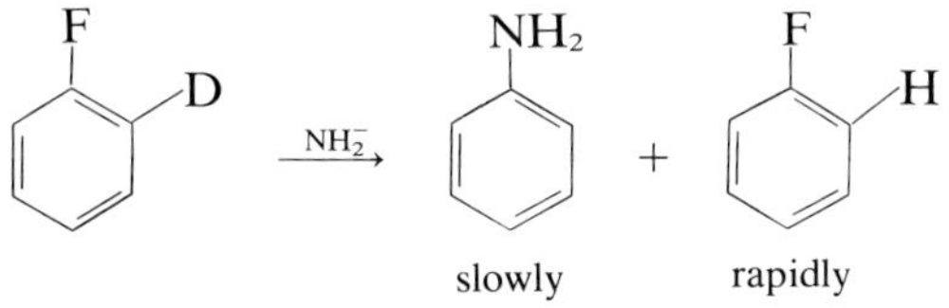

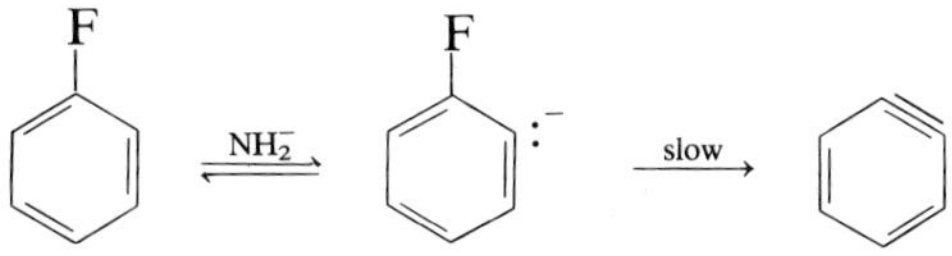

by rapid and reversible carbanion formation; the concerted mechanism

THE S_N2 MECHANISM

The benzyne mechanism can continue to operate with bases weaker than NH_2^- (e.g., OH^-) if temperatures in the 200–300°C range are used. For example, *cine* substitution occurs for reaction of ^{14}C-labeled chlorobenzene with sodium hydroxide at 340°C:[62]

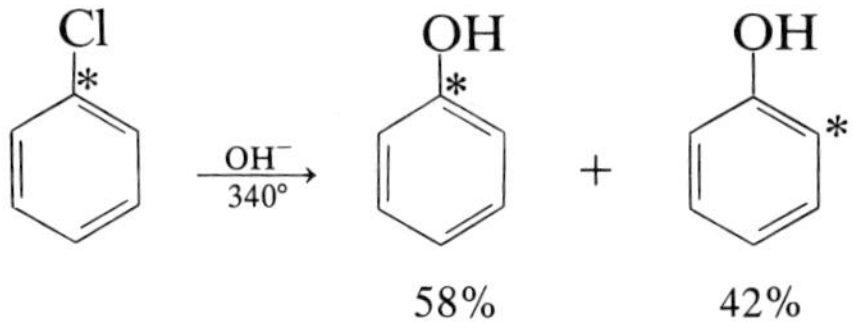

It is important to note that the two products from this reaction are not formed in equal amounts; substitution at the site of leaving group attachment exceeds substitution *ortho* to this position. Such a result is consistent with competition between substitution by a benzyne mechanism and a S_N2 displacement mechanism:

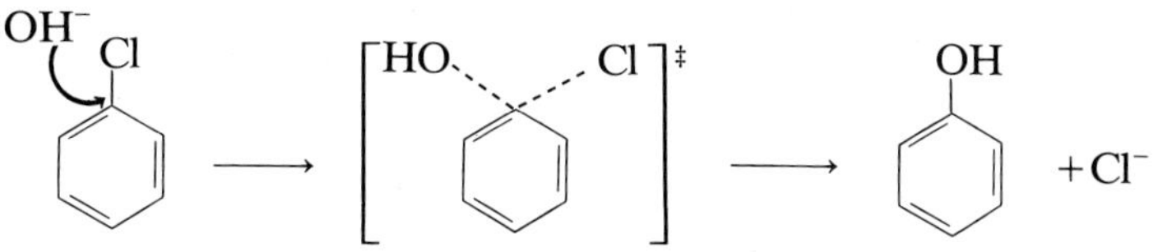

Support for this interpretation comes from the observation of an increase in the amount of substitution without rearrangement for better leaving groups; in this case, the S_N2 mechanism simply becomes more favorable as the leaving group improves:[62]

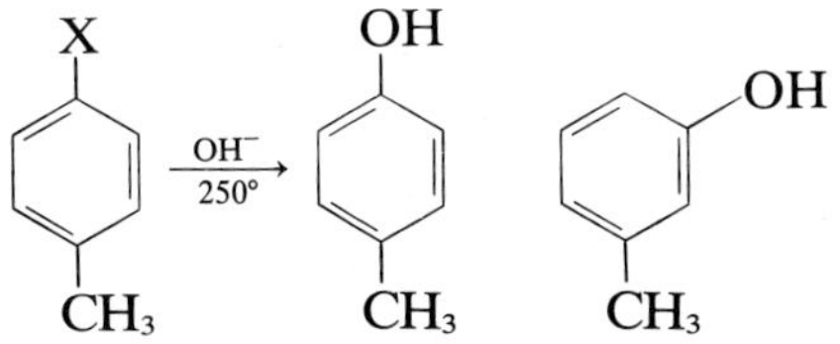

X = Cl	58%	42%
X = Br	78.5%	21.5
X = I	~97%	~3%

THE S_N1 MECHANISM

Despite the instability of the phenyl cation there is evidence that some reactions of aryl diazonium ions proceed by an S_N1 mechanism to give this cation:[51]

It should be appreciated that decomposition of the diazonium ion is facilitated by formation of the stable N_2 molecule; in other words, nitrogen is a very good leaving group. In fact, nitrogen is such a good leaving group that aliphatic diazonium ions are generally too reactive to even be identified as transient intermediates in solution. The 2,2,2-trifluoroethyl diazonium ion is an exception, being detectable by nmr for 1 hr at $-60°C$ in FSO_3H, presumably because of the great instability of the trifluoroethyl cation.[63]

As a measure of the stability of the phenyl cation relative to the trifluoroethyl cation and simple alkyl cations, phenyldiazonium fluoroborates can actually be isolated as stable crystalline salts; despite the tremendous leaving group ability of nitrogen, there remains a significant activation energy for formation of the phenyl cation from the phenyldiazonium ion.

There are three pieces of evidence for occurrence of the S_N1 mechanism for decomposition of aryldiazonium ions: (a) the reaction rates are first-order, depending only on diazonium concentration;[64] (b) electron donating substituents facilitate reaction;[65] and (c) strong base or electron withdrawing groups are not required as in the addition–elimination or benzyne mechanisms.[66]

It is interesting that labeling one nitrogen of the diazonium salt as ^{15}N shows that some nitrogen scrambling or equilibration results during reaction:[67]

While this result may derive from return from the aryl cation

it has been suggested that a cyclic diazirine cation may be formed in which both nitrogens are equivalent:[64,67]

We stated previously that diazonium reactions were first-order reactions. Actually, in the presence of relatively good nucleophiles such as hydroxide, the reactions are second-order, and in this case, an intermediate azo compound is formed in the rate-determining step:[68]

$$Ar—N_2^+ + OH^- \xrightarrow{slow} Ar—N=N—OH \xrightarrow{fast} Ar—OH + N_2$$

References

1. R. L. Letsinger, *J. Amer. Chem. Soc.*, **72,** 4842 (1950).
2. (a) L. D. McKeever, "Ions and Ion Pairs in Organic Reactions," Vol. 1, M. Szwarc, Ed., Wiley-Interscience, New York, 1972, Chapter 6; (b) J. March, "Advanced Organic Chemistry," McGraw-Hill, New York, 1968, p. 148; (c) E. C. Ashby, J. Laemmle, and H. M. Newmann, *Acc. Chem. Res.*, **7,** 272 (1974); (d) J. Smid, "Ions and Ion Pairs in Organic Reactions," Vol. 1, M. Szwarc, Ed., Wiley-Interscience, New York, 1972, Chapter 3.
3. D. E. Applequist and A. H. Peterson, *J. Amer. Chem. Soc.*, **83,** 862 (1969).
4. D. J. Cram and R. D. Partos, *J. Amer. Chem. Soc.*, **85,** 1093 (1963).
5. H. M. Walborsky, A. A. Youssef, and J. M. Notes, *J. Amer. Chem. Soc.*, **84,** 2465 (1962).
6. S. Winstein and T. G. Traylor, *J. Amer. Chem. Soc.*, **78,** 2597 (1956).
7. D. J. Cram, "Fundamentals of Carbanion Chemistry," Academic Press, New York, 1965.
8. J. R. Jones, *Quart. Rev. (London)*, **25,** 365 (1971).
9. A. Streitwieser, Jr., and J. H. Hammons, *Progr. Phys. Org. Chem.*, **3,** 41 (1965).
10. M. Szwarc, A. Streitwieser, and P. C. Mowery, "Ions and Ion Pairs in Organic Reactions," Vol. 2, M. Szwarc, Ed., Wiley-Interscience, New York, 1974, Chapter 2.
11. E. M. Kaiser and D. W. Slocum, "Organic Reactive Intermediates," S. P. McManus, Ed., Academic Press, New York, 1973, Chapter 5.
12. D. E. Applequist and D. F. Obrien, *J. Amer. Chem. Soc.*, **85,** 743 (1963).
13. L. P. Hammett, "Physical Organic Chemistry," 2nd ed., McGraw-Hill, New York, 1970.
14. J. Hine, "Physical Organic Chemistry," 2nd ed., McGraw-Hill, New York, 1962.
15. R. P. Bell, "The Proton in Chemistry," 2nd ed., Methuen, London, 1973.
16. C. H. Rochester, "Acidity Functions," Academic Press, New York, 1970.
17. R. H. Boyd, "Solute-Solvent Interaction," J. F. Coetzee and C. D. Ritchie, Eds., Dekker, New York, 1969.
18. R. Stewart and D. Dolman, *Can. J. Chem.*, **45,** 911 (1967).
19. F. G. Bordwell and W. S. Matthews, *J. Amer. Chem. Soc.*, **96,** 1214, 1216 (1974).
20. J. M. Emsley, J. Feeney, and L. H. Sutcliffe, "High Resolution Nuclear Magnetic Resonance Spectroscopy," Vol. 2, Pergamon, Oxford, 1966, pp. 988–1031.
21. A. Streitwieser, Jr., P. H. Owens, R. A. Wolf, and J. E. Williams, Jr., *J. Amer. Chem. Soc.*, **96,** 5448 (1974).
22. G. L. Closs and L. E. Closs, *J. Amer. Chem. Soc.*, **85,** 2022 (1963).
23. W. von E. Doering and A. K. Hoffman, *J. Amer. Chem. Soc.*, **77,** 521 (1955).
24. S. Andreades, *J. Amer. Chem. Soc.*, **86,** 2003 (1964).
25. J. I. Brauman and L. K. Blair, *J. Amer. Chem. Soc.*, **92,** 5986 (1970).
26. T. B. McMahon and P. Kebarle, *J. Amer. Chem. Soc.*, **96,** 5940 (1974).
27. D. J. Cram, B. Rickborn, C. A. Kingsbury, and P. Haberfield, *J. Amer. Chem. Soc.*, **83,** 3678 (1961).
28. D. J. Pederson and H. F. Frensdorff, *Angew. Chem. Int. Ed. Engl.*, **11,** 16 (1972).

29. D. J. Sam and H. E. Simmons, *J. Amer. Chem. Soc.*, **96,** 2252 (1974).
30. F. R. Jensen and B. Rickborn, "Electrophilic Substitution of Organomercurials," McGraw-Hill, New York, 1968.
31. J. March, "Advanced Organic Chemistry," McGraw-Hill, New York, 1968, Chapter 12.
32. C. D. Gutsche, "The Chemistry of Carbonyl Compounds," Prentice-Hall, Englewood Cliffs, N.J., 1967.
33. A. A. Frost and R. G. Pearson, "Kinetics and Mechanism," 2nd ed., Wiley, New York, 1961.
34. W. P. Jencks, *Progr. Phys. Org. Chem.*, **2,** 63 (1964).
35. H. O. House, "Modern Synthetic Reactions," Benjamin, New York, 1965.
36. M. L. Bender, *Chem. Rev.*, **60,** 53 (1960).
37. S. L. Johnson, *Adv. Phys. Org. Chem.*, **5,** 237 (1967).
38. C. K. Ingold, "Structure and Mechanism in Organic Chemistry," 2nd ed., Cornell Univ. Press, New York, 1969.
39. H. M. Walborsky and A. E. Young, *J. Amer. Chem. Soc.*, **86,** 3288 (1968).
40. M. J. S. Dewar and J. M. Harris, *J. Amer. Chem. Soc.*, **91,** 3652 (1969).
41. Ref. 30, p. 32.
42. L. H. Gale, F. R. Jensen, and J. A. Landgrebe, *Chem. and Ind.*, **118** (1960).
43. F. R. Jensen and L. H. Gale, *J. Amer. Chem. Soc.*, **81,** 6337 (1959).
44. R. D. Brown, A. S. Buchanon, and A. A. Humffray, *Aust. J. Chem.*, **18,** 1507, 1513 (1965).
45. F. R. Jensen and L. H. Gale, *J. Amer. Chem. Soc.*, **82,** 145, 148 (1960).
46. F. R. Jensen, L. D. Whipple, D. K. Wedegaertner, and J. A. Landgrebe, *J. Amer. Chem. Soc.*, **82,** 2466 (1960).
47. Ref. 30, p. 82.
48. I. P. Beletskaya, T. P. Fetisova, and O. A. Reutov, *Proc. Acad. Sci. USSR Chem. Sect.*, **55,** 347 (1964).
49. Ref. 30, p. 117.
50. H. B. Charman, E. D. Hughes, and C. K. Ingold, *J. Chem. Soc.*, 2523, 2530 (1959).
51. J. Miller, "Aromatic Nucleophilic Substitution," Elsevier, Amsterdam, 1968.
52. M. R. Crampton, *Adv. Phys. Org. Chem.*, **7,** 211 (1969).
53. F. Pietra, *Quart. Rev. (London)*, **23,** 504 (1969).
54. R. W. Hoffman, "Dehydrobenzene and Cycloalkynes," Academic Press, New York, 1967.
55. J. F. Bunnett, *Quart. Rev. (London)*, **12,** 1 (1958).
56. J. F. Bunnett, E. W. Garbisch, and K. M. Pruitt, *J. Amer. Chem. Soc.*, **79,** 385 (1957).
57. J. D. Roberts, H. E. Simmons, L. A. Carlsmith, and C. W. Vaughan, *J. Amer. Chem. Soc.*, **75,** 3290 (1953).
58. F. A. Hart, *J. Chem. Soc.*, 3324 (1960).
59. M. Stiles and R. G. Miller, *J. Amer. Chem. Soc.*, **82,** 3802 (1960).
60. G. Wittig and R. W. Hoffmann, *Chem. Ber.*, **95,** 2718 (1962).
61. J. D. Roberts, D. A. Semenow, H. E. Simmons, and L. S. Carlsmith, *J. Amer. Chem. Soc.*, **78,** 601, 611 (1956).
62. A. T. Bottini and J. D. Roberts, *J. Amer. Chem. Soc.*, **79,** 1458 (1957).
63. J. R. Mohrig and K. Keegstra, *J. Amer. Chem. Soc.*, **89,** 5492 (1967).
64. C. G. Swain, J. E. Sheats, and K. G. Harbison, *J. Amer. Chem. Soc.*, **97,** 796 (1975).
65. M. L. Crossley, R. H. Kienle, and C. H. Benbrook, *J. Amer. Chem. Soc.*, **62,** 1400 (1940); see also ref. 51, p. 33.
66. Ref. 51, p. 38.
67. E. S. Lewis, R. E. Holliday, and L. D. Hartung, *J. Amer. Chem. Soc.*, **91,** 430 (1969).
68. C. D. Ritchie, *Acc. Chem. Res.*, **5,** 348 (1972).

Bibliography

General

D. J. Cram, "Fundamentals of Carbanion Chemistry," Academic Press, New York, 1965.

E. M. Kaiser and D. W. Slocum, "Organic Reactive Intermediates," S. P. McManus, Ed., Academic Press, New York, 1973, Chapter 5.

Structure and Stability

J. R. Jones, *Quart. Rev. (London)*, **25**, 365 (1971).

A. Streitwieser, Jr., and J. H. Hammons, *Progr. Phys. Org. Chem.*, **3**, 41 (1965).

Acidity Functions and Acid-Base Catalysis

J. Hine, "Physical Organic Chemistry," 2nd ed., McGraw-Hill, New York, 1962.

R. P. Bell, "The Proton in Chemistry," Methuen, London, 1959.

C. H. Rochester, "Acidity Functions," Academic Press, New York, 1970.

L. P. Hammett, "Physical Organic Chemistry," 2nd ed., McGraw-Hill, New York, 1970.

Aldol Condensation and Ester Hydrolysis

C. D. Gutsche, "The Chemistry of Carbonyl Compounds," Prentice-Hall, Englewood Cliffs, N.J., 1967.

W. P. Jencks, *Progr. Phys. Org. Chem.*, **2**, 63 (1964).

M. L. Bender, *Chem. Rev.*, **60**, 53 (1960).

S. L. Johnson, *Adv. Phys. Org. Chem.*, **5**, 237 (1967).

H. O. House, "Modern Synthetic Reactions," Benjamin, New York, 1965.

C. K. Ingold, "Structure and Mechanism in Organic Chemistry," 2nd ed., Cornell Univ. Press, New York, 1969.

Substitutions of Organometallics

F. R. Jensen and B. Rickborn, "Electrophilic Substitution of Organomercurials," McGraw-Hill, New York, 1968.

E. C. Ashby, J. Laemmle, and H. M. Neumann, *Acc. Chem. Res.*, **7**, 272 (1974).

J. March, "Advanced Organic Chemistry," McGraw-Hill, New York, 1968, Chapter 12.

Nucleophilic Aromatic Substitution

J. Miller, "Aromatic Nucleophilic Substitution," Elsevier, Amsterdam, 1968.

M. R. Crampton, *Adv. Phys. Org. Chem.*, **7**, 211 (1969).

F. Pietra, *Quart. Rev. (London)*, **23**, 504 (1969).

R. W. Hoffmann, "Dehydrobenzene and Cycloalkynes," Academic Press, New York, 1967.

Problems

1. Arrange the following compounds in order of acidity and justify your answer: cyclopropane, methane, cyclohexane, and ethylene.
2. Explain the following order of carbon acidities: acetone $<$ cyclohexanone $<$ ethyl malonate $<$ acetylacetone.

3. Give a mechanism for the malonic ester synthesis (p. 226).

4. Give methods for performing the following transformations:

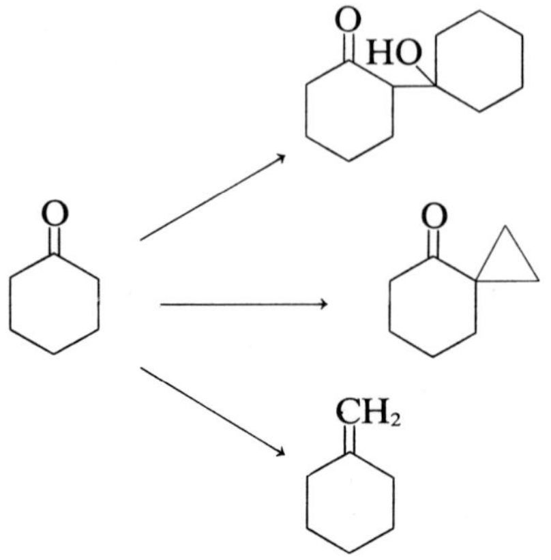

5. The benzoin condensation is catalyzed by cyanide ion, but not by stronger bases. Give a mechanism for this reaction.

$$2C_6H_5CHO \xrightarrow{CN^-} C_6H_5-\overset{O}{\underset{\|}{C}}-\overset{OH}{\underset{|}{C}}H-C_6H_5$$

6. Provide mechanisms for the following rearrangements:

$$C_6H_5-\overset{O}{\underset{\|}{C}}-\overset{O}{\underset{\|}{C}}-C_6H_5 \xrightarrow{OH^-} (C_6H_5)_2\overset{OH}{\underset{|}{C}}-COO^-$$

(Benzilic acid rearrangement)

$$C_6H_5\overset{O}{\underset{\|}{C}}CH_2-\overset{CH_3}{\underset{\underset{CH_3}{|}}{\overset{\oplus}{N}}}-CH_2C_6H_5 \xrightarrow{OH^-} C_6H_5\overset{O}{\underset{\|}{C}}CHCH_2C_6H_5 \underset{N(CH_3)_2}{}$$

(Stevens rearrangement)

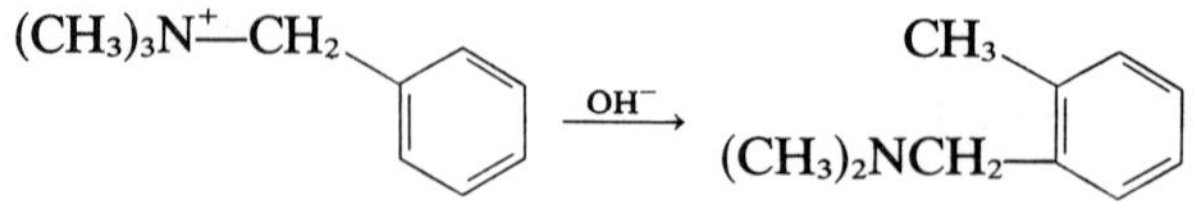

$$(CH_3)_3\overset{+}{N}-CH_2 \text{ } \xrightarrow{OH^-} (CH_3)_2NCH_2- \text{ } CH_3$$

(Sommelet–Hauser rearrangement)

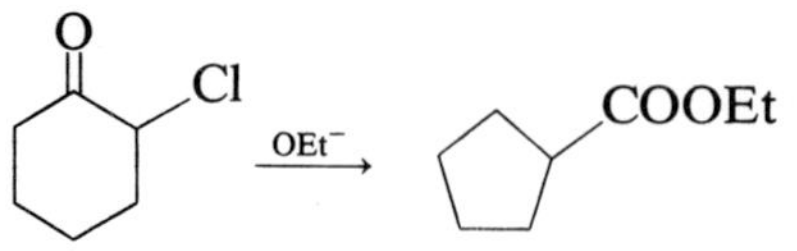

$\xrightarrow{OEt^-}$ COOEt

(Favorskii rearrangement)

$$(CH_3)_2\overset{OH}{\underset{|}{C}}-\overset{O}{\underset{\|}{C}}-CH(CH_3)_2 \xrightarrow{H^+} CH_3\overset{O}{\underset{\|}{C}}-\overset{OH}{\underset{\underset{CH_3}{|}}{C}}-CH(CH_3)_2$$

7. The Cannizzaro reaction involves an intermolecular hydride shift. Provide a mechanism.

$$2C_6H_5CHO \xrightarrow{OH^-} C_6H_5COO^- + C_6H_5CH_2OH$$

8. The decomposition of nitramide is base catalyzed. Use the following data to determine whether the catalysis is general or specific:

$$H_2N-NO_2 \xrightarrow{B:} H_2O + N_2O$$

$C_6H_5CO_2Na$	$C_6H_5CO_2H$	10^5k
0.0225	0.0125	7.76
0.0167	0.0083	5.92
0.01125	0.00625	4.30
0.0075	0.00375	2.90

9. Considering acid-catalyzed esterification to be the microscopic reverse of ester hydrolysis, rationalize the observation that **I** is resistant to esterification:

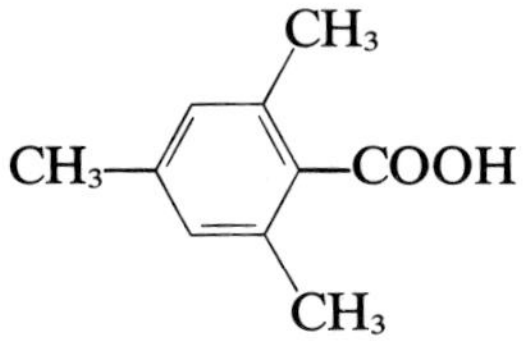

I

10. The base-catalyzed hydrolysis of ring-substituted methyl benzoates exhibits a positive Hammett ρ value. What conclusion can be drawn from this result?

11. Explain the product ratio observed for the following reaction:

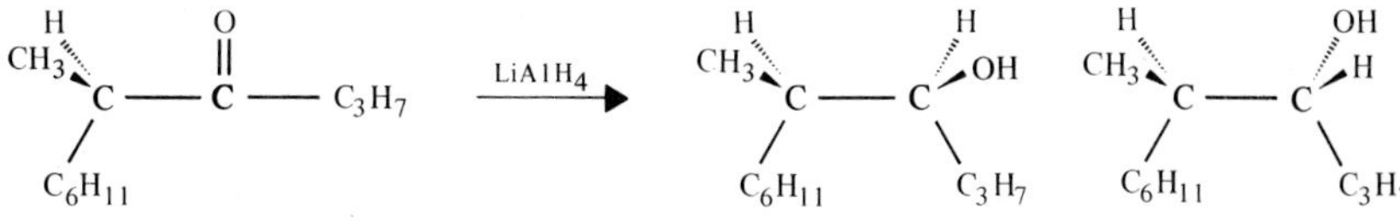

Major product   Minor product

12. ·Compound **II** gives only *meta* substitution upon reaction with NH_2^-. Explain.

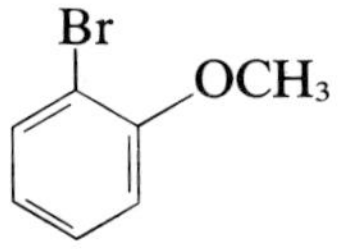

II

6

Free Radicals

6.1 History

The term "radical" was first introduced by Lavoisier in 1789. Its use through the nineteenth century was as an organizing concept, to refer to commonly encountered groups, such as the methyl radical. The concept of a "free" radical had been postulated for compounds like cyanogen (CN), but gas density molecular weight determinations indicated these to be dimers. In 1896, the authoritative Ostwald had written: "It took a long time before it was finally recognized that the very nature of organic radicals is inherently such as to preclude the possibility of isolating them." Thus, the report by Moses Gomberg in 1900[1] that he had prepared triphenylmethyl, a stable organic radical, was met with much suspicion and outright disbelief. For many years, the majority of chemists considered Gomberg's compound to be the dimer, hexaphenylethane, which happened to have some very unusual properties.

Gomberg had, in fact, been attempting to synthesize hexaphenylethane, using what was expected to be a routine coupling reaction:

$$2\ \phi_3CBr + 2\ Ag \rightarrow \phi_3C\!-\!C\phi_3 + 2\ AgBr$$

The white solid product obtained initially was not hexaphenylethane, however, since elemental analysis for carbon and hydrogen gave results distinctly too low. Repetition of the experiment under oxygen-free conditions avoided the white solid product and instead gave a yellow solution that reverted to the initial product upon exposure to air. Based upon the unexpected color and the high reactivity towards oxygen and halogens, Gomberg postulated that he had prepared a solution of triphenylmethyl radicals.

$$2\ \phi_3C\!\cdot + O_2 \rightarrow \phi_3C\!-\!O\!-\!O\!-\!C\phi_3$$

(yellow)   (white solid)

$$2\ \phi_3C\!\cdot + I_2 \rightarrow 2\ \phi_3C\!-\!I$$

In attempting to prove that the yellow molecule in question was the triphenylmethyl radical and not the dimer, Gomberg performed cryoscopic molecular weight determinations. These results, however, pointed to an apparent molecular weight much closer to dimer than to monomer. Gomberg was then forced to postulate an equilibrium between monomer and dimer, favoring the dimer.

$$\phi_3C\!-\!C\phi_3 \rightleftharpoons 2\ \phi_3C\!\cdot$$

The equilibrium concept proved to be highly successful. The yellow color, due to the radical, deepened with an increase in temperature, suggesting an increase in dissociation. Furthermore, the absorption did not obey Beer's law (see page 326)—that is, the intensity did not decrease in proportion to the dilution. Finally, the chemical reactivity was also suggestive of an equilibrium. If a small amount of oxygen was admitted to the yellow solution, the color quickly disappeared but in time the color was regenerated.

During the remainder of his life (1866–1947), Gomberg worked mainly on various aspects of free radical chemistry, solidifying his position and gradually winning the support and esteem of the chemistry establishment. Interestingly, a significant chapter in the triphenylmethyl story has been written as recently as 1968. By nmr studies, the actual structure of the triphenylmethyl dimer, always assumed to be hexaphenylethane, was actually shown to be:[2]

A free radical is now usually defined as a molecule or atom with an unpaired electron. Most molecules are composed of an even number of electrons, all paired; however, a molecule containing an odd number of electrons, such as nitric oxide or triphenylmethyl, will necessarily be a free radical by the above definition. Furthermore, some molecules, such as the oxygen molecule, exist with two unpaired electrons and these are considered biradicals.

The simplest organic free radical, the methyl radical, displays the features we will use to characterize organic radicals—a trivalent carbon atom with seven valence electrons. The structure of the methyl radical is planar or very nearly planar, utilizing sp^2 hybrids for the three covalent bonds, with the unpaired electron occupying a p orbital perpendicular to the plane of the molecule. (Compare with the structure of simple carbocations and carbanions, Chapters 4 and 5.)

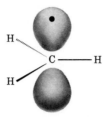

6.2   Detection of Free Radicals

The unique feature of a free radical is the unpaired electron; the spin magnetic moment of this unpaired electron makes the free radical a paramagnetic species. Thus, free radicals may be attracted by a magnetic field. This is the principle of the Gouy balance, which measures the effect of a magnetic field upon the apparent weight of a sample—its magnetic susceptibility. Measurements of the magnetic susceptibility of triphenylmethyl radicals in a 5% benzene solution at 20°C indicate that there are only 2% free radicals with the remainder being undissociated dimer.[3]

Electron Spin Resonance

By far the most useful and widely used technique for detection of free radicals is electron spin resonance (esr), also called electron paramagnetic resonance (epr). The theory and experimental aspects of esr are very similar to those of nmr, except that esr deals with the magnetic moment of an unpaired electron, where nmr deals with magnetic moments of hydrogen or other nuclei.

In an applied magnetic field, the magnetic moment of an unpaired electron has only two allowed quantum states, which we may consider to be aligned either with or against the applied field. These two orientations are of different energies, and, therefore, a resonance absorption may take place when electromagnetic radiation is applied with just the right energy to promote the transition (Figure 6.1).

For a typical esr experiment, a magnetic field of about 3200 gauss is applied, and resonance absorption occurs at around 9000 MHz, in the microwave region. (Compare this with a typical nmr experiment that uses a stronger applied field of around 14,000 gauss but a less energetic radiofrequency of 60 MHz.)

Like an nmr spectrum, an esr spectrum displays certain characteristic features. In nmr, we may obtain information from the location of the resonance absorptions, the intensities of the absorptions, and the specific splitting patterns of the absorptions. An esr spectrum typically consists of a single absorption. The location of the esr absorption is measured by the

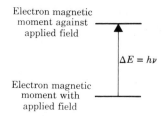

Fig. 6.1 *Electron spin resonance absorption where* $\Delta E = h\nu = g\beta H$, *and* g *is the gyromagnetic ratio,* β *is the Bohr magneton (constant), and* H *is the applied magnetic field.*

gyromagnetic ratio, or g *value*. For a hypothetical free electron g = 2.0023, and for most organic radicals g values from 2.002 to 2.006 are observed. These g values provide some information about the environment of the unpaired electron, but the very narrow range of observed g values makes this information much less useful than an nmr chemical shift. Splitting patterns, however, are fully as prominent and as useful as they are in nmr spectroscopy. The so-called *hyperfine splitting* of an esr absorption is caused by interactions (coupling) of the electron magnetic moment with other nearby magnetic moments, usually hydrogen nuclei. The familiar $(n+1)$ rule applies as well to esr spectroscopy as to nmr spectroscopy: coupling to n equivalent protons produces a splitting pattern of $(n+1)$ lines. Thus, the methyl radical shows an esr spectrum that is a quartet in the intensity ratio of $1:3:3:1$. Typically, esr spectra are plotted as derivative curves for greater sensitivity, so the methyl radical spectrum would appear as in Figure 6.2.

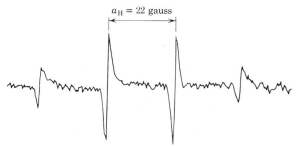

Fig. 6.2 *Electron spin resonance spectrum of the methyl radical.*[4]

Besides hydrogen nuclei, other elements that have nonzero nuclear spin will split an esr absorption. These include D, ^{13}C, ^{14}N, ^{19}F, ^{31}P, and ^{35}Cl.

The magnitude of the hyperfine splitting constant, usually designated a_H, can be particularly informative. The esr spectrum of the isopropyl radical, for example, shows $a_{H\alpha} = 22$ gauss and $a_{H\beta} = 25$ gauss. Thus, the spectrum appears as two overlapping septets:[5]

$$a_{H\alpha} = 22 \text{ gauss}$$

$$a_{H\beta} = 25 \text{ gauss}$$

Although it may appear surprising that the coupling to the β hydrogens is greater than that to the α hydrogen, this is a general phenomenon. The coupling between the unpaired electron and the α hydrogen is probably minimized because the C–H bond and the p orbital are orthogonal. The coupling with the β hydrogens is not subject to such a restriction; in fact, very strong interactions between the unpaired electron and the β hydrogens are possible by hyperconjugation:

Hyperconjugative delocalization

   The special value of esr spectroscopy lies in its selectivity and its sensitivity. It detects only unpaired electrons; therefore, the vast majority of molecules, having all paired electrons, are completely transparent to esr. Furthermore, the esr technique is extremely sensitive, being capable of detecting free radical concentrations as low as 10^{-9} M.

   Despite this sensitivity, however, special experimental problems are frequently encountered in attempting to generate sufficient radical concentrations. Most free radicals are highly reactive species that disappear by recombination very quickly. To achieve a sufficient steady-state concentration, two approaches may be taken: rapid generation of radicals to counterbalance their rapid decay or adjustment of conditions so the decay of the radicals is slowed or halted. The first approach requires a flow system in which a constant supply of the radical precursor is flowed through the esr sample area (cavity), where it is photolyzed or thermolyzed to produce radicals. This approach has made possible the observation of the esr spectra of many radical intermediates and offers the additional possibility of determining the rates of radical decay processes. The second approach involves generation of radicals under conditions where they are stable and may be studied at leisure. The normal conditions for such an experiment involve a frozen solvent acting as a matrix for the radical precursor. Photolysis of the precursor then produces radicals frozen in the matrix where they are unable to diffuse together to recombine. The esr spectrum of the isopropyl radical, discussed earlier, was in fact obtained by this method, although in a slightly indirect way:[5]

$$(CH_3)_3COOC(CH_3)_3 \xrightarrow[\substack{-162°C \\ \text{in solid} \\ C_3H_8}]{h\nu} 2(CH_3)_3CO\cdot \xrightarrow{C_3H_8} (CH_3)_3COH + (CH_3)_2CH\cdot$$

$$\text{(stable at } -162°C)$$

   An alternative method of stabilizing free radicals involves trapping the radical by some chemical reaction. If the trapping reaction leads to a stable free radical, it may then be studied to identify or study the initial radical. 2-Methyl-2-nitrosopropane and phenyl-*tert*-butyl nitrone have been used in this manner, both giving stable nitroxide radicals for which the hyperfine splitting constants provide information about the trapped radical:

$$(CH_3)_3C-N{=}O + R\cdot \longrightarrow (CH_3)_3C-\overset{\overset{\displaystyle\dot{O}}{|}}{N}-R$$

$$(CH_3)_3C-\overset{\overset{O}{\uparrow}}{N}=CH-\phi + R\cdot \longrightarrow (CH_3)_3C-\overset{\overset{\dot{O}}{|}}{N}-\underset{\underset{R}{|}}{C}H-\phi$$

Spin trapping reactions

Chemically Induced Dynamic Nuclear Polarization

The very recent technique called chemically induced dynamic nuclear polarization (CIDNP) has found increasing utility in detecting the intermediacy of free radicals in reaction mechanisms. Chemically induced dynamic nuclear polarization is based upon nmr spectroscopic analysis of the *products* of reactions involving radical intermediates. Normally, nmr spectra of free radicals are so broadened by interactions of the protons with the unpaired electron that the spectra are not very useful. However, it was discovered that the nmr spectra of the products of radical reactions may show the effects of having passed through a free radical stage. The CIDNP effect may be quite startling: the nmr spectrum of a product may show nmr emission (inverted absorption peaks) or enhanced absorption (unusually strong absorption peaks) immediately after it has existed as a free radical.

The CIDNP phenomenon is caused by the effect of the unpaired electron on the nuclear spin orientations. Normal nmr absorption spectroscopy is based upon the usual Boltzmann distribution of nuclear spins—a slightly greater number of protons will have spins aligned with rather than against the applied magnetic field. The nmr absorption intensity depends upon the difference in population between the upper and lower energy levels. In a free radical, however, the spin of the unpaired electron may interact with the spin of a proton. This may result in *polarization* of the proton spins: either excess population of the lower energy spin state or excess population of the upper state. Even after the radical has formed a diamagnetic product, the polarization of the proton spins may remain for up to several minutes. The nmr spectrum of a molecule with its proton spins polarized in a non-Boltzmann distribution will then display some unusual intensity features. If there are more than the normal number of protons in the lower energy state, the absorption will be more intense than normal. If there are excess protons in the higher energy state, irradiation at the resonance frequency may stimulate some of those protons to drop to the lower state with simultaneous *emission* of that same frequency.

A good example of the CIDNP effect can be seen in the exchange reaction between *n*-butyllithium and *sec*-butyl iodide:

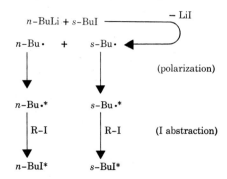

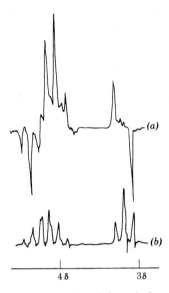

Fig. 6.3 *Nuclear magnetic resonance spectra of the α-hydrogens of n-BuI and s-BuI:*[6] *(a) during the exchange reaction of n-BuLi and s-BuI; and (b) reference spectrum of n-BuI and s-BuI.*

Figure 6.3 indicates the downfield portion of the nmr spectra of *n*-butyl iodide and *sec*-butyl iodide (the α-hydrogens), clearly indicating the enhanced absorption and emission characteristics during the reaction.[6]

6.3 Homolysis Reactions

Bond Dissociation Energies

Since nearly all stable organic molecules have all their electrons paired, free radicals are nearly always generated in pairs by homolytic cleavage of a bond. The heat of reaction for such a homolysis is defined as the bond dissociation energy D:

$$A\!\!-\!\!B \rightarrow A\cdot + \cdot B \qquad \Delta H = D_{A\text{-}B}$$

Table 6.1 *Representative Bond Dissociation Energies (kcal/mole)*

H–H	104	H–F	136
F–F	38	H–Cl	103
Cl–Cl	58	H–Br	88
Br–Br	46	H–I	71
I–I	36		
CH_3–H	104	CH_3–F	108
1° C–H	98	CH_3–Cl	84
2° C–H	95	CH_3–Br	70
3° C–H	91	CH_3–I	56
Ar–H	112		
$ArCH_2$–H	85		

Bond dissociation energies can be used very simply and effectively for an analysis of the energetics involved in free radical reactions. In general for a radical reaction, ΔH may be obtained as the difference between the bond dissociation energies of the bonds that are broken and the bonds that are formed. Using Table 6.1, the following heats of reaction are readily calculated:

$$Cl\cdot + CH_3CH_2CH_2\!-\!H \rightarrow H\!-\!Cl + CH_3CH_2CH_2\cdot \qquad \Delta H = -5 \text{ kcal/mole}$$
$$(98)(103)$$

$$Cl\cdot + CH_3\underset{\underset{H}{|}}{C}HCH_3 \rightarrow H\!-\!Cl + CH_3\dot{C}HCH_3 \qquad\qquad \Delta H = -8 \text{ kcal/mole}$$
$$(95)(103)$$

A comparison of the energetics of these two reactions shows that abstraction of a secondary hydrogen is 3 kcal/mole more favorable than abstraction of a primary hydrogen (this would be true regardless of the abstracting species). In the case of abstraction by a chlorine atom, both reactions are moderately exothermic, and chlorine shows only a slight preference for abstraction of a secondary hydrogen over a primary hydrogen.

The situation for bromine atom abstraction of hydrogen is somewhat different:

$$Br\cdot + CH_3CH_2CH_2\!-\!H \rightarrow H\!-\!Br + CH_3CH_2CH_2\cdot \qquad \Delta H = +10 \text{ kcal/mole}$$
$$(98)(88)$$

$$Br\cdot + CH_3\underset{\underset{H}{|}}{C}HCH_3 \rightarrow H\!-\!Br + CH_3\dot{C}HCH_3 \qquad\qquad \Delta H = +7 \text{ kcal/mole}$$
$$(95)(88)$$

Again, abstraction of the 2° hydrogen is favored over 1° abstraction by 3 kcal/mole, but both reactions are endothermic. Applying Hammond's postulate (p. 124), the transition state for these endothermic reactions will resemble products (free radicals), and therefore the differences between product stabilities will be more significant than it was for the chlorine abstraction. Bromine abstraction of hydrogen is much more *selective* than chlorine atom abstraction of hydrogen (Figure 6.4) (see also Figure 3.8).

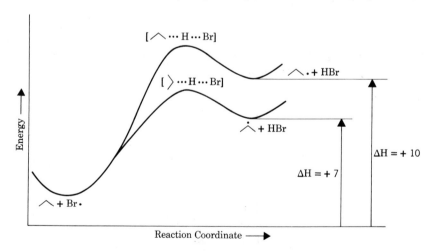

Fig. 6.4 *Bromine atom reaction with propane.*

In a similar manner, bond dissociation energy data can be used to order the various types of hydrogen in their reactivity towards radical abstraction (benzyl > 3° > 2° > 1° > CH_3 > aryl); furthermore, we may rank the halogens in their ability to abstract hydrogen (F > Cl > Br > I). In the analysis of free radical mechanisms, we will frequently find it useful to work out the energetics of individual mechanistic steps in order to determine the feasibility of potential reaction pathways.

Relative Radical Stability

Figure 6.4 indicates that the isopropyl radical is 3 kcal/mole more stable than the n-propyl radical, compared to propane as a common reference. In general, a greater number of alkyl substituents stabilizes a radical center. Since free radicals are electron-deficient species, the alkyl groups act as electron donors, perhaps by hyperconjugation (see p. 132):

Stabilization by hyperconjugation

Other factors leading to radical stabilization are π conjugation and steric hindrance.

Allylic and benzylic radicals are greatly stabilized by resonance delocalization of the unpaired electron:

Resonance delocalization

Free radicals are also stabilized if they are sterically prevented from undergoing usual radical reactions such as recombination. An example is 2,4,6-tri-*tert*-butylphenoxy, a stable radical, which is stabilized by conjugation as well as steric hindrance:

Stabilization by steric hindrance

6.4 Chain Reactions

Mechanism

One of the most characteristic aspects of free radicals is their tendency to undergo chain reactions. Since free radicals have an unpaired electron, their reaction with a molecule that has only paired electrons must again yield another free radical. Thus, radical–molecule reactions are self-sustaining in that there is no net disappearance of radicals. In general, a free radical chain reaction consists of: (*a*) *initiation*, in which free radicals are generated from some source; (*b*), *propagation*, in which radical–molecule reactions take place; and (*c*), *termination*, in which radicals disappear, usually by recombination with one another.

The free radical chlorination of alkanes is a classic chain reaction that we shall consider in some detail. Overall, the reaction amounts to a substitution process, and either heat or light is required:

$$RH + Cl_2 \xrightarrow[\text{or } \Delta]{h\nu} RCl + HCl$$

The heat or light serves to dissociate Cl_2 to generate chlorine atoms in the initiation step. Thermally, the energy required is the bond dissociation energy of 58 kcal/mole. Photochemically, any light absorbed by Cl_2 in the visible or ultraviolet region ($\lambda < 500$ nm) has sufficient energy to cause dissociation.

Initiation
$$Cl_2 \xrightarrow[\text{or } \Delta]{h\nu} 2Cl\cdot$$

The chlorine atoms may then abstract hydrogen from the alkane forming an alkyl radical. It is in this step that any selectivity is displayed and the products are determined. We have already analyzed the energetics of this step. The second propagation step involves reaction of the alkyl radical with a chlorine molecule to generate alkyl chloride and another chlorine atom. Using Table 6.1, it can be readily shown that this reaction is strongly exothermic for all C–Cl bonds. Thus, the pair of propagation reactions accomplishes the overall conversion of alkane to alkyl chloride, and most importantly, the necessary chlorine atom is regenerated so that the propagation steps may be repeated indefinitely.

Propagation
$$\begin{cases} Cl\cdot + RH \rightarrow R\cdot + HCl \\ R\cdot + Cl_2 \rightarrow RCl + Cl\cdot \end{cases}$$

Overall,
$$RH + Cl_2 \rightarrow RCl + HCl$$

A cyclic reaction diagram similar to the type commonly used to illustrate biochemical reactions is shown in Figure 6.5. Each complete turnover of the radical cycle converts $RH + Cl_2$ to $RCl + HCl$. All that is needed is an entry into the cycle (initiation).

From the considerations thus far, it would seem that a free radical chain reaction could be infinitely efficient; a single chlorine atom could convert any amount of alkane to alkyl chloride. In fact, chain reactions can be highly efficient but are never totally efficient. This is due to termination reactions in which the chain-carrying free radicals disappear. Normally, radical concentrations are at a very low steady-state value, so that radical–radical encounters are rare compared to radical–molecule encounters.

When they do occur, however, each termination reaction leads to the disappearance of two radicals and the ending of two chains.

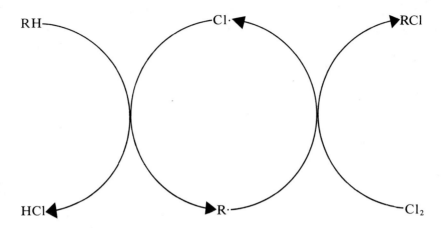

Fig. 6.5 *Cyclic diagram for a chain mechanism.*

Termination $\begin{cases} 2Cl \cdot \rightarrow Cl_2 \\ 2R \cdot \rightarrow RR \\ R \cdot + Cl \cdot \rightarrow RCl \end{cases}$

Some of the evidence that a free radical chain mechanism is operating in the chlorination of alkanes includes the following: (*a*) the quantum yield for photochlorination can be as high as 10^6—that is, for each photon absorbed, up to a million molecules are converted to product; (*b*) in some instances, small quantities of the termination product R–R can be isolated (usually less than 0.1%); (*c*) radical inhibitors (species such as O_2 that react with free radicals to give unreactive products) shut off the reaction far more effectively than would have been expected stoichiometrically.

Kinetics

The kinetic behavior of a chain reaction can be very distinctive, since fractional kinetic orders often appear in the rate law. We will consider the kinetics of chlorination of an alkane as a typical example:

$$Cl_2 \xrightarrow{k_1} 2Cl \cdot \tag{1}$$

$$Cl \cdot + RH \xrightarrow{k_2} HCl + R \cdot \tag{2}$$

$$R \cdot + Cl_2 \xrightarrow{k_3} RCl + Cl \cdot \tag{3}$$

$$2Cl \cdot \xrightarrow{k_4} Cl_2 \tag{4}$$

$$2R \cdot \xrightarrow{k_5} RR \tag{5}$$

$$R \cdot + Cl \cdot \xrightarrow{k_6} RCl \tag{6}$$

The overall rate of a chain reaction will depend upon two factors: (*a*) the rate of initiation—how fast are radicals being made and (*b*) the rate of propagation relative to the rate of termination—how many propagation steps are accomplished before termination ends the chain. For the foregoing chain reaction, there are two propagation steps (steps (2) and (3)), either of which may be rate-determining. We will consider these two limiting cases, where either propagation step (2) or (3) is much faster than the other. Between these two limiting cases, the kinetics are substantially more complex.

CASE I—PROPAGATION STEP (2) RATE-DETERMINING

If propagation step (2) is much slower than propagation step (3), Cl atoms react slowly (step (2)) and alkyl radicals react quickly (step (3)). Effectively, the only radicals existing in the system will be Cl atoms. Therefore, the termination of chains will occur predominantly by combination of Cl atoms (step (4)). Since step (2) is the rate-determining step, the rate law is as follows:

$$\text{Rate} = k_2[\text{RH}][\text{Cl}\cdot]$$

The steady-state assumption for Cl atoms gives

$$2k_1[\text{Cl}_2] + k_3[\text{R}\cdot][\text{Cl}_2] = k_2[\text{Cl}\cdot][\text{RH}] + 2k_4[\text{Cl}\cdot]^2$$

The steady-state assumption for R· radicals gives

$$k_2[\text{Cl}\cdot][\text{RH}] = k_3[\text{R}\cdot][\text{Cl}_2]$$

Combining the two equations, we obtain

$$k_1[\text{Cl}_2] = k_4[\text{Cl}\cdot]^2$$

This result is general; that is, the rate of initiation will always equal the rate of termination under steady-state conditions. In effect, it amounts to a steady-state condition for total radicals, since there is only a net generation of radicals in initiation and net disappearance of radicals in termination. The steady-state concentration for Cl atoms is, therefore,

$$[\text{Cl}\cdot] = \frac{k_1^{1/2}}{k_4^{1/2}}[\text{Cl}_2]^{1/2}$$

If step (2) is the rate-determining step, we obtain the final rate law as follows:

$$\text{Rate} = k_2[\text{RH}][\text{Cl}\cdot]$$

$$\text{Rate} = \frac{k_1^{1/2}k_2}{k_4^{1/2}}[\text{RH}][\text{Cl}_2]^{1/2}$$

CASE II—PROPAGATION STEP (3) RATE-DETERMINING

If propagation step (3) is much slower than propagation step (2), then alkyl radicals will be generated quickly (step (2)) but react slowly (step (3)). Then R· will be essentially the only radical present in the system, and termination will consist of step (5) rather than step (4) or (6). Consideration of the steady-state equations for R· and Cl· again leads to the equating of initiation and termination rates:

$$k_1[Cl_2] = k_5[R\cdot]^2$$

$$[R\cdot] = \frac{k_1^{1/2}}{k_5^{1/2}}[Cl_2]^{1/2}$$

$$\text{Rate} = k_3[R\cdot][Cl_2]$$

$$\text{Rate} = \frac{k_1^{1/2}k_3}{k_5^{1/2}}[Cl_2]^{3/2}$$

The appearance of half-order dependence upon initiator concentration is characteristic of radical chain reactions. The specific form of the rate law can also be used to distinguish the slower of the two propagation steps.

An additional kinetic characteristic that may be derived for chain reactions is the *kinetic chain length*. This is a measure of how long the chain reactions are, or in other words, how many propagation reactions occur for each initiation. The kinetic chain length ν may be defined as the ratio of the rate of the overall reaction compared to the rate of initiation:

$$\nu = \frac{\text{Rate (overall)}}{\text{Rate (initiation)}}$$

The kinetic analysis of chain reactions has been mainly utilized in the study of polymerization reactions. In these cases, the kinetic chain length can be directly related to the size of the polymer, and experimental conditions may be adjusted to achieve polymers of the desired average length. The most readily adjusted experimental parameter, the rate of initiation, is inversely related to chain length. Thus, a slow initiation, at lower temperatures for example, will lead to longer chains. This is because slow initiation leads to a lower overall radical concentration, and termination will be minimized.

6.5 Radical Initiators

The most characteristic feature of free radical reactions is the propensity for chain reactions. A corollary to this is the observation that a given chain reaction may be initiated by any of a large variety of radical sources. For example, the chlorination of an alkane may be initiated by any free radical that is capable of either abstracting hydrogen from the alkane or a chlorine atom from Cl_2:

$$\text{Initiator} \longrightarrow 2A\cdot$$

$$A\cdot + Cl_2 \longrightarrow ACl + Cl\cdot$$

(followed by propagation steps (2),(3),(2),(3), and so on)

or

$$A\cdot + RH \longrightarrow AH + R\cdot$$

(followed by propagation steps (3),(2),(3),(2), and so on)

Overall,

$$RH + Cl_2 \xrightarrow[\text{initiator}]{\text{radical}} RCl + HCl$$

A large and varied collection of convenient free radical initiators is available. Most initiators involve the peroxide functional group because the weak O–O bond is readily cleaved thermally. Most peroxides decompose with activation energies between 30 and 40 kcal/mole (the O–O bond strength), making them useful sources of radicals at temperatures between 80 and 150°C. Some of the more commonly used peroxide initiators are *tert*-butyl peroxide, benzoyl peroxide, and peroxydisulfate:

$$(CH_3)_3COOC(CH_3)_3 \longrightarrow 2(CH_3)_3CO\cdot$$

$$\phi-\overset{\overset{\displaystyle O}{\|}}{C}-O-O-\overset{\overset{\displaystyle O}{\|}}{C}-\phi \longrightarrow 2\phi-\overset{\overset{\displaystyle O}{\|}}{C}-O\cdot \longrightarrow 2\phi\cdot + 2CO_2$$

$$^-O-\overset{\overset{\displaystyle O}{\|}}{\underset{\underset{\displaystyle O}{\|}}{S}}-O-O-\overset{\overset{\displaystyle O}{\|}}{\underset{\underset{\displaystyle O}{\|}}{S}}-O^- \longrightarrow 2\ ^-O-\overset{\overset{\displaystyle O}{\|}}{\underset{\underset{\displaystyle O}{\|}}{S}}-O\cdot$$

Homolytic cleavage of all peroxides leads to oxygen radicals, which are highly reactive because of the electronegativity of oxygen. Occasionally, peroxides may be used in conjunction with redox reactions to generate free radicals:

$$Fe^{2+} + ROOH \rightarrow Fe^{3+} + RO\cdot + OH^-$$

The other major class of compounds used as free radical initiators are the azo compounds, which decompose thermally to expel molecular nitrogen:

$$R-N\!\!=\!\!N-R \xrightarrow[\text{or } h\nu]{\Delta} 2R\cdot + N_2$$

In many cases, azo compounds and peroxides may be decomposed photolytically as well as thermally. The advantage of photolysis is that the temperature may be adjusted independently of the need for initiation. Numerous other photoreactions such as the photoreduction of benzophenone, involve free radical intermediates and are suitable for radical initiation.

$$\phi_2CO + (CH_3)_2CHOH \xrightarrow{h\nu} \phi_2\dot{C}OH + (CH_3)_2\dot{C}OH$$

Cage Effects

From organic initiators, free radicals are necessarily generated in pairs, usually in close proximity to one another. One possible fate of such a radical pair is simple recombination with one another before they can become "free" radicals. This geminate recombination occurs frequently in solution because the surrounding solvent molecules create a "cage" that slows the diffusion of the two radicals away from one another. Hence these effects are usually called cage effects. In the case of alkyl peroxides, cage recombination simply regenerates the peroxide initiator. However, in the case of azo compounds or acyl peroxides, intervening molecules of N_2 or CO_2 may be lost such that recombination yields a stable product. Thus, the effectiveness of a radical initiator as a source of "free" radicals depends upon the extent of these cage effects:

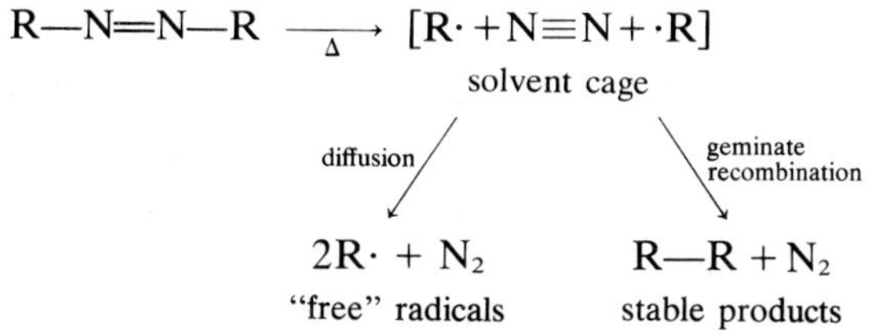

$$R—N=N—R \xrightarrow{\Delta} [R·+N≡N+·R]$$

solvent cage

diffusion / \ geminate recombination

$$2R· + N_2 \qquad R—R + N_2$$

"free" radicals   stable products

Quantitatively, the cage effect is defined as the fraction of total radical pairs generated that react directly in a cage reaction. An evaluation of the extent of cage reaction is usually made by trapping all the radicals that escape from the cage and assuming that any radicals which are not trappable have undergone geminate recombination. Thiols and other good hydrogen donors are often used as radical traps, but the most efficient radical scavengers are stable free radicals:

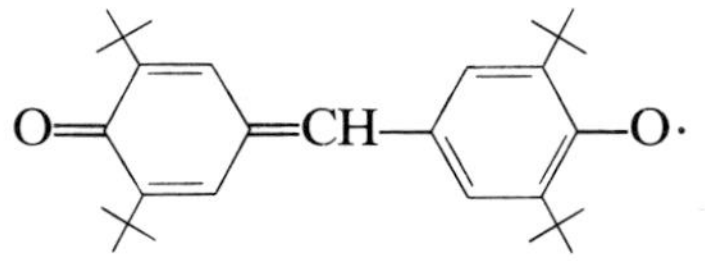

Galvinoxyl

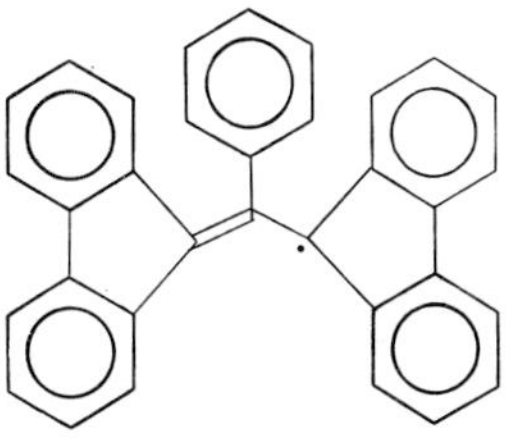

Koelsch radical

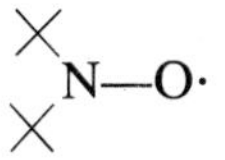

Di-*tert*-butyl nitroxide

Typical cage effects account for a substantial part of the disappearance of initiator. For example, the decomposition of azocumene in benzene at 40°C shows a 27% cage effect.[7] This result indicates that diffusion of the geminate radical pair out of the cage is about three times faster than recombination:

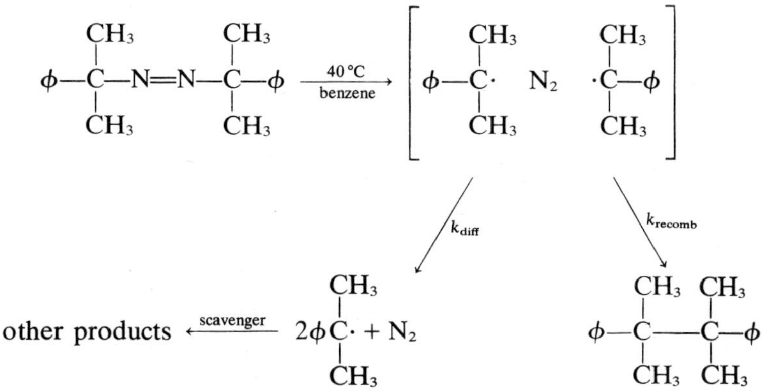

other products $\xleftarrow{\text{scavenger}}$ 2φ$\overset{CH_3}{\underset{CH_3}{C}}$· + N_2

$$\frac{k_{\text{diff}}}{k_{\text{recomb}}} = 3$$

As might be expected, temperature and viscosity play an important role in determining the extent of cage recombination. At higher temperatures or at lower solvent viscosities, the radical pair more readily escapes from the solvent cage. A greater number of intervening molecules also minimizes cage recombination; for example, diacyl peroxides (which eliminate $2CO_2$) usually show smaller cage effects than azo compounds which eliminate N_2.

Cage effects have been used in many cases to determine relative rates of some fundamental molecular processes. For example, the three allylic azo isomers that follow yield the same product mixture upon cage recombination:[8]

This indicates that the radical pair quickly equilibrates or randomizes with respect to its orientation within the solvent cage relative to the rate of bond formation.

In the vapor phase, cage effects are nonexistent, except at very high pressures, simply because the concept of a solvent cage is not appropriate. In the solid phase, the rigidity of the medium leads to some very pronounced cage effects. The photodecomposition of crystalline acetyl benzoyl peroxide, specifically labeled at the peroxide oxygens, produces, among other cage products, methyl benzoate with 60% of the labeled oxygen at the methoxy position.[9] Thus, the methyl radical combines with the benzoyloxy radical before complete equilibration of the two oxygens takes place. The comparable experiment in solution leads to complete scrambling of the oxygen label:

$$CH_3-\overset{\overset{\displaystyle O}{\|}}{C}-O^{18}-O^{18}-\overset{\overset{\displaystyle O}{\|}}{C}-\phi \xrightarrow[-70\,°C]{h\nu}$$

$$\left[CH_3-\overset{\overset{\displaystyle O}{\|}}{C}-O^{18}\cdot\quad\cdot O^{18}-\overset{\overset{\displaystyle O}{\|}}{C}-\phi\right]\longrightarrow\left[CH_3-\overset{\overset{\displaystyle O}{\|}}{C}-O^{18}\cdot+CO_2+\cdot\phi\right]$$

$$\downarrow\qquad\qquad\qquad\qquad\qquad\downarrow$$

$$\left[CH_3\cdot+CO_2+\cdot O^{18}-\overset{\overset{\displaystyle O}{\|}}{C}-\phi\right]\longrightarrow\left[CH_3\cdot+2CO_2+\cdot\phi\right]$$

$$\downarrow$$

$$\underset{(60\%)}{CH_3-O^{18}-\overset{\overset{\displaystyle O}{\|}}{C}-\phi}\ +\underset{(40\%)}{CH_3-O-\overset{\overset{\displaystyle O^{18}}{\|}}{C}-\phi}\qquad (+\text{other cage products})$$

Biradicals

In the analysis of radical-radical interactions, such as cage effects, some of the most interesting correlations come from radical pairs contained within the same molecule. A molecule with two radical centers is called a biradical. Biradicals are normally classified according to the location of the two radical sites. 1,1-Biradicals have both unpaired electrons on the same carbon atom and are carbenes (Chapter 7). 1,2-Biradicals are a representation of an excited state of a π bond (Chapter 8). 1,5-Biradicals and biradicals separated by even more atoms behave as relatively independent radical centers. The 1,3- and 1,4-biradicals are the classes we shall deal with here.

A few organic molecules have been predicted to exist as biradicals in their ground state. Trimethylenemethane, which we have seen in Chapter 1 is of some theoretical interest, is calculated to have a triplet ground state[*] in its planar geometry.[10] The alternative geometry of a methylenecyclopropane structure, of course, is still lower in energy. This 1,3-biradical has now been made and its triplet ground state confirmed by esr:[11]

By a similar synthesis, tetramethyleneethane has been made and has a 1,4-triplet biradical ground state in its planar geometry:[12]

[*] Triplet state will be defined and described in detail in Chapters 7 and 8. A triplet state is an electronic state of a molecule that has two unpaired electrons with parallel spins. Most organic molecules have singlet ground states with all electrons paired and antiparallel.

We will be concerned primarily with simpler biradicals without such extensive delocalization—for example, trimethylene $\cdot CH_2CH_2CH_2\cdot$. Trimethylene also is not truly a ground state but is an excited state of cyclopropane. Theoretical calculations for trimethylene suggest that it has a stable conformation (relative potential energy minimum) with the two p orbitals parallel,[13,14] a structure that has been called π cyclopropane:

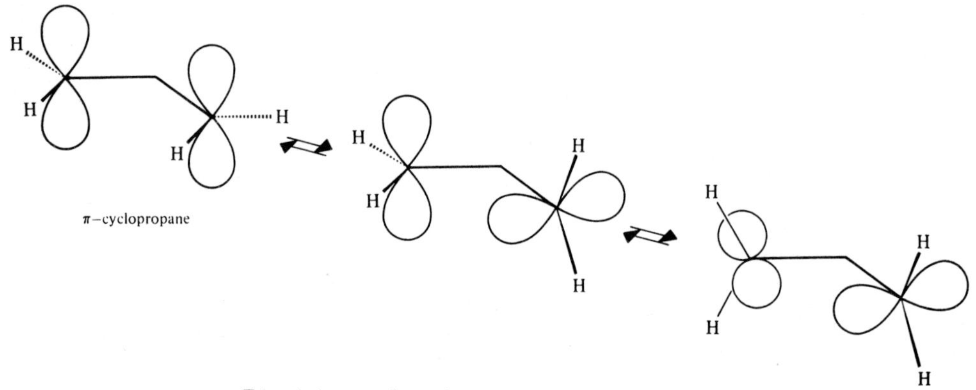

Trimethylene conformations

"Stretched" cyclopropane
(leads to σ -bonding)

One of the most important aims in the study of biradical reactions has been *spin correlation* studies—a correlation of the different reactivities of singlet and triplet biradicals. The six-membered ring azo compound shown below was prepared in *meso* (*cis*) and in D,L (*trans*) forms and subjected to decomposition by thermolysis, direct photolysis, and triplet photosensitization. The stereochemistry of the product depends markedly upon whether a singlet or triplet biradical is involved (Table 6.2).

Table 6.2 *Spin Correlation in Azo Decomposition*[15]

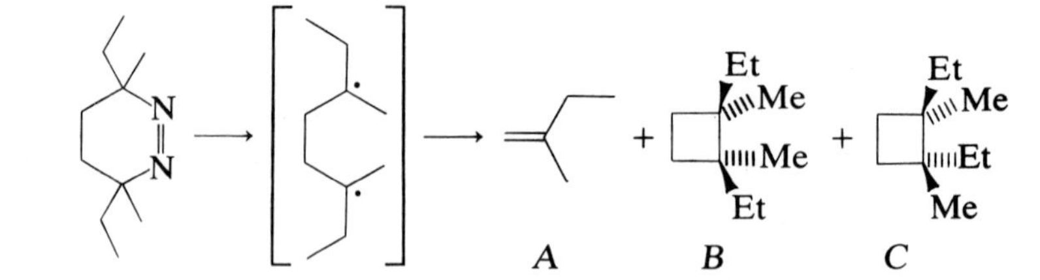

(either *meso* or D, L)

		%A	%B	%C	% retention
Thermolysis	*meso*	49	43	2.5	98
	D,L	51	3.5	42	98
Direct photolysis	*meso*	61	35	3.5	95
	D,L	60	4	33	97
Triplet photosensitization	*meso*	77	11.5	8	61
	D,L	75	8	12	65

The spin correlation explanation of these results follows. Thermolysis or direct photolysis produces a singlet biradical with antiparallel electron spins from which direct bond formation may take place. Hence the singlet biradical reacts quickly to give ring closure with nearly complete retention of configuration. Substantial β cleavage to alkenes is a common characteristic of 1,4-biradicals. The triplet biradical that is formed by photosensitization (see Chapter 8) necessarily has a longer lifetime because bond formation cannot take place until the two electrons have converted to antiparallel spins. Thus, less ring closure takes place and the ring closure that does occur takes place with scrambling of the geometry.

Comparable spin correlation effects are observed for the 1,3-biradicals obtained by addition of singlet and triplet carbenes to olefins (see p. 304–305).

Biradicals have also been implicated as reaction intermediates, most notably in cycloaddition reactions. The products observed from the thermal [2+2] cycloaddition of 1,1-dichloro-2,2-difluoroethene to the geometric isomers of 2,4-hexadiene indicate biradical intermediates; the orientation of addition is always such that the most stable biradical is formed (an allyl radical on the one hand and $-CCl_2\cdot$ rather than $-CF_2\cdot$ on the other hand), and the stereochemistry is such that partial rotation is observed about the bond not involved in the allylic radical.[16] Note that this nonconcerted [2+2] thermal cycloaddition is consistent with the orbital symmetry rules that predict that the most favorable suprafacial orientation is forbidden (Chapter 2).

Biradical intermediates in [2+2] cycloadditions[16]

6.6 Characteristic Radical Reactions

The reactions that organic free radicals undergo may be broadly classified into the following categories:

1. Unimolecular rearrangements and fragmentations,
2. Bimolecular radical–radical (termination) reactions,
3. Bimolecular electron transfer (redox) reactions,
4. Bimolecular radical–molecule reactions, including abstractions, additions, and substitutions.

Unimolecular Rearrangements and Fragmentations

In direct contrast to carbocation chemistry free radical rearrangements are relatively rare. With a strong driving force and a good migrating group such as aryl, 1,2 migrations are occasionally observed. Rearrangement of aryl groups is apparently facilitated by electron withdrawing groups (Hammett ρ value $\sim +1$);[17] however, the ability to delocalize the odd electron may be the dominant factor in the substituent effects:

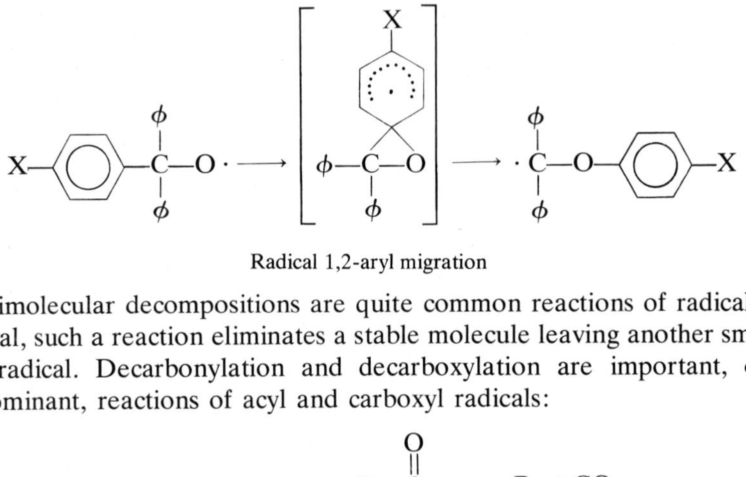

Radical 1,2-aryl migration

Unimolecular decompositions are quite common reactions of radicals. In general, such a reaction eliminates a stable molecule leaving another smaller free radical. Decarbonylation and decarboxylation are important, often predominant, reactions of acyl and carboxyl radicals:

decarbonylation $$R\text{—}\overset{\overset{\displaystyle O}{\|}}{C}\cdot \longrightarrow R\cdot + CO$$

decarboxylation $$R\text{—}\overset{\overset{\displaystyle O}{\|}}{C}\text{—}O\cdot \longrightarrow R\cdot + CO_2$$

For alkoxy radicals, β-scission is a common fragmentation pathway. The preferred cleavage, where there is a choice, produces the most stable radical:[18]

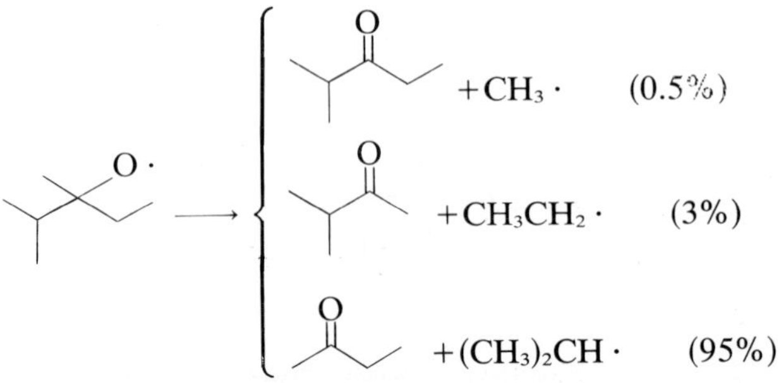

β-Scission

Radical–Radical Reactions

Radical–radical reactions differ from other radical reactions in that they necessarily lead to a net destruction of radicals. Hence, these are the termination steps that end a free radical chain reaction. The two main types of termination reaction observed for organic free radicals are combination and disproportionation. Combination may be thought of as the reverse of homolysis, leading to direct bond formation between the two radical sites. Disproportionation, on the other hand, involves transfer of a β-hydrogen from one radical to the other leading to alkane and olefin product. For most alkyl radicals, combination is the favored termination reaction, unless there is steric hindrance or other constraint:

$$2CH_3CH_2CH_2\cdot \xrightarrow[\text{combination}]{} CH_3CH_2CH_2CH_2CH_2CH_3$$

$$2CH_3CH_2CH_2\cdot \xrightarrow[\text{disproportionation}]{} CH_3CH_2CH_3 + CH_3\text{—}CH\text{=}CH_2$$

$$\frac{k_{\text{comb}}}{k_{\text{dis}}} = 7 \text{ for } n\text{-propyl radicals}^{19}$$

However,

$$\frac{k_{\text{comb}}}{k_{\text{dis}}} = 0.2 \text{ for } t\text{-butyl radicals}^{20}$$

Since radical–radical combination is simply the reverse of a bond homolysis, the combination of radicals is usually considered to be an unactivated process (Figure 6.6). Most radical–radical reactions have been found to have zero activation energies. In cyclohexane solution, methyl radicals combine at a rate of $9 \times 10^9 \, M^{-1} \, \text{sec}^{-1}$, which is very close to the diffusion-controlled limit.* Other radicals combine with slightly lower rate constants, such as t-butyl radical recombination that has a rate constant of $2 \times 10^9 \, M^{-1} \, \text{sec}^{-1}$. The difference is thought to be due to orientation factors that make larger radicals less likely to couple during a given encounter.[21]

The impact that these very rapid combination reactions have upon chain reactions is quite significant. Since termination steps are so efficient, an

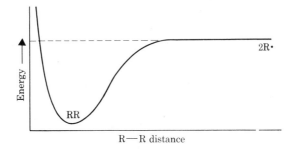

Fig. 6.6   *Potential energy diagram for bond homolysis or radical–radical combination.*

---

* A bimolecular reaction that proceeds with the diffusion-controlled rate constant is a reaction that occurs upon every encounter of the two reactants; that is, the rate-determining step is diffusion of the two reactants together. For typical solvents around room temperature, the maximum, diffusion-controlled rate constant is about $10^{10} \, M^{-1} \, \text{sec}^{-1}$.

effective chain reaction can be achieved only if the propagation steps are rapid enough to compete. Thus, chain reactions are favored with very reactive radicals and are also favored under low steady-state radical concentrations.

Radical Oxidation–Reduction Reactions

In the presence of suitable oxidants or reductants, radicals may be oxidized to cations or reduced to anions. The ferrous ion catalysis of peroxide decomposition that was mentioned previously as a source of free radicals may be followed by a subsequent reduction to alkoxide ion; thus, low initiator concentrations are desirable:

$$ROOH + Fe^{2+} \rightarrow RO\cdot + OH^- + Fe^{3+}$$

$$RO\cdot + Fe^{2+} \rightarrow RO^- + Fe^{3+}$$

Oxidants such as cupric ion are often capable of converting radical intermediates to cations. The differentiation between radical and carbocation intermediates is strikingly displayed in the cyclobutyl system.[22] Cyclobutyl radicals do not rearrange, yet cyclobutyl cations undergo extensive rearrangement.

(2%)   (44%)   (54%)

Copper ion catalysis of peroxide reactions has been developed into a highly useful synthetic scheme. Typically, cuprous ion is used as reductant to catalyze decomposition, after which the cupric ion oxidizes the radical intermediate, regenerating cuprous ion:[22]

$$CH_3COOtBu + Cu^+ \longrightarrow Cu^{2+} + CH_3COO^- + tBuO\cdot$$

$$tBuO\cdot + RH \longrightarrow tBuOH + R\cdot$$

$$R\cdot + Cu^{2+} \longrightarrow R^+ + Cu^+$$

$$R^+ + CH_3COO^- \longrightarrow ROOCCH_3$$

Overall,

$$RH + CH_3COOtBu \xrightarrow{Cu^+} ROOCCH_3 + tBuOH$$

Radical–Molecule Reactions

By far the most common type of radical reaction is the reaction of a radical with a molecule. In the following sections, we will be considering radical reactions with diamagnetic molecules, which have all electrons paired; hence, such a radical–molecule reaction will necessarily generate another radical.

These are the propagation reactions that determine the course of radical chain reactions. We will consider in detail some of the most important of these reactions: abstractions, in which an atom is transferred to the radical; additions, in which the radical adds to a π bond (this includes polymerizations); and substitutions, in which the radical displaces a leaving group.

6.7   Radical Abstraction Reactions

Abstraction reactions correspond to the removal of an atom by a free radical. In general, the univalent atoms are those which are subject to radical abstraction; hydrogen and halogen abstractions are virtually the only types of radical abstractions observed:

$$B{-}A + R\cdot \rightarrow B\cdot + R{-}A$$

Using bond dissociation energies, heats of reaction for abstraction reactions may be obtained, and from these, estimates of the relative rates may be made. For example, we have already discussed the selectivity that a chlorine atom exhibits in abstraction of alkyl hydrogen atoms. By assuming that the relative activation energies roughly parallel the relative exothermicities, we can readily demonstrate that chlorine will prefer to abstract a 3° hydrogen, followed by 2°, 1°, and methyl (Table 6.3).

Table 6.3 *Hydrogen Abstraction by Chlorine Atoms*

| R | $RH + Cl\cdot \rightarrow R\cdot + HCl$ | |
	ΔH_{calc}(kcal/mole)	$E_{a\,(exptl)}$(kcal/mole)[23]
3°	−12	0.1
2°	−8	0.5
1°	−5	1
CH$_3$	+1	4

Alkyl hydrogen abstraction by bromine atoms is always endothermic with correspondingly larger activation energies and hence a greater selectivity than abstraction by chlorine. This is the usual reactivity–selectivity correlation: the less reactive species will be more selective (Table 6.4).

Table 6.4 *Hydrogen Abstraction by Bromine Atoms*

| R | $RH + Br\cdot \rightarrow R\cdot + HBr$ | |
	ΔH_{calc}(kcal/mole)	$E_{a\,(exptl)}$(kcal/mole)[23]
3°	+3	7.5
2°	+7	10
1°	+10	13
CH$_3$	+16	18

Aside from these energetic effects, polar effects have been shown to be important in radical reactions. We have considered the electron distribution in transition states to be relatively symmetrical without much charge separation. In fact, hydrogen transfer from carbon to a halogen appears to involve a substantial polarization in the transition state, with partial carbocation character on the developing alkyl radical:

$$\left[\overset{\delta+}{\underset{/}{\overset{\backslash}{C}}} \cdots\cdots H \cdots\cdots \overset{\delta-}{Cl} \right]^{\ddagger}$$

Since carbocation stability parallels radical stability, either criterion would predict the correct order of selectivity for hydrogen abstraction from simple alkanes. However, with polar or polarizable substituents, polarization effects in the transition state may lead to different selectivities. The abstraction of hydrogen from propionic acid follows different courses depending upon the abstracting species. The methyl radical abstracts primarily from the α position where the incipient radical may be delocalized to the carboxyl group. However, a chlorine atom abstracts from the β position, farther removed from the electron-withdrawing carboxyl substituent:[24]

$$CH_3CH_2COOH + CH_3\cdot \longrightarrow \left[\begin{array}{c} \overset{\delta-}{CH_3-CH-COOH} \\ | \\ H \\ | \\ \overset{\delta+}{CH_3} \end{array} \right]^{\ddagger} \longrightarrow CH_3\dot{C}HCOOH$$

(85% α abstraction)

$$CH_3CH_2COOH + Cl\cdot \longrightarrow \left[\begin{array}{c} \overset{\delta+}{CH_2-CH_2-COOH} \\ | \\ H \\ | \\ Cl^{\delta-} \end{array} \right]^{\ddagger} \longrightarrow \dot{C}H_2CH_2COOH$$

(98% β abstraction)

Furthermore, the bromination of substituted toluenes indicates that abstraction is favored by substituents that stabilize benzyl cations. The kinetic substituent effects correlate with σ^+ values and yield a ρ value of -1.39:[25]

In addition to hydrogen abstractions, halogen atoms may also be abstracted by free radicals. Abstraction from a dihalogen molecule, as in the halogenation chain reaction, is by far the most common instance of halogen abstraction. However, hypochlorites and N-bromosuccinimide provide alternative sources for radical chlorination or bromination, respectively. *Tert*-butyl hypochlorite is a convenient chlorination agent which works through the following chain mechanism:

$$tBuOCl \xrightarrow{\Delta} tBuO\cdot + Cl\cdot$$

$$tBuO\cdot + RH \longrightarrow tBuOH + R\cdot$$

$$R\cdot + tBuOCl \longrightarrow RCl + tBuO\cdot$$

Occasionally, hypochlorite chlorination will provide a synthetic route not available with molecular chlorine:

$$\text{(major)} \qquad \text{(minor)}$$

The use of *N*-bromosuccinimide (NBS) provides an exceptionally useful method for specific bromination of allylic positions. The special value of NBS for such reactions is due to the fact that molecular halogens would be unable to accomplish a substitution reaction in the presence of an olefinic bond and would give addition instead. The mechanism of NBS substitution does not involve NBS as a bromine atom donor, as had long been thought, but rather NBS acts as a source of very low concentrations of molecular bromine. It has been confirmed that the selectivity observed in NBS brominations is the same as the selectivity of molecular bromine. Furthermore, very low concentrations of Br_2 have been shown to give allylic bromination, rather than addition to the π bond.[26] The key question remains unsolved, however: why do low concentrations of Br_2 favor radical substitution over ionic addition?

1. $Br\cdot + RH \longrightarrow R\cdot + HBr$

2.

3. $Br_2 + R\cdot \longrightarrow RBr + Br\cdot$

There are two significant advantages of NBS over Br_2 as a bromination agent: the NBS keeps a low concentration of Br_2, enhancing radical rather than ionic processes, and at the same time the NBS removes the HBr that would inhibit the substitution by acting as a good hydrogen donor (essentially reversing the abstraction step).

Propagation

$$Br\cdot + RH \xrightarrow{\Delta H = +10 \text{ kcal/mole}} R\cdot + HBr$$

Inhibition

$$R\cdot + HBr \xrightarrow{\Delta H = -10 \text{ kcal mole}} RH + Br\cdot$$

6.8 Radical Addition Reactions

Free radicals may also undergo addition to carbon–carbon double bonds by chain mechanisms. By far the most thoroughly studied and the most useful of these reactions is polymerization, which we shall consider after dealing with some of the simpler addition reactions:

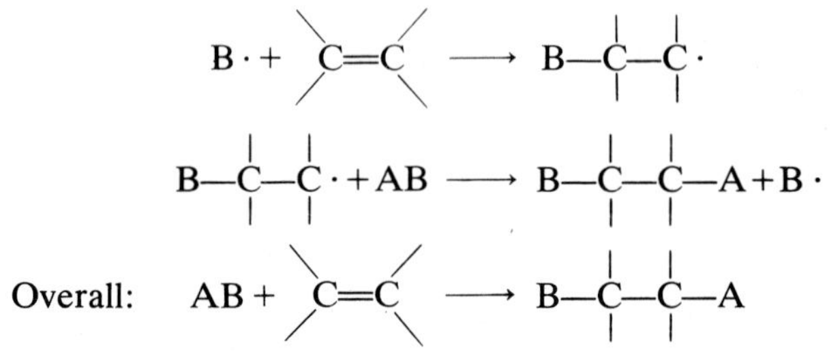

If the initial radical addition is reversible, this provides a route for *cis–trans* equilibration of geometric isomers. This occurs in the case of I_2 photolysis in the presence of the 2-butenes:[27]

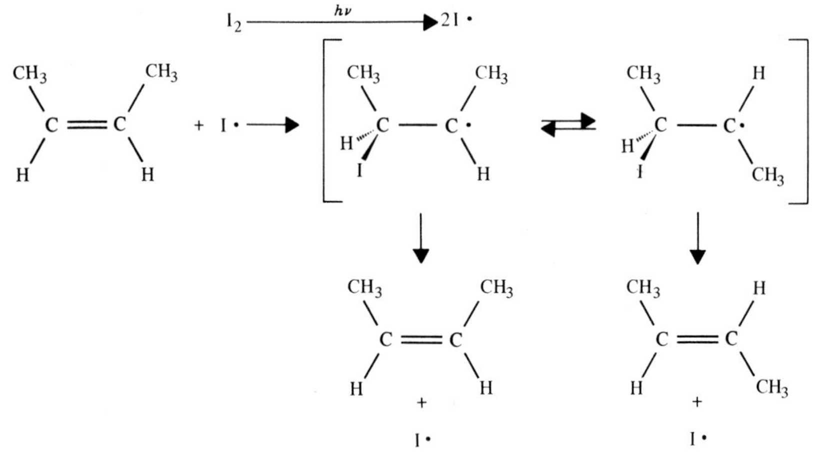

If the initial radical addition is not readily reversible, and the molecule AB is reactive towards abstraction, simple addition will occur. This is the case with the anti-Markovnikov addition of HBr to olefins.

Markovnikov (ionic) addition:

$$R—CH{=}CH_2 \xrightarrow{H^+} R—\overset{+}{C}H—CH_3 \xrightarrow{Br^-} R—\underset{\underset{Br}{|}}{C}H—CH_3$$

Anti-Markovnikov (radical) addition:

$$R—CH{=}CH_2 \xrightarrow{Br\cdot} R—\underset{\cdot}{C}H—CH_2Br \xrightarrow{HBr} R—CH_2—CH_2—Br$$

In both the ionic and radical additions, the orientation of addition is such as to give the most stable intermediate. (This, of course, would be the unambiguous way to state an orientation "rule.") The different orientations result from the different species which undergo the initial addition: H^+ in the ionic addition and $Br\cdot$ in the radical addition. For synthetic purposes, the anti-Markovnikov addition may be accomplished by inclusion of peroxides or other radical initiators with the olefin–HBr mixture.

None of the other hydrogen halides add to olefins by a radical chain process. Only with HBr are *both* of the propagation steps sufficiently favorable for an efficient chain reaction (Table 6.5).

Table 6.5 *Energetics of Radical Addition of HX*[28]

1. $X \cdot + CH_2 = CH_2 \rightarrow X - CH_2 - CH_2 \cdot$

2. $X - CH_2 - CH_2 \cdot + HX \rightarrow X - CH_2 - CH_3 + X \cdot$

X	ΔH_1(kcal/mole)	ΔH_2(kcal/mole)
F	~−40	+37
Cl	−26	+5
Br	−5	−11
I	+7	−27

Many other species may be added to olefins by radical mechanisms. Thiols represent one important class, since the S—H bond is rather easily broken homolytically:

$$RS \cdot + \ \text{\Large$>$}C{=}C\text{\Large$<$} \ \longrightarrow \ RS{-}\overset{|}{\underset{|}{C}}{-}\overset{|}{\underset{|}{C}}\cdot$$

$$RS{-}\overset{|}{\underset{|}{C}}{-}\overset{|}{\underset{|}{C}}\cdot + RSH \ \longrightarrow \ RS{-}\overset{|}{\underset{|}{C}}{-}\overset{|}{C}H + RS \cdot$$

Halomethanes also may be added:

$$+ CCl_4 \xrightarrow{(CH_3COO)_2} CCl_3 \text{ ... (85\% yield)}$$

Autoxidation

Autoxidation may be considered as a special case of radical addition, specifically addition of a radical to a molecule of oxygen. Oxygen, being a triplet (biradical) itself, is highly reactive towards most free radicals. Overall, the reaction corresponds to formation of a hydroperoxide at a reactive C–H bond. The radical chain mechanism is exemplified by the autoxidation of cumene (isopropylbenzene):

$$\phi CH(CH_3)_2 + R \cdot \rightarrow \phi \dot{C}(CH_3)_2 + RH$$

$$\phi \dot{C}(CH_3)_2 + O_2 \longrightarrow \phi - \overset{CH_3}{\underset{CH_3}{\overset{|}{\underset{|}{C}}}} - OO \cdot$$

$$\phi - \overset{CH_3}{\underset{CH_3}{\overset{|}{\underset{|}{C}}}} - OO \cdot + \phi CH(CH_3)_2 \longrightarrow \phi \overset{CH_3}{\underset{CH_3}{\overset{|}{\underset{|}{C}}}} - OOH + \phi \dot{C}(CH_3)_2$$

Overall: $\phi CH(CH_3)_2 + O_2 \longrightarrow \phi \overset{CH_3}{\underset{CH_3}{\overset{|}{\underset{|}{C}}}} - OOH$

The air oxidation of cumene is an industrially important reaction, since acid-catalyzed rearrangement of the hydroperoxide leads to phenol and acetone in good yield:

$$\phi-\underset{\underset{CH_3}{|}}{\overset{\overset{CH_3}{|}}{C}}-O-O-H \xrightarrow{H^+} \phi-\underset{\underset{CH_3}{|}}{\overset{\overset{CH_3}{|}}{C}}-O-\overset{+}{\underset{\underset{H}{}}{\overset{H}{O}}} \xrightarrow[\phi\sim]{-H_2O} +\underset{\underset{CH_3}{|}}{\overset{\overset{CH_3}{|}}{C}}-O\phi \xrightarrow[-H^+]{H_2O}$$

$$HO-\underset{\underset{CH_3}{|}}{\overset{\overset{CH_3}{|}}{C}}-O\phi \longrightarrow O=\underset{\underset{CH_3}{|}}{\overset{\overset{CH_3}{|}}{C}} + HO\phi$$

Autoxidation also proceeds readily with aldehydes with the final product being a carboxylic acid. The initially produced peracid is capable of oxidizing another molecule of aldehyde. This process accounts for the great susceptibility of most aldehydes to air oxidation:

$$\phi\overset{O}{\overset{||}{C}}-H+R\cdot \longrightarrow \phi-\overset{O}{\overset{||}{C}}\cdot +RH$$

$$\phi\overset{O}{\overset{||}{C}}\cdot +O_2 \longrightarrow \phi-\overset{O}{\overset{||}{C}}-OO\cdot$$

$$\phi-\overset{O}{\overset{||}{C}}-OO\cdot +\phi\overset{O}{\overset{||}{C}}H \longrightarrow \phi-\overset{O}{\overset{||}{C}}-OOH+\phi\overset{O}{\overset{||}{C}}\cdot$$

$$\phi-\overset{O}{\overset{||}{C}}-OOH+\phi\overset{O}{\overset{||}{C}}H \longrightarrow 2\phi\overset{O}{\overset{||}{C}}-OH$$

$$\text{Overall:} \ \ 2\phi\overset{O}{\overset{||}{C}}H+O_2 \longrightarrow 2\phi\overset{O}{\overset{||}{C}}-OH$$

In general, autoxidation proceeds efficiently only with highly reactive C–H bonds. If the hydrogen abstraction step is not very fast, then termination predominates and alcohols or alkyl peroxides rather than hydroperoxides result:

$$2ROO\cdot \rightarrow ROOOOR \rightarrow 2RO\cdot +O_2$$

$$2RO\cdot \rightarrow ROOR$$

$$RO\cdot +RH \rightarrow ROH+R\cdot$$

The best evidence for the intermediacy of a tetroxide is an elegant oxygen-labeling experiment. By performing the autoxidation with a mixture of $^{16}O-^{16}O$ and $^{18}O-^{18}O$, it was found that some $^{16}O-^{18}O$ was formed as the reaction proceeded.[29] Other conceivable mechanisms, such as a direct displacement of O_2 from a peroxy radical, would not be expected to give any mixing of ^{16}O and ^{18}O.

The termination of secondary peroxy radicals has generated substantial interest because it can be accompanied by chemiluminescence. The bimolecular termination of secondary peroxy radicals yields an alcohol, a ketone that is formed at least partially in an electronically excited state, and molecular oxygen. The termination apparently proceeds through intramolecular disproportionation of the tetroxide intermediate:[30,31]

$$2R_2CHOO \cdot \longrightarrow R_2CHOH + (R_2CO)^* + O_2$$

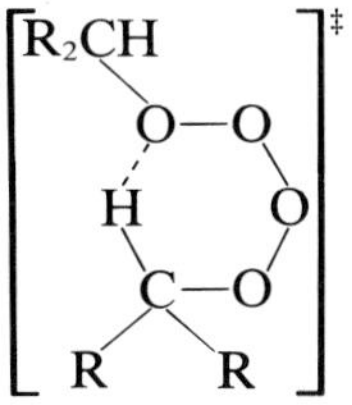

Autoxidation is also the pathway followed in combustion reactions. The products formed in autoxidation are peroxides, which themselves could be radical initiators at elevated temperatures. Thus, autoxidations can be self-perpetuating under combustion conditions leading to complete oxidation.

Arylations

Additions to an aromatic ring are also possible reactions of free radicals, although usually only for the most reactive radicals. The most common example of aromatic substitution by a radical mechanism is phenylation by phenyl radicals that provides some of the best synthetic routes to biaryls. The mechanism involves a resonance-stabilized radical intermediate, similar to the intermediates in electrophilic or nucleophilic aromatic substitution (see Chapters 4 and 5). In this case, however, the radical intermediate reacts either by combination or by disproportionation (hydrogen donation or abstraction):

$$(\phi COO)_2 \longrightarrow 2\phi CO_2 \cdot \longrightarrow 2\phi \cdot + 2CO_2$$

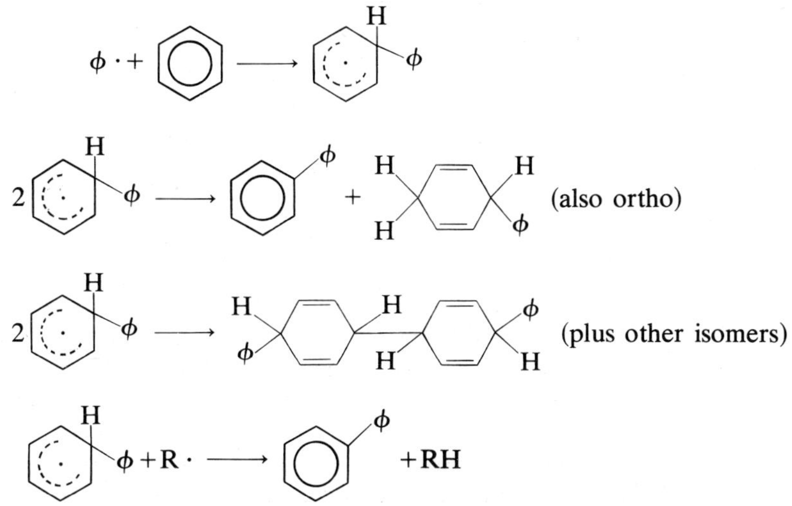

(also ortho)

(plus other isomers)

+RH

The combination and disproportionation reactions have been demonstrated by isolation of dihydrobiphenyls and tetraphenyl derivatives. For synthetic purposes, biphenyls are best prepared by including an oxidant, usually a relatively stable free radical, to abstract hydrogen from the intermediate. In the case of the Gomberg reaction, arenediazonium ions in the presence of strong base yield aryl radicals and a relatively stable radical that serves to remove a hydrogen atom from the arylated radical intermediate:[32]

$$ArN_2^+ + OH^- \longrightarrow Ar—N=N—OH$$

$$Ar—N=N—OH + OH^- \rightleftharpoons Ar—N=N—O^- + H_2O$$

$$Ar—N=N—O^- + ArN_2^+ \longrightarrow Ar—N=N—O—N=N—Ar$$

$$Ar—N=N—O—N=N—Ar \longrightarrow Ar\cdot + N_2 + \cdot O—N=N—Ar$$

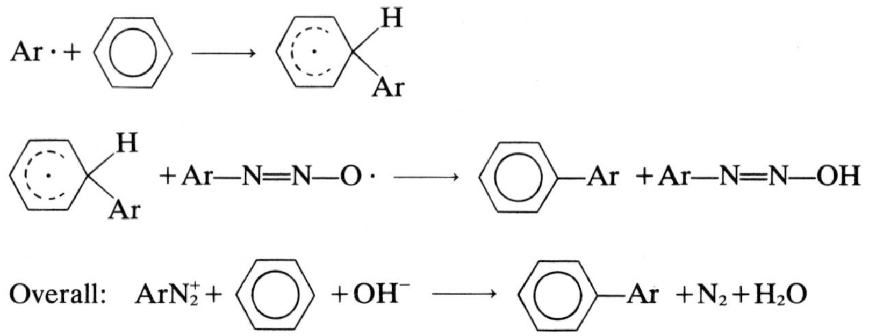

Overall:

Polymerizations

Catalysis of olefin polymerization is one of the most important reactions of free radicals. This may be considered to be a special case of an addition reaction in which the radical generated by addition to the double bond then acts as the radical to add to another double bond and so on:

$$R\cdot + \phi CH=CH_2 \longrightarrow R—CH_2—\dot{C}H\phi$$

$$R—CH_2—\dot{C}H\phi + \phi CH=CH_2 \longrightarrow R—CH_2CH—CH_2\dot{C}H\phi$$
$$\qquad\qquad\qquad\qquad\qquad\qquad\qquad\quad \phi$$

$$R\text{-}\!\left(CH_2—\underset{\underset{\phi}{|}}{CH}\right)_{\!n-1}\!\!CH_2—\dot{C}H\phi + \phi CH=CH_2 \longrightarrow R\text{-}\!\left(CH_2—\underset{\underset{\phi}{|}}{CH}\right)_{\!n}\!\!CH_2—\dot{C}H\phi$$

Polymers formed in radical chain reactions normally consist of 10^3 to 10^5 olefin (monomer) units. Thus, the initiating radicals that appear as the end groups of the polymer are normally insignificant in considering the properties of polymers. Depending upon the olefin monomer chosen, polymers with huge varieties of properties may be constructed. The obvious commercial applications have led to very rapid progress in polymer chemistry over the last 40 years. The correlation of polymer structures with properties has been one of the most important aspects of polymer research. Some of the varied polymers produced by radical chain processes are shown in Table 6.6.

Table 6.6 *Some Common Polymers*

Monomer	Name	Polymer	Common name
$CH_2{=}CH_2$	Ethylene	$\{CH_2{-}CH_2\}_n$	Polyethylene
$CF_2{=}CF_2$	Tetrafluoroethylene	$\{CF_2{-}CF_2\}_n$	Teflon
$CH_2{=}CHCl$	Vinyl chloride	$\{CH_2{-}CHCl\}_n$	PVC, Tygon
$CH_2{=}C\overset{\displaystyle CH_3}{\underset{\displaystyle COOCH_3}{\big\langle}}$	Methyl methacrylate	$\left(CH_2{-}\underset{COOCH_3}{\overset{CH_3}{C}}{-}\right)_n$	Plexiglas or Lucite
$CH_2{=}CCl_2$	Vinylidene chloride	$\{CH_2{-}CCl_2\}_n$	Saran
$CH_2{=}CH{-}\underset{Cl}{C}{=}CH_2$	Chloroprene	$\left(CH_2{-}CH{=}\underset{Cl}{C}{-}CH_2\right)_n$	Neoprene
$CH_2{=}CH{-}\underset{CH_3}{C}{=}CH_2$	Isoprene	$\left(CH_2{-}\overset{H}{C}{=}\overset{CH_3}{C}{-}CH_2\right)_n$	
		(all *cis*)	Natural rubber
		(all *trans*)	Gutta percha

The polymerization of unsymmetrical monomers always leads to regular polymers. For example, polystyrene will have phenyl groups on exactly every second carbon of the polymer. This is due to the stability of the benzyl radical that will be formed preferentially upon every addition of styrene to the growing chain. This regularity is general for all polymers, since the preferred orientation of radical addition will be observed upon every addition of a monomer unit.

Copolymerization, the use of two different monomer units to construct mixed polymers, is an important method of providing even greater variation, and greater control, in polymer properties. For example, polystyrene is a rigid, rather brittle polymer used mainly to make foamed insulation. Styrene copolymerized with 20 to 30% of 1,3-butadiene is much tougher and is used in making bowling balls or helmets. Polymerization of 25% styrene with 75% 1,3-butadiene gives a good elastomer (elastic polymer) which is the major synthetic rubber used in automobile tires. Styrene copolymerized with equimolar maleic anhydride gives, upon hydrolysis, a water-soluble polymer used as a dispersant or sizing agent.

One of the most important aspects of copolymerization is the strong tendency for alternation of monomers. The extreme case is the copolymer of stilbene and maleic anhydride, which is always an alternating 1:1 copolymer regardless of the relative concentrations of the two monomers:

$$n\phi CH{=}CH\,\phi + n\,CH{=}CH \longrightarrow \left(\underset{\phi}{CH}{-}\underset{\phi}{CH}{-}\underset{O{=}C}{CH}{-}\underset{C{=}O}{CH}\right)_n$$

The strong preference for alternating copolymers that is shown by most monomer pairs may be ascribed to polar effects. Using the copolymerization of styrene with methylmethacrylate as an example, polar effects direct that a terminal styryl radical will always prefer to add to a methyl methacrylate double bond, while a terminal acrylic radical will always prefer to add to styrene:

$$\begin{array}{c}\dashv CH_2-CH\cdot\\|\\\phi\end{array}$$

$$CH_2\!=\!C\begin{array}{c}CH_3\\\\COOCH_3\end{array}$$

$$\left[\begin{array}{c}\dashv CH_2-\overset{\delta^+}{CH}----CH_2-C\begin{array}{c}CH_3\\\\\overset{\delta^-}{\raise2pt{:}}C-OCH_3\\\|\\O\end{array}\\|\\\phi\end{array}\right]^{\ddagger}$$

$$\dashv CH_2-CH-CH_2-\overset{CH_3}{\underset{COOCH_3}{C}}\cdot\atop\phi$$

$$\phi CH\!=\!CH_2$$

$$\left[\begin{array}{c}CH_3\\\dashv CH_2-C----CH_2-\overset{\delta^+}{CH}\\\overset{\delta^-}{\raise2pt{:}}C-OCH_3\quad\phi\\O\end{array}\right]^{\ddagger}$$

$$\dashv CH_2-\overset{CH_3}{\underset{COOCH_3}{C}}-CH_2-\underset{\phi}{CH}\cdot$$

This copolymerization has been utilized in an ingenious scheme for differentiation between cation, radical, and anion intermediates. Olefin polymerization in principle may be initiated by any of these intermediates. In fact, however, the specific monomer structure will determine which types of intermediates are able to effect its polymerization. For example, methyl methacrylate can very effectively stabilize a negative charge on its α carbon but cannot stabilize a positive charge. Hence methyl methacrylate is polymerized by anions or radicals but not by cations. Similarly, styrene is polymerized by radicals or cations but not by anions:

$$R\!:\!^{\ominus}+CH_2\!=\!C\begin{array}{c}CH_3\\\\COOCH_3\end{array}\longrightarrow R-CH_2-\overset{CH_3}{\underset{\underset{O}{\overset{\|}{C}}-OCH_3}{C:^{\ominus}}}\longleftrightarrow R-CH_2-\overset{CH_3}{\underset{C-OCH_3}{C}}$$

$$R^{\oplus}+CH_2\!=\!CH\phi\longrightarrow R-CH_2-\underset{\phi}{CH^{\oplus}}\longleftrightarrow R-CH_2-CH\longleftrightarrow$$

Table 6.7 *Polymerization of Styrene–Methyl Methacrylate*[33]

Initiator	Polymer	%C	%H
Cations	Polystyrene	92.3	7.7
Anions	Poly(methyl methacrylate)	71.4	9.6
Radicals	1:1 copolymer	82.9	8.6

Given this information, it is possible to use a mixture of styrene and methyl methacrylate as a probe for the types of intermediates present in a particular reaction. In the presence of cations, only polystyrene will be produced. In the presence of anions, only poly(methyl methacrylate) results. In the presence of radicals, the 1:1 copolymer is produced. The polymer that is produced is easily collected and purified, and the three possible polymers are readily distinguished by elemental analysis (Table 6.7).

In the synthesis of polymers, one of the most important goals is control of the polymer chain length. The length of any chain reaction depends upon the competition between chain propagation steps and termination. In general, the more reactive the monomer, the greater the likelihood that the growing chain will continue reacting with monomer in propagation steps before termination occurs. However, the chain initiator can also be important in determining chain length, since the growing chain may react with the initiator molecule. This is known as chain transfer, and although it may generate another radical suitable for starting another chain, the initial chain is ended:

Initiation

$$A—B \longrightarrow A\cdot + B\cdot$$

Propagation

$$B\cdot + CH_2{=}CH—\phi \longrightarrow B—CH_2—\dot{C}H\phi$$

Chain transfer

$$\text{-}\!\!\!\!\text{+}CH_2—\dot{C}H\phi + AB \longrightarrow \text{-}\!\!\!\!\text{+}CH_2—\underset{\phi}{CH}—A + B\cdot$$

Usually, high polymers are formed readily with peroxides as initiators, since peroxides are poor transfer agents. However, with good transfer agents, very short chains, called telomers, are formed. Carbon tetrachloride is a good transfer agent, and polymerizations in the presence of CCl_4 lead to relatively short chain lengths. Depending on the concentration of CCl_4, average chain lengths may be controlled quite well.

Quantitatively, the measure of the effectiveness of a transfer agent is how well it competes with monomer for reaction with the growing chain. The ratio of rate constants for chain transfer and chain propagation is defined as the *transfer constant*. For styrene polymerization at 60°C, CCl_4 has a transfer constant of 10^{-2} and CBr_4 has a transfer constant of about 2:[34]

$$\text{transfer constant} = \frac{k_{transfer}}{k_{propagation}}$$

Thus, we would expect an equimolar mixture of styrene and CCl_4 to give polymerization only to an extent of about 100 monomer units. In the limit, as with styrene in excess CBr_4 solvent, polymer chains of length one are expected. This, of course, is simply a case of radical addition to a double bond:

$$R \cdot + CBr_4 \rightarrow RBr + \cdot CBr_3$$

$$\cdot CBr_3 + \phi CH {=} CH_2 \rightarrow \phi \dot{C}H {-} CH_2 {-} CBr_3$$

$$\phi \dot{C}HCH_2CBr_3 + CBr_4 \rightarrow \underset{\underset{Br}{|}}{\phi CH} {-} CH_2 {-} CBr_3 + \cdot CBr_3$$

A final factor governing the length of any chain reaction is the presence of inhibitors. A radical inhibitor may be defined as any species that reacts with a reactive radical (one involved in chain propagation) to produce an unreactive radical (one incapable of chain propagation). Commercial styrene normally includes a small amount of hydroquinone or other inhibitor, in order to prevent the styrene from polymerizing in the presence of traces of radicals. Oxygen is also a good inhibitor since peroxy radicals are quite unreactive in chain mechanisms (except in autoxidation of very reactive C–H bonds).

Reactive Radical	+	*Inhibitor*		*Unreactive Species*
R ·	+	HO—⬡—OH	⟶	HO—⬡—O · + RH
R ·	+	O=⬡=O	⟶	RO—⬡—O ·
R ·	+	O_2	⟶	R—O—O ·
R ·	+	Fe^{3+}	⟶	$R^+ + Fe^{2+}$

While traces of inhibitors are useful to avoid undesired chain reactions, their presence in a monomer to be polymerized is counterproductive and makes control of the polymerization impossible. Monomers for use in commercial polymerizations are among the purest organic compounds made.

6.9 Radical Displacement Reactions

One subject of continued interest in free radical chemistry has been the question of a direct radical displacement reaction that would be analogous to the S_N2 nucleophilic displacement mechanism or the S_E2 electrophilic displacement mechanism. Such a radical mechanism has been called the S_H2 mechanism (substitution, homolytic, bimolecular); conceptually, it would be electronically intermediate between the S_N2 mechanism, which involves inversion of stereochemistry, and the S_E2 mechanism, which involves retention of stereochemistry. The only reactions we have encountered so far that might be considered radical displacement reactions are the abstraction reactions. However, since these abstractions involve hydrogen or the halogens, the specific stereochemistry of the displacement is unknown.

Some of the evidence that the S_H2 mechanism is stereochemically analogous to the S_N2 mechanism comes from kinetic analyses. As a rule, S_N2 reactions are extremely sensitive to the steric environment about the substrate. The steric effects upon the rates of S_N2 substitutions on disulfides by anion nucleophiles, as well as the rates of radical displacements by the corresponding radicals, were shown to display similar dependencies upon the substituents. This was taken as an indication that both displacements follow similar courses (back-side attack):[35,36]

$$Y:^- + R-S-S-R \xrightarrow{k(S_N2)} YSR + {}^-SR$$

$$Y\cdot + R-S-S-R \xrightarrow{k(S_H2)} YSR + \cdot SR$$

By far the most convincing evidence for inversion in the S_H2 mechanism, as it is with the S_N2 mechanism, is the observation of specific stereochemistry about an asymmetric carbon atom. Such evidence had long been sought and has only recently been clearly demonstrated in the case of radical addition of bromine to cyclopropanes. Using photoinitiation at $-78°C$, bromine adds to cyclopropane in a radical chain process to give 1,3-dibromopropane. By consideration of many alkylated cyclopropanes, the generalization could be made that bromine atoms preferentially attack the least substituted carbon and displace the most stable alkyl radical.[37] Both of these criteria, of course, are consistent with an S_H2 mechanism: greatest reactivity at the least hindered site and displacement of the best leaving group:

$$Br_2 \xrightarrow{h\nu} 2Br\cdot$$

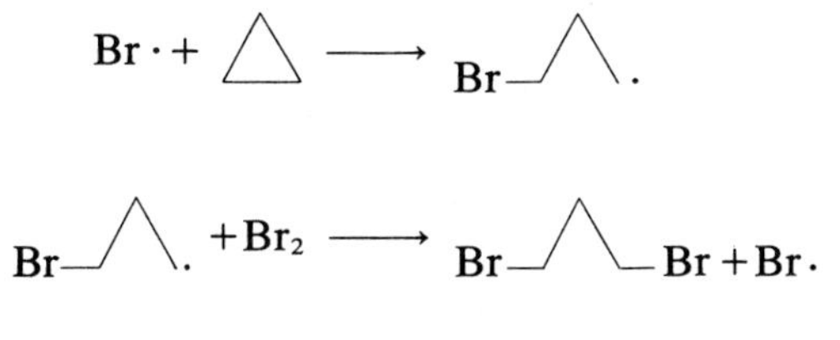

Overall:

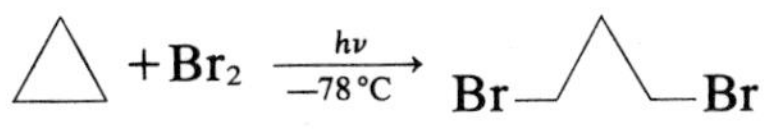

For example:

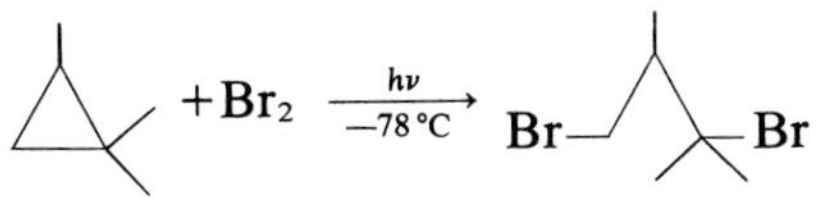

The addition of bromine to *cis*-1,2,3-trimethylcyclopropane gave products that could clearly be considered to result from inversion of configuration at one center and racemization at the other center:[38]

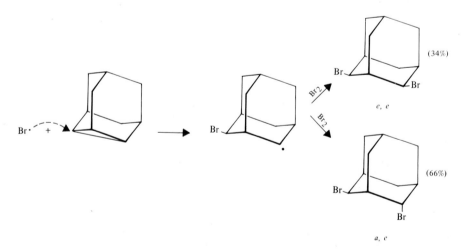

A similar conclusion was reached from photobromination of 2,4-dehydroadamantane, which gives diequatorial and axial,equatorial dibromoadamantanes but no diaxial product, indicating inversion at one center and random attack at the other center:[37]

While the S_H2 mechanism has been demonstrated to proceed with inversion of configuration, the difficulty in finding a clearcut example of an S_H2 mechanism indicates that radical displacements on carbon are not common reactions of free radicals. Like the S_N2 mechanism, which also requires inversion, the S_H2 mechanism is subject to strong steric hindrance. Thus, radical displacements have so far been observed only on the bent bonds of cyclopropane or other strained rings or on univalent atoms (abstractions).

6.10 Nonclassical Free Radicals

By analogy to the extensive studies of rearrangements and neighboring group assistance in carbocation chemistry, determination of the significance of nonclassical structures and rearrangements in radical chemistry has been a major area of interest. As we noted earlier, rearrangements of free radicals are much less common than cationic rearrangements. In general, the differences between radical and cation rearrangements may be understood in terms of the molecular orbitals involved. A nonclassical structure (or the transition state for an alkyl shift) is essentially a three-center bond (see Chapter 4). For a cation, this is a two-electron, three-center bond, and for a radical, there are three electrons to be accommodated in the three-center bond. The situation is similar to the cyclopropenyl cation and radical—a three-center cyclic system has only one bonding molecular orbital (see Chapter 1). Thus, the two-electron cation structure is more stable than the three-electron radical structure:

MOs for an alkyl shift

Classical	Nonclassical	Classical

Aryl migrations need not involve three-center bonding since the radical may be delocalized into the ring. Thus, radical aryl migrations are most common. Additionally, evidence supportive of bridged radical intermediates has been accumulated in the case of bromine as the bridging atom. Free radical bromination of bromocyclohexane leads to 94% *trans*-1,2-dibromocyclohexane.[39] This has been attributed to anchimeric assistance by the neighboring bromine during the hydrogen abstraction process leading to a stereochemistry like that from bromonium cation intermediates. The predominance of 1,2 products indicates that bromine facilitates hydrogen abstraction at the adjacent position. The predominance of *trans*-1,2 dibromide over the *cis*, however, could be due to steric effects which make the *trans* dibromide much more stable than the *cis* dibromide, thus weakening the argument for a bridged radical:

94% *trans* racemic

Careful stereochemical studies of bromination reactions provide further evidence for the significance of bromine bridging in radical reactions. In free radical bromination, *erythro*-3-deuterio-2-bromobutane yields undeuterated *meso* dibromide and monodeuterated racemic dibromide. The *threo* stereoisomer yields monodeuterated *meso* and undeuterated racemic product.[40] These results strongly indicate that bromine provides neighboring group assistance as the hydrogen (or deuterium) is abstracted and leads to a bridged radical that determines the stereochemistry of the product as either *meso* or racemic:

Racemic–d_1 *Meso*–d_0

erythro produces racemic – d_1 and *meso* – d_0

threo produces racemic – d_0 and *meso* – d_1

Furthermore, optically active (S)-2-bromobutane yields both *meso* and racemic 2,3-dibromobutane. A classical radical would have retained stereochemistry about the initial chiral center and, therefore, would have produced *meso* and only one enantiomer. Radioactive labeling experiments also show that the initial bromine atom becomes statistically scrambled between the two locations of the dibromo product.[40]

Although neighboring group participation by bromine in radical reactions has been convincingly demonstrated, the existence of nonclassical hydrocarbon radicals has been more difficult to establish. In the norbornyl system, 2-norbornyl radicals do not racemize, unlike the facile cationic rearrangement. A nonclassical structure had been suggested for the 7-norbornenyl radical; however, recent calculations and esr analyses of the 7-norbornenyl radical have determined that structure **1** is more stable than either structure **2** or **3**, which might have made use of neighboring group participation with the π bond.[41] This reemphasizes the molecular orbital explanation pointed out earlier—a three-center bond is unstable for three electrons.

Most stable

6.11 Radical Ions

Another highly important class of organic reaction intermediates that contain an unpaired electron are the radical ions. Radical ions are simply free radicals that also bear a positive or negative charge, and are accordingly called radical cations or radical anions. In nearly all cases, radical ions may be considered to be derived from a parent organic molecule by gain or loss of one electron. The nomenclature often reflects this:

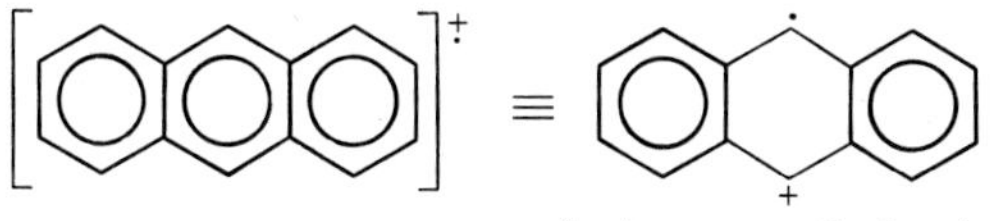

Anthracene radical cation

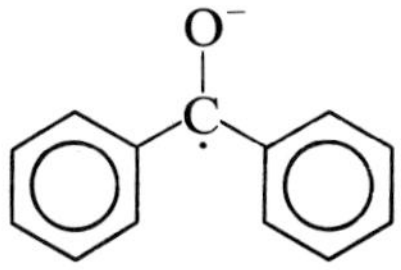

Benzophenone radical anion (or benzophenone ketyl)

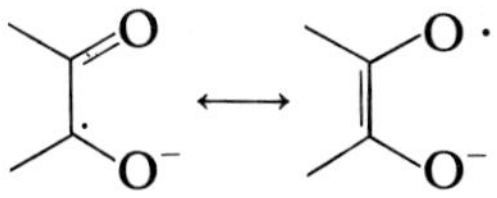

Semidione

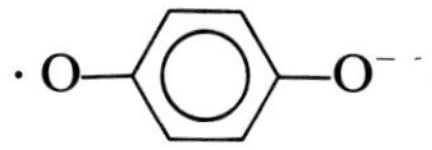

Semiquinone

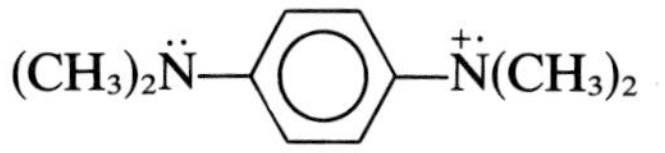

$(CH_3)_2\ddot{N}$—⟨ ⟩—$\overset{+\cdot}{N}(CH_3)_2$ Wurster's salt

As would be expected, radical anions may be intermediates in reduction reactions and radical cations intermediates in oxidation reactions. Ketones or aldehydes are reduced to pinacols by the action of sodium, magnesium, or aluminum:

$$(CH_3)_2C{=}O \xrightarrow{\text{Na(Hg)}} (CH_3)_2\dot{C}{-}O^-$$

$$\underset{\overset{|}{OH}\ \overset{|}{OH}}{(CH_3)_2C{-}C(CH_3)_2} \longleftarrow \underset{\overset{|}{O_-}\ \overset{|}{O_-}}{(CH_3)_2C{-}C(CH_3)_2}$$

The synthetically valuable Birch reduction accomplishes a partial reduction of an aromatic ring by electron transfer from sodium (or more correctly, a solvated electron) in liquid ammonia:[42]

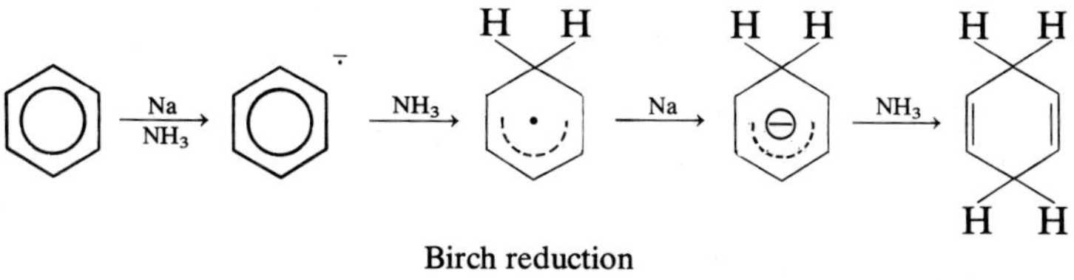

Birch reduction

A particularly interesting feature of the Birch reduction is the predominance of 1,4 reduction rather than formation of the more stable conjugated 1,3 diene. This phenomenon has been explained on the basis of "the principle of least motion," which states that those elementary reactions will be favored that involve the least change in atomic and electronic configuration. The selective formation of the 1,4 product occurs at the last step, protonation of the cyclohexadienyl anion. Considering all the resonance forms of this anion, it is clear that not all of the C–C bonds have equivalent double-bond character. On the average, we may consider the structure of the cyclohexadienyl anion to be as follows:

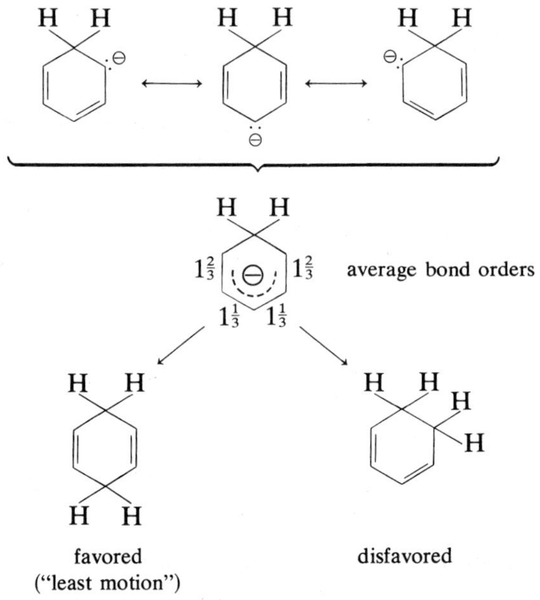

Thus, there will be a lesser change of bonding characteristics (atomic and electronic configuration) if the 1,4 product is formed rather than the 1,3 product.

The principle of least motion pertains to transition state energies, and, therefore it may be applied to predict the kinetically-controlled product. Thermodynamically, the 1,3-cyclohexadiene is a more stable product than the 1,4-cyclohexadiene.

Radical ions may also be encountered in organic electrochemical processes in which electron transfer is accomplished by an electrode process. In particular, aromatic compounds have relatively low-lying π^* molecular orbitals and relatively high-energy filled π molecular orbitals, so they may be readily oxidized or reduced by electron transfer. Using an alternating current, radical cations and anions are alternately produced in the same immediate vicinity. This process leads to *electrogenerated chemiluminescence* caused by the mutual annihilation of the oppositely-charged species.[43,44] The electron transfer processes which occur are exemplified for naphthalene. Only the HOMO and LUMO are shown in order to keep track of the electrons.

Reduction

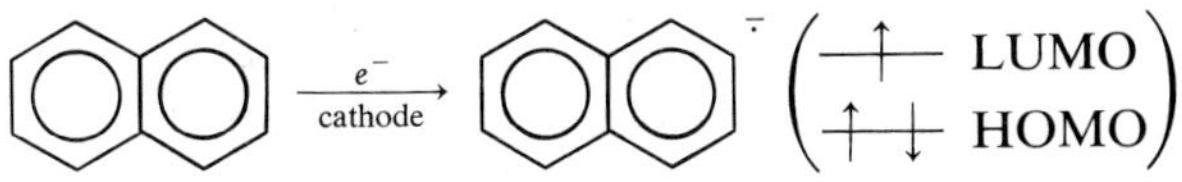

Oxidation

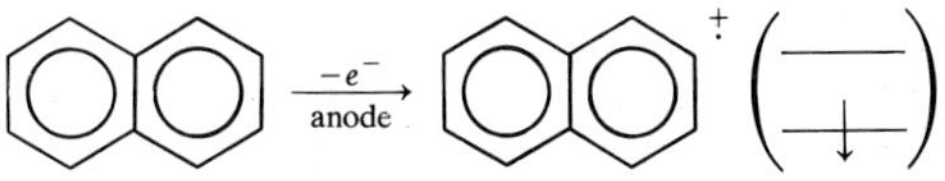

Annihilation

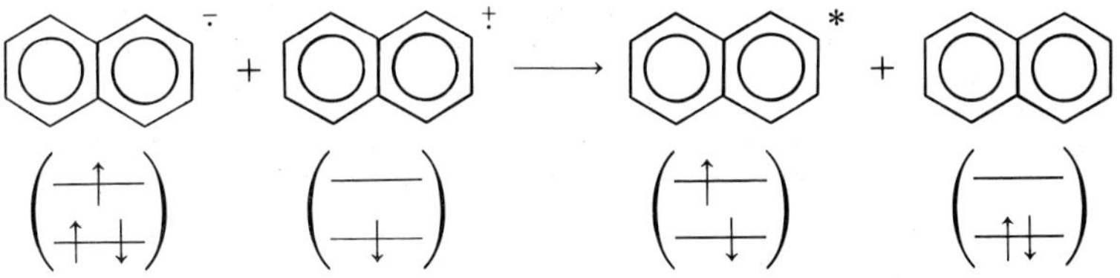

Luminescence

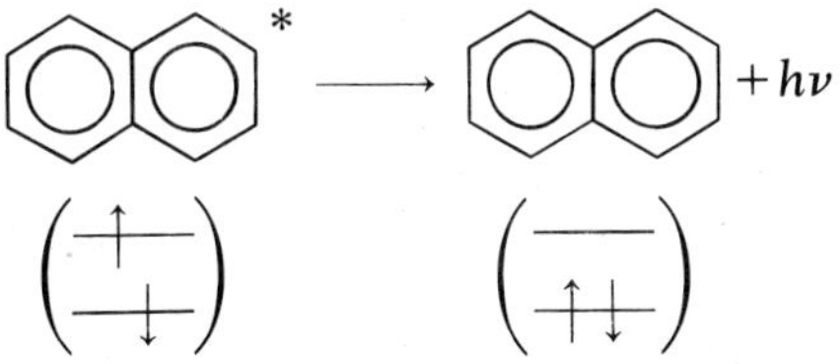

Electrogenerated chemiluminescence

References

1. M. Gomberg, *J. Amer. Chem. Soc.*, **22,** 757 (1900).
2. H. Laukamp, W. T. Nanta, and C. MacLean, *Tetrahedron Lett.*, 249 (1968).
3. M. F. Roy and C. S. Marvel, *J. Amer. Chem. Soc.*, **59,** 2622 (1937).
4. J. E. Wertz and J. R. Bolton, "Electron Spin Resonance: Elementary Theory and Practical Applications," McGraw-Hill, New York, 1972, p. 53.
5. P. J. Krusic and J. K. Kochi, *J. Amer. Chem. Soc.*, **90,** 7155 (1968).
6. H. R. Ward, R. G. Lawler, and R. A. Cooper, *J. Amer. Chem. Soc.*, **91,** 746 (1969).
7. S. F. Nelsen and P. D. Bartlett, *J. Amer. Chem. Soc.*, **88,** 143 (1966).
8. P. S. Engel and D. J. Bishop, *J. Amer. Chem. Soc.*, **94,** 2148 (1972).
9. N. J. Karch and J. M. McBride, *J. Amer. Chem. Soc.*, **94,** 5092 (1972).
10. D. R. Yarkony and H. F. Schaefer III, *J. Amer. Chem. Soc.*, **96,** 3754 (1974).
11. P. Dowd, A. Gold, and K. Sachdev, *J. Amer. Chem. Soc.*, **90,** 2715 (1968).
12. P. Dowd, *J. Amer. Chem. Soc.*, **92,** 1066 (1970).
13. R. Hoffmann, *J. Amer. Chem. Soc.*, **90,** 1475 (1968).
14. P. J. Hay, W. J. Hunt, and W. A. Goddard, *J. Amer. Chem. Soc.*, **94,** 638 (1972).
15. P. D. Bartlett and N. A. Porter, *J. Amer. Chem. Soc.*, **90,** 5317 (1968).
16. L. K. Montgomery, K. Schueller, and P. D. Bartlett, *J. Amer. Chem. Soc.*, **86,** 622 (1964).
17. M. S. Kharasch, A. C. Poshkus, A. Fono, and W. Nudenberg, *J. Org. Chem.*, **16,** 1458 (1951).

18. F. D. Greene, M. L. Savitz, F. D. Osterholtz, H. H. Lau, W. N. Smith, and P. M. Zanet, *J. Org. Chem.*, **28**, 55 (1963).

19. J. Grotewald and J. A. Kerr, *J. Chem. Soc.*, 4337 (1963).

20. A. F. Trotman-Dickenson, "Free Radicals," Methuen, London, 1959, p. 53.

21. D. J. Carlsson and K. U. Ingold, *J. Amer. Chem. Soc.*, **90**, 7047 (1968).

22. J. K. Kochi and A. Bemis, *J. Amer. Chem. Soc.*, **90**, 4038 (1968).

23. A. F. Trotman-Dickenson, *Adv. Free-Radical Chem.*, **1**, 1 (1965); R. S. Davidson, *Quart. Rev. (London)*, **21**, 249 (1967).

24. J. M. Tedder, *Quart. Rev. (London)*, **14**, 340 (1960).

25. R. E. Pearson and J. C. Martin, *J. Amer. Chem. Soc.*, **85**, 354 (1963).

26. W. A. Thaler, *Methods Free-Radical Chem.*, **2**, 189 (1969).

27. M. H. Back and R. J. Cvetanovic, *Can. J. Chem.*, **41**, 1396 (1963).

28. C. Walling, "Free Radicals in Solution," Wiley, New York, 1957, p. 241.

29. J. E. Bennett and J. A. Howard, *J. Amer. Chem. Soc.*, **95**, 4008 (1973).

30. V. A. Belyakov and R. F. Vassil'ev, *Photochem. Photobiol.*, **11**, 179 (1970).

31. R. E. Kellogg, *J. Amer. Chem. Soc.*, **91**, 5433 (1969).

32. W. E. Bachmann and R. A. Hoffman, *Org. React.*, **2**, 244 (1944).

33. C. Walling, E. R. Briggs, W. Cummings, and F. R. Mayo, *J. Amer. Chem. Soc.*, **72**, 48 (1950).

34. G. Henrici-Olivé and S. Olivé, *Fortschr. Hochpolymer. Forsch.*, **2**, 496 (1961).

35. W. A. Pryor and T. L. Pickering, *J. Amer. Chem. Soc.*, **84**, 2705 (1962).

36. W. A. Pryor and H. Guard, *J. Amer. Chem. Soc.*, **86**, 1150 (1964).

37. K. J. Shea and P. S. Skell, *J. Amer. Chem. Soc.*, **95**, 6728 (1973).

38. G. G. Maynes and D. E. Applequist, *J. Amer. Chem. Soc.*, **95**, 856 (1973).

39. W. Thaler, *J. Amer. Chem. Soc.*, **85**, 2607 (1963).

40. P. S. Skell, R. R. Pavlis, D. C. Lewis, and K. J. Shea, *J. Amer. Chem. Soc.*, **95**, 6735 (1973).

41. J. K. Kochi, P. Bakuzis, and P. J. Krusic, *J. Amer. Chem. Soc.*, **95**, 1516 (1973).

42. A. J. Birch and H. Smith, *Quart. Rev (London)*, **4**, 69 (1958).

43. D. M. Hercules, *Science*, **145**, 808 (1964).

44. A. Zweig, *Adv. Photochem.*, **6**, 425 (1968).

Bibliography

General References

W. A. Pryor, "Free Radicals," McGraw-Hill, New York, 1966.

C. Walling, "Free Radicals in Solution," Wiley, New York, 1957.

J. K. Kochi, ed., "Free Radicals," Vol. I and succeeding volumes, Wiley-Interscience, New York, 1973.

E. S. Huyser, "Free-Radical Chain Reactions," Wiley-Interscience, New York, 1970.

D. C. Nonhebel and J. C. Walton, "Free-Radical Chemistry," Cambridge Univ. Press, Cambridge, 1974.

Electron Spin Resonance

J. E. Wertz and J. R. Bolton, "Electron Spin Resonance: Elementary Theory and Practical Applications," McGraw-Hill, New York, 1972.

E. G. Janzen, *Acc. Chem. Res.*, **44**, 31 (1971).

CIDNP

H. R. Ward, *Acc. Chem. Res.*, **5**, 18 (1972).

R. G. Lawler, *Acc. Chem. Res.*, **5**, 25 (1972).

Biradicals

G. Jones, *J. Chem. Educ.*, **51**, 175 (1974).

L. Salem and C. Rowland, *Angew. Chem. Int. Ed. Engl.*, **11**, 92 (1972).

S$_H$2 Reactions

K. U. Ingold and B. P. Roberts, "Free-Radical Substitution Reactions," Wiley-Interscience, New York, 1971.

Biochemistry

W. A. Pryor, "Free Radicals in Biological Systems," *Scientific American*, August (1970).

Problems

1. Using bond dissociation energies, calculate ΔH for each of the following reactions, and predict which of each pair of reactions is more likely to occur. For reactions involving X, do the calculations for each halogen.

   (*a*)

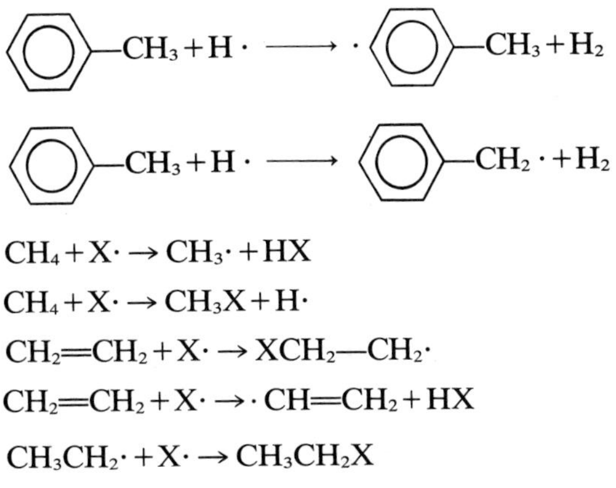

   (*b*) $CH_4 + X\cdot \rightarrow CH_3\cdot + HX$

   $CH_4 + X\cdot \rightarrow CH_3X + H\cdot$

   (*c*) $CH_2{=}CH_2 + X\cdot \rightarrow XCH_2{-}CH_2\cdot$

   $CH_2{=}CH_2 + X\cdot \rightarrow \cdot CH{=}CH_2 + HX$

   (*d*) $CH_3CH_2\cdot + X\cdot \rightarrow CH_3CH_2X$

   $CH_3CH_2\cdot + X\cdot \rightarrow CH_2{=}CH_2 + HX$

   (*e*) $2CH_3CH_2\cdot \rightarrow CH_3CH_2CH_2CH_3$

   $2CH_3CH_2\cdot \rightarrow CH_3CH_3 + CH_2{=}CH_2$

2. The esr spectrum of the isopropyl radical shows splittings of $a_{H\alpha} = 22$ gauss and $a_{H\beta} = 25$ gauss. Roughly sketch what the spectrum would look like, indicating where the different splitting constants would be measured.

3. Write all the resonance forms for diphenylpicrylhydrazyl (DPPH), a stable free radical:

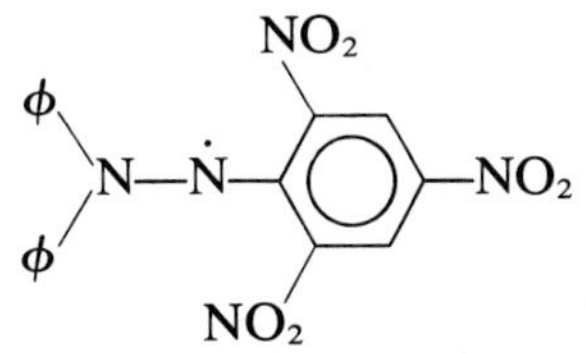

4. Diethyl ether that has been left open to the air for long periods becomes dangerously explosive. Write the complete mechanism for this autoxidation.

5. Predict the products and write radical chain mechanisms for the following reactions. Assume suitable radical initiation is available in each case:

    (a) $(CH_3)_3CCHO \rightarrow$

    (b) ⟨O⟩—$CH_2CH_3 + Br_2 \longrightarrow$

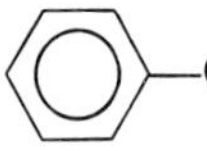

    (c) [naphthalene] $+ O_2 \longrightarrow$

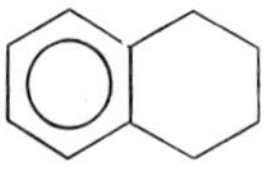

    (d) $CH_3CHO + EtOOC—CH=CH—COOEt \rightarrow$

    (e) [cyclohexene] $+$ [N-bromosuccinimide] N—Br $\longrightarrow$

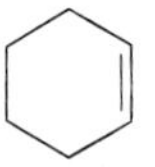

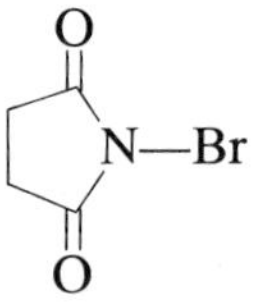

    (f) [cyclohexene] $+$ $CBr_4 \longrightarrow$

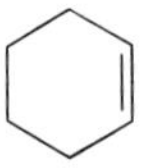

    (g) $CH_3OOCCH=CHCH_3 \rightarrow$

6. Explain what conclusions may be drawn from the following cage effects experiments:

    (a) $tBuON=NOtBu \xrightarrow[\Delta]{\text{hexane}} 6\%$ cage recombination

    $tBuON=NOtBu \xrightarrow[\Delta]{\text{nujol}} 68\%$ cage recombination

    (b) $\begin{matrix} CH_3 \\ | \\ \phi—CH—N=N—CH_3 \end{matrix} \xrightarrow[\Delta]{\text{nujol}}$ racemization of starting material

    (optically active)

    (c) $tBuOOtBu \xrightarrow[45°]{\text{nujol}} 76\%$ cage effect

    $tBuO—N=N—OtBu \xrightarrow[45°]{\text{nujol}} 68\%$ cage effect

    $tBuO—\overset{O}{\overset{||}{C}}—O—O—\overset{O}{\overset{||}{C}}—OtBu \xrightarrow[45°]{\text{nujol}} 54\%$ cage effect

7. Suggest as many methods as possible which would indicate that phenyl radicals are intermediates in the following radical-initiated reaction:

$$\phi I + NaOCH_3 \xrightarrow[\text{CH}_3\text{OH}]{\text{initiator}} \phi H + CH_2O + NaI$$

Propose a reasonable mechanism.

Hint: Iodobenzene radical anion is another key intermediate.

8. Vitamin E, α-tocopherol, acts as an effective biological antioxidant, apparently by acting as a free radical inhibitor. Write a mechanism for its action as a radical inhibitor and predict its ultimate oxidation product:

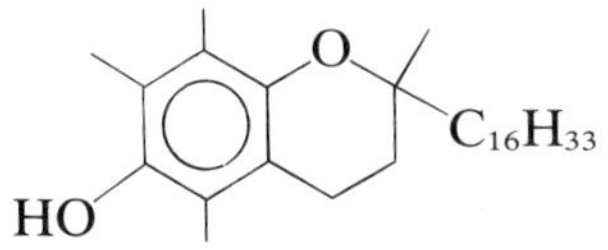

(α-tocopherol)

9. Acrylonitrile (A) and 1,3-butadiene (B) form an alternating copolymer with radical initiation. Write a mechanism that explains the alternating effect in the copolymerization.

   Toward a radical with B terminus, A is 2.5 times as reactive as B, but toward a radical with A terminus, B is 20 times as reactive as A. Explain this contrast.

10. Allyl acetate does not polymerize well under normal radical initiation conditions. However, perdeuterated allyl acetate polymerizes more rapidly to yield longer polymer chains. Explain these data.

11. Occasionally, radical termination will show a preference for cross-termination rather than symmetrical termination. Draw the appropriate transition states for all possible termination reactions in the autoxidation of cumene.

    Explain the preference for cross-termination.

12. Decomposition of *tert*-butyl peroxide in toluene leads to both *tert*-butyl alcohol and acetone as products. Write a mechanism and predict the other products of the decomposition. In cumene solvent, how would the *tert*-butyl alcohol : acetone ratio compare to toluene solvent?

13. Use the principle of least motion to explain why product A is kinetically favored although it is thermodynamically less stable than B:

14. Abstraction of iodine from aryl iodides by phenyl radical follows a Hammett plot with $\rho = +0.57$:

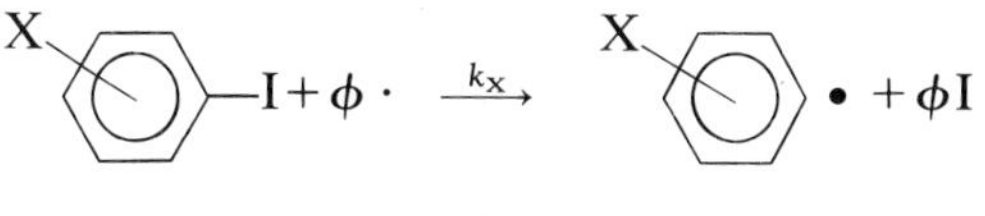

The observation of a positive ρ is highly unusual for radical abstraction reactions. Explain why most radical abstractions (of H or Cl, for example) show a negative ρ, and why abstraction of iodine shows a positive ρ.

7

Carbenes

7.1 History and Nomenclature

Carbenes may be defined as compounds that contain a divalent carbon atom having only six valence electrons, four involved in the two covalent bonds and two nonbonding electrons. The simplest member of the carbene family is methylene, $:CH_2$. Substituted carbenes RCH: and $R_2C:$ are now known for a large variety of substituents R from aryl and alkyl groups to halogens and acyl groups.

The name "carbene" was first conceived by R. B. Woodward, W. von E. Doering, and S. Winstein "in a nocturnal Chicago taxi" in 1951. The carbene nomenclature system is by far the most commonly used, even though it is in conflict with the IUPAC system, which employs the ending "-ylidene" (Table 7.1). In the carbene nomenclature system, the two substituents on the divalent carbon atom are simply listed before the name carbene.

Although carbenes have been the subject of active research only since 1950, the intermediacy of "carbon dichloride" had been postulated in the basic hydrolysis of chloroform as early as 1862.[1] In the intervening years, carbenes suffered from the same prejudices as Gomberg's free radicals and did not become a respectable research field until the pioneering work of

Table 7.1 *Carbene Nomenclature Systems*

	"carbene"	IUPAC
CH_2:	carbene	methylidene
CH_3CH:	methylcarbene	ethylidene
$CH_2{=}CH{-}CH$:	vinylcarbene	propenylidene
$(C_6H_5)_2C$:	diphenylcarbene	—

Hine and Doering in the 1950s. Prior to this era of respectability, it is reported that carbene was utilized in the chemical laboratories of D. Duck, in an unsuccessful synthesis.[2]

Walt Disney Productions

Because carbenes have only six electrons about the central carbon atom, they are highly electron-deficient and normally undergo reactions as electrophiles. Carbenes react readily with π bonds and even with C–H bonds. The discovery of the reaction of methylene with benzene to produce toluene and cycloheptatriene prompted Doering to call methylene "the most indiscriminate reagent known in organic chemistry."

When the carbene has the possibility of strong π donations from an adjacent atom, the electrophilic nature is diminished. Carbon monoxide, for example, may be viewed as a carbene with strong π interactions from the adjacent oxygen. Bond strength and bond length data indicate that the ionic resonance form makes a much more important contribution than the carbene form. Thus, carbon monoxide is not carbenelike, but more carbanion-like; it is nucleophilic and a very good ligand:

$$:C{=}\ddot{O}: \leftrightarrow :\bar{C}{\equiv}\overset{+}{O}:$$

We will now consider in some detail the structure of carbenes, some methods for the generation of carbenes, and some of the typical reactions of carbenes.

7.2 Structure of Carbenes

Carbon has four available atomic orbitals capable of containing up to eight valence electrons. Since carbenes utilize only two bonding molecular orbitals, the two nonbonding electrons have two remaining atomic orbitals in which to accommodate themselves. Two possibilities immediately present themselves. The two electrons may occupy the same orbital, in which case they must have opposite spins; this situation represents a singlet carbene. Alternatively, the two electrons may occupy different orbitals, in which case they would be free to have parallel spins; this is a triplet carbene. According to Hund's rule, the triplet should be the lower energy species, since it minimizes the electron–electron repulsions.

The simplest description of the structure of triplet methylene would utilize *sp* hybridization for the central carbon atom. The two *sp* hybrid orbitals bond to the two hydrogen atoms, and the two unhybridized *p* orbitals contain one electron each. This is the most effective manner of making two σ bonds and minimizing electron–electron repulsions for a total of six electrons.

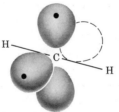

Triplet methylene (*sp* hybridized)

The molecular geometry of triplet methylene, thus, should be linear. Triplet methylene, because of its singly-occupied *p* orbitals, can be considered to be a biradical, and we shall see later that it does indeed behave as a biradical in many respects.

Singlet methylene would be most simply constructed with sp^2 hybridization at the central carbon atom. Two of the sp^2 hybrid orbitals bond to the two hydrogen atoms and the third contains the lone pair of electrons. The unhybridized *p* orbital remains empty. This is the most effective manner of minimizing electron–electron repulsions for three electron pairs, analogous to the structure of simple carbocations.

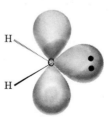

Singlet methylene
(sp^2 hybridized)

The geometry of singlet methylene should, thus, be bent with a bond angle of 120° or perhaps somewhat smaller due to repulsions of the C–H bonds by the lone pair. Singlet methylene simultaneously displays the lone pair of a carbanion and the empty p orbital of a carbocation—a schizophrenia that is particularly intriguing to organic chemists.

In actuality, the structures we have just considered for singlet and triplet methylene are oversimplified. Determination of the exact structure of singlet and triplet methylene has prompted an enormous amount of theoretical and experimental work. The most recent calculations and measurements arrive at bond angles of about 103° for singlet methylene[3] and 136° for triplet methylene.[4,5] The magnitude of the energy difference between singlet and triplet methylene has also been the subject of intense study, and the data now seem to point to the triplet state as 8 to 9 kcal/mole more stable than the singlet state.[6–8]

7.3 Electron Spin Resonance Studies

Since triplet carbenes have unpaired electrons, they give an electron spin resonance signal (see p. 255). As with free radicals, there are special problems associated with obtaining an esr spectrum of very reactive species, mainly the attainment of a sufficiently large concentration of the paramagnetic molecule. This problem is most readily solved by generating the carbenes by photolysis in the solid state. If the medium is a solid, rigid matrix, such as a frozen glassy hydrocarbon at 77°K, the carbenes are stable indefinitely. The key to this stability is not only the low temperature, but also the rigid structure that prevents the carbenes from diffusing together and reacting.

Electron spin resonance spectroscopy has demonstrated that the ground states for arylcarbenes are triplets. The initial photolysis of the precursor leads to the singlet carbene, with the triplet carbene formed thereafter by spin inversion. The triplet is shown to be the ground state by the long lifetime of the esr signal—up to several hours at 77°K:

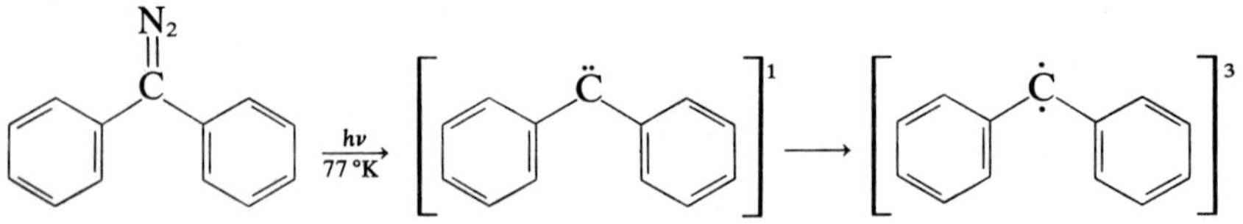

The complete analysis of the esr spectrum of an arylcarbene can give detailed structural information. Such an analysis has shown that arylcarbenes are bent and have one unpaired electron conjugated with the aromatic π system and one unpaired electron perpendicular to it.[9]

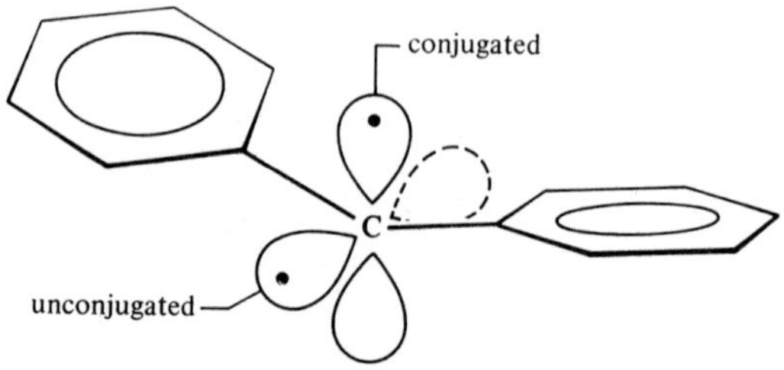

Electron spin resonance experiments have demonstrated that quintet states (four unpaired spins) are the ground states for some dicarbenes.[10]

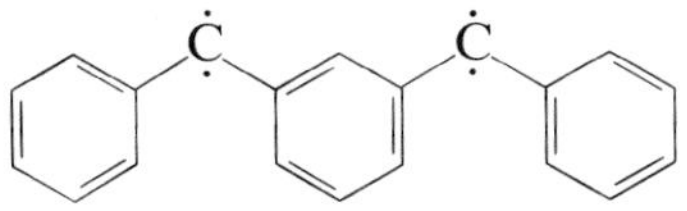

Meta—quintet ground state[10] 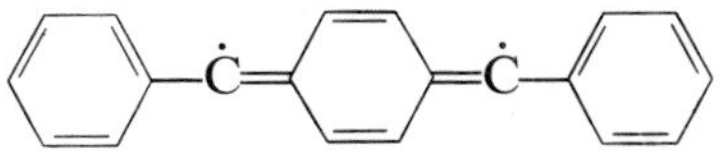   Para—triplet ground state[11]

Finally, geometric isomers of arylcarbenes have been shown to exist by esr experiments. This is possible only because the aryl carbenes are bent rather than linear. The overlap of one of the p orbitals with the aromatic system apparently provides sufficient double-bond character for hindered rotation at 77°K.[12]

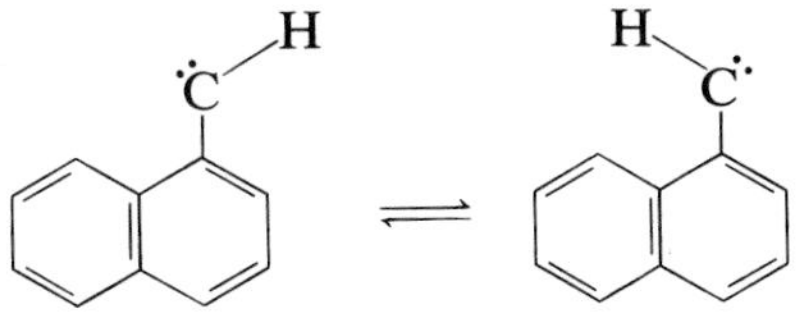

anti   syn

7.4 Differentiation Between Singlet and Triplet Carbenes

The overwhelming majority of carbene reactions may be classified into two general types: insertion reactions into σ bonds and cycloaddition reactions with π bonds.

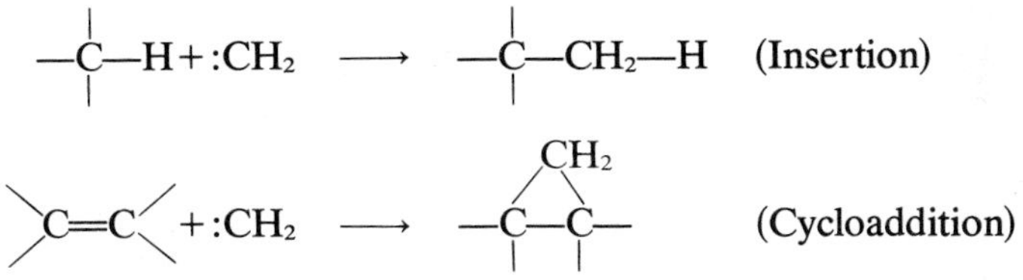

$$-\overset{|}{\underset{|}{C}}-H+:CH_2 \longrightarrow -\overset{|}{\underset{|}{C}}-CH_2-H \quad \text{(Insertion)}$$

$$C=C + :CH_2 \longrightarrow -\overset{|}{C}-\overset{|}{C}- \quad \text{(Cycloaddition)}$$

For each of these two classes of reactions, both singlet and triplet carbenes may lead to the same product. Thus, a crucial and recurrent problem in carbene chemistry is the determination of the spin state of the carbene involved in a given reaction. Even though the triplet state has been established as the ground state for most carbenes, this is not an assurance that only the triplet state will participate in chemical reactions. In fact, most methods that generate carbenes produce initially the singlet state. Furthermore, the reactivity of carbenes is so extreme that the singlet carbene may very well react before it has an opportunity to convert to the more stable triplet state. Thus, a complete mechanism of a carbene reaction must specify whether the singlet or triplet state, or both, is involved in the reaction.

Although both singlet and triplet carbenes normally lead to the same products, some of the more subtle mechanistic techniques, such as selectivity and stereochemistry, can be utilized to differentiate between the two forms. The differentiation may be made with a single unifying postulate: a triplet carbene, because it contains two unpaired electrons, must react in a stepwise manner; whereas a singlet carbene, with all electrons paired, may give product in a concerted reaction. Thus, the insertion of singlet methylene into the C–H bonds of an alkane is considered to be a concerted process:

$$-\underset{|}{\overset{|}{C}} \cdots \overset{CH_2}{\underset{H}{\diagup}} \longrightarrow -\underset{|}{\overset{|}{C}}-CH_2-H$$

Triplet methylene is unable to react in a concerted manner, since one spin inversion is necessary before ground state product can be obtained. Thus, triplet carbenes react by an abstraction–recombination mechanism that is what would be envisioned for a biradical-like species:

$$-\underset{|}{\overset{|}{C}}-H + \uparrow\cdot\overset{\uparrow}{\dot{C}H_2} \longrightarrow \underbrace{-\underset{|}{\overset{|}{C}}\cdot\uparrow + \uparrow\cdot CH_3}$$

$$\downarrow \text{spin inversion}$$

$$-\underset{|}{\overset{|}{C}}-CH_3 \longleftarrow -\underset{|}{\overset{|}{C}}\cdot\downarrow + \uparrow\cdot CH_3$$

Stereochemistry of Carbene Reactions

A more definitive characterization appears in the comparative cycloaddition reactions of singlet and triplet carbenes. In the concerted reaction of singlet methylene with an olefin, the stereochemistry of the olefin is retained in the cyclopropane product.

$$^1CH_2 + \underset{H}{\overset{CH_3}{\diagdown}}C=C\underset{H}{\overset{CH_3}{\diagup}} \longrightarrow \underset{CH_3 \ CH_3}{H^{\prime\prime\prime\prime}\triangle^{\prime\prime\prime\prime}H} \qquad \text{(plus insertion products)}$$

In contrast, triplet methylene reaction with either *cis*- or *trans*-2-butene yields a mixture of the *cis*- and *trans*-1,2-dimethylcyclopropanes.[13] The loss of stereochemistry is most readily explained in terms of the Skell–Woodworth hypothesis.[13] This explanation considers that triplet methylene can initially make only one new bond, due to spin considerations and, therefore, yields an intermediate triplet 1,3-biradical. There is then competition between bond rotation that produces the loss of stereochemistry and spin inversion that is necessary for formation of the second bond. Spin correlation studies for such biradicals (see p. 270) indicate that bond rotation is indeed faster than ring closure (spin inversion) for triplet biradicals but not for singlet biradicals.

An alternative view of the different stereoselectivity of singlet and triplet carbenes in cycloaddition reactions has been given in terms of orbital symmetries.[14] The orbitals of singlet methylene plus ethylene correlate with ground state cyclopropane, but the orbitals of triplet methylene plus ethylene correlate with an excited state of the trimethylene biradical in which rotation is facilitated. Operationally, the two theories are difficult to distinguish, and most practicing carbene chemists still refer to the Skell–Woodworth hypothesis for its ease of application.

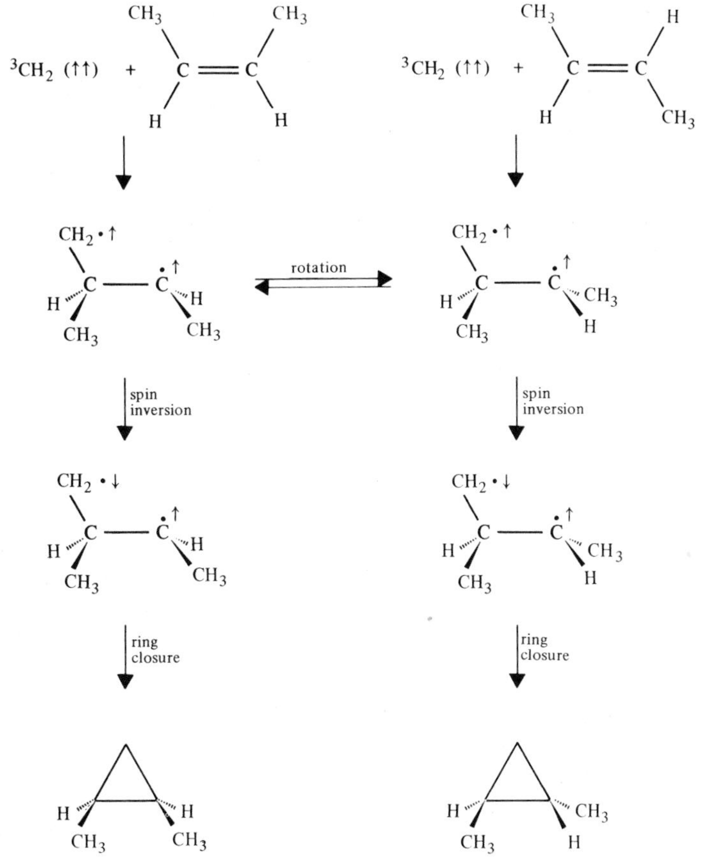

$^3CH_2$ (↑↑) + [structure: CH_3–CH=C–CH_3/H] $^3CH_2$ (↑↑) + [structure: CH_3–CH=C–H/CH_3]

$CH_2 \cdot ↑$ [biradical structure]   ⇌ rotation ⇌ $CH_2 \cdot ↑$ [biradical structure]

spin inversion   spin inversion

$CH_2 \cdot ↓$ [biradical structure] $CH_2 \cdot ↓$ [biradical structure]

ring closure   ring closure

[cyclopropane: H, CH_3, H, CH_3]   [cyclopropane: H, CH_3, CH_3, H]

We will now examine some of the key experimental evidence that has led to the general belief in concerted reactions for singlet carbenes and stepwise biradical mechanisms for triplet carbenes. Establishment of a concerted reaction mechanism always requires substantial negative evidence to rule out potential intermediates, as well as observation of specific stereochemistry that suggests a single-step reaction. We have already mentioned the retention of stereochemistry in the reaction forming cyclopropanes. A few studies have been made regarding the stereochemistry of singlet carbene insertion reactions, and all indicate that the insertion proceeds with retention of configuration:[15]

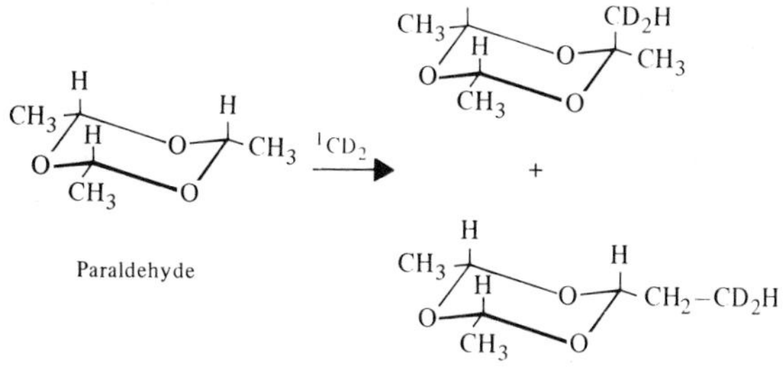

[reaction scheme: Paraldehyde + $^1CD_2$ → products]

Paraldehyde

Rather than a concerted reaction, it is also plausible that a singlet biradical of very short lifetime could be an intermediate in either the cycloaddition or insertion reactions. However, very convincing evidence against this interpretation was provided by a ^{14}C labeling experiment:[16]

$$^1CH_2 + \underset{CH_3}{\overset{CH_3}{\diagdown}}C=\overset{*}{C}H_2 \longrightarrow \underset{CH_3}{\overset{CH_3CH_2}{\diagdown}}C=\overset{*}{C}H_2 \quad \text{(plus other products)}$$

(>99% maintenance of $1-^{14}C$)

The experiment thus rules out the intermediacy of a singlet radical pair, since it would be expected to undergo some allylic rearrangement:

$$^1CH_2(\uparrow\downarrow) + \underset{CH_3}{\overset{CH_3}{\diagdown}}C=\overset{*}{C}H_2 \longrightarrow \left[\begin{array}{c} CH_3 \cdot (\uparrow) \\ (\downarrow) \quad \overset{*}{C}H_2 \\ CH_2 \cdots \cdots \\ \underset{CH_3}{\overset{|}{C}} \end{array}\right]$$

$$\underset{CH_3}{\overset{CH_2}{\diagdown}}C-\overset{*}{C}H_2CH_3 \qquad \underset{CH_3}{\overset{CH_2}{\diagdown}}C=\overset{*}{C}H_2$$

Finally, singlet carbenes are distinguished from triplet carbenes by their lack of selectivity. In C–H insertion reactions, there is only a slight selectivity in the order $3° > 2° > 1°$ with typical ratios of $1.5 : 1.2 : 1.0$ (for reaction with isopentane).[17] In alkene cycloadditions, there are only small changes in rates depending on substituents. Thus singlet methylene reacts with ethylene, mono-, di-, tri-, or tetrasubstituted alkenes, or with butadiene all within a reactivity range of a factor of three.[18] These selectivity data suggest that the singlet carbene reactions are not sensitive to radical stability.

Contrasting results are observed for triplet carbene reactions. Triplet methylene reacts rather selectively with different classes of C–H bonds, preferring $3° > 2° > 1°$ by ratios of about $7:2:1$.[19,20] Reactivities towards alkenes also show a wider range—for example, butadiene reacts 19 times faster than ethylene towards triplet methylene in the vapor phase.[18] In both cases, the preferred reactions are readily explained on the basis of formation of the most stable biradical or radical pair.

Furthermore, aside from the scrambling of stereochemistry observed in both cycloaddition and insertion reactions, triplet carbene insertion reactions give side products characteristic of recombinations of free radicals that have escaped from the solvent cage:[21]

$$^3CH_2 + -\overset{|}{\underset{|}{C}}-H \longrightarrow \left[CH_3 \cdot + -\overset{|}{\underset{|}{C}}\cdot\right] \longrightarrow CH_3-\overset{|}{\underset{|}{C}}-$$

$$CH_3 \cdot + -\overset{|}{\underset{|}{C}}\cdot \longrightarrow CH_3-\overset{|}{\underset{|}{C}}-$$

$$CH_3-CH_3 + -\overset{|}{\underset{|}{C}}-\overset{|}{\underset{|}{C}}-$$

The stereochemistry of carbene insertions into C–H bonds on an asymmetric carbon has been the subject of only a few mechanistic studies primarily because of experimental difficulties. The most thorough study of C—H insertion has been done using singlet dichlorocarbene from a mercurial precursor and optically active 2-phenylbutane.[22] The observed product showed predominant retention of stereochemistry. A primary deuterium isotope effect of 2.5 was measured.

$$\phi HgCCl_2Br \xrightarrow{\Delta} \phi HgBr + {}^1CCl_2$$

$$\phi\!-\!\underset{\underset{CH_3}{|}}{CH}\!-\!CH_2CH_3 \xrightarrow{{}^1CCl_2} \phi\!-\!\underset{\underset{CCl_2H}{|}}{\overset{\overset{CH_3}{|}}{C}}\!-\!CH_2CH_3$$

(retention)

Furthermore, a Hammett study using the same carbene and substituted cumenes gave relative rates that led to a linear Hammett plot with slope of -1.19. The negative ρ value is consistent with the electrophilicity expected for carbenes. The magnitude of the ρ value indicates that $:CCl_2$ is rather selective, unlike $:CH_2$. (In general, $:CCl_2$ is selective enough to be synthetically useful, especially in cycloaddition reactions.)

$$Z\!-\!\!\left\langle\!\!\bigcirc\!\!\right\rangle\!\!-\!CH(CH_3)_2 \xrightarrow{{}^1CCl_2} Z\!-\!\!\left\langle\!\!\bigcirc\!\!\right\rangle\!\!-\!\underset{\underset{CH_3}{|}}{\overset{\overset{CH_3}{|}}{C}}\!-\!CCl_2H$$

All the foregoing data may be combined to provide an accurate picture of the transition state for C–H insertion by $:CCl_2$. The process is viewed as a concerted front-side attack (to give retention) with substantial C—H bond cleavage in the transition state (moderate isotope effect) and with some positive charge developed on the benzylic carbon in the transition state (negative Hammett ρ value):

$$\underset{/}{\overset{\backslash}{}}C\!-\!\!-\!\!-H + :CCl_2 \longrightarrow \left[\underset{\underset{Cl}{}\;\;\underset{Cl}{}}{\overset{\overset{\delta+}{\backslash}}{\underset{/}{C}}\!-\!\!-\!\!-H}\right]^{\ddagger} \longrightarrow \underset{/}{\overset{\backslash}{}}C\!-\!\!-\!\!-CCl_2H$$

Vapor Phase Carbene Chemistry

In the vapor phase, there are two additional considerations that are very important for an understanding of carbene chemistry. The first point reflects the fact that carbene reactions are normally highly exothermic (about 90 kcal/mole for insertions or additions). Thus, a product molecule is frequently produced with a large amount of excess internal energy. In the vapor phase without solvent molecules to help dissipate the excess vibrational energy, the molecule may be subject to further reactions. Such reactions are often called "hot molecule" reactions. Cyclopropanes, from cycloaddition reactions, are particularly susceptible to hot molecule decomposition to the thermodynamically more stable olefin, since E_a for cyclopropane isomerization is only 64 kcal/mole. The following experiments illustrate the significance of this point.

At pressures over 400 Torr, ethylene reacts with singlet methylene to produce 80% cyclopropane by cycloaddition and 20% propene by insertion. As the pressure is lowered, the cyclopropane yield drops and the propene yield rises. Apparently at pressures below 400 Torr, the hot cyclopropane product does not undergo enough collisions to completely cool it to ground state cyclopropane, and some propene is formed by isomerization, presumably via a trimethylene biradical. At increasingly lower pressures, more and more hot cyclopropane is converted to propene. Thus, the product distribution varies with pressure, and only above 400 Torr, where the product composition has leveled off, can the relative amounts of cycloaddition and insertion be ascertained:[23]

The effects of pressure on the stereochemistry of methylene reactions can also be observed. The stereoselective reaction of singlet methylene with cis-2-butene is observed only at high pressure; at low pressure, the products include trans- and cis-1,2-dimethylcyclopropane as well as insertion products. In this case, the hot cis cyclopropane can be isomerized to the trans cyclopropane, thus obscuring the stereoselectivity of the initial cycloaddition.[24]

In both of the previous cases, the effects are not observed in solution phase studies. Hot molecule reactions are rarely encountered in solution work since the surrounding molecules very rapidly and efficiently remove any excess vibrational energy (over a Boltzmann distribution).

These effects of pressure in vapor phase reactions are not limited to carbene chemistry. Vibrationally activated hot molecules may also be encountered in photochemical or radical reactions. For example, the following sequence has been demonstrated to occur in the gas-phase photolysis of 2,3-diazabicyclo [2.2.1]hept-2-ene:[25]

Since the initial reaction is exothermic by 14 kcal/mole, plus the energy of the photon (about 90 kcal/mole), the initial bicyclo[2.1.0]pentane product has internal energy much in excess of E_a for the subsequent rearrangements. The rearranged products diminish in proportion at higher pressures of inert gas M.

A second complication of vapor phase carbene chemistry also relates to the effects of pressure. The transition of a singlet carbene to a triplet carbene, called intersystem crossing, also requires a dissipation of energy and is, therefore, facilitated by increased numbers of collisions. Reaction of methylene with a given constant pressure of cis-2-butene, such that hot cyclopropane reactions are negligible, with the addition of increasing pressures of an inert gas, such as N_2, CO, Ar, or CF_4, results in the formation of increasing amounts of trans-1,2-dimethylcyclopropane.[26] This arises from the increased intersystem crossing to triplet methylene and its nonstereoselective addition to cis-2-butene. It is, therefore, possible to observe a nonstereoselective cycloaddition at very low pressures, due to scrambling of a hot cyclopropane product, or at high pressures (of inert gas), due to an enhancement of intersystem crossing followed by a nonstereoselective triplet reaction.

Control of the Triplet Versus Singlet Nature of a Carbene Reaction

The concept of enhanced intersystem crossing by addition of an inert diluent should bring to mind the idea that the singlet versus triplet character of a carbene reaction might be controlled in this manner. In fact, this effect is readily observable in solution using an inert solvent for dilution rather than an inert gas. The photolysis of diazofluorene **1** in neat liquid cis-2-butene gives predominantly the cis adduct **2**, while in neat liquid trans-2-butene the trans adduct **3** is the major product. Dilution of the reaction mixtures with hexafluorobenzene, which is inert towards carbene reactions, leads to decreasing stereoselectivity. The increased number of collisions between the singlet carbene and the inert solvent tend to enhance crossing to the more stable triplet state, and therefore the characteristic nonstereoselectivity of a triplet carbene is observed.[27]

Furthermore, upon extrapolation of the data from each series of experiments to infinite dilution, a limiting ratio of 14% **2** and 86% **3** is obtained regardless of the cis or trans stereochemistry of the starting alkene. This indicates that both *cis*- and *trans*-2-butene yield the same product mixture with triplet fluorenylidene, which suggests that the biradical intermediate in the cycloaddition reaction has fully equilibrated rotationally before ring closure takes place.

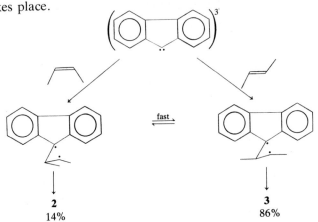

2	**3**
14%	86%

Aside from the use of inert diluents, several other methods are often used to control the nature of the carbene involved in a reaction. Triplet carbene, being somewhat selective, can frequently be scavenged successfully leaving only singlet carbene reactions. Oxygen, having a triplet ground state, is a very efficient scavenger for this purpose. For example, the reaction of fluorenylidene in neat *cis*-2-butene is not completely stereoselective, since the ratio of **2/3** is 1.95.[27] Apparently, some intersystem crossing occurs even in the absence of inert diluent. The stereoselectivity is improved by the addition of oxygen, however, since it readily scavenges the triplet carbene and leaves the singlet carbene.

In general, a carbene reaction can be rather well constrained to a singlet state reaction by running the reaction in the presence of high concentrations of substrate to allow minimum time for intersystem crossing and by including a specific triplet carbene scavenger, such as oxygen. Control of a carbene reaction in a triplet state pathway would appear to be more difficult, since the carbene must initially pass through the singlet state in any pyrolytic or photolytic generation. However, the technique of photosensitization allows the complete bypassing of the singlet carbene by generating the carbene in a triplet state directly. The technique is actually an indirect photolysis in which the light is absorbed by a sensitizer molecule that itself has a high efficiency of intersystem crossing. In its electronically excited triplet state, the sensitizer molecule can then transfer its excitation energy and its triplet character to diazomethane, which will now decompose directly to triplet methylene (see Chapter 8 for a full discussion of energy transfer and photosensitization).

$$CH_2{=}N_2 \xrightarrow{h\nu} (CH_2{=}N_2)^{*1} \longrightarrow {}^1CH_2 + N_2$$

Direct photolysis of diazomethane

$$\phi_2CO \xrightarrow{h\nu} (\phi_2CO)^{*1} \xrightarrow{fast} (\phi_2CO)^{*3}$$

$$(\phi_2CO)^{*3} + CH_2{=}N_2 \rightarrow \phi_2CO + (CH_2{=}N_2)^{*3}$$

$$(CH_2{=}N_2)^{*3} \rightarrow {}^3CH_2 + N_2$$

Photosensitization of diazomethane

7.5 Carbene Complexes—Carbenoids

Now that we have established the different reactivity patterns for singlet and triplet carbenes, we must recognize another significant complexity in carbene reactions. The reactivity of carbenes is so great that in many instances they do not exist as free divalent carbenes during the course of a reaction; rather, they are complexes of a carbene with some other molecule, often the molecule from which the carbene was to be generated. Such complexes are called carbenoids. Thus, the generalized dissociation reaction that follows does involve free carbene:

$$R_2C\big\langle{}^A_B \longrightarrow AB + R_2C: \xrightarrow{XY} R_2C\big\langle{}^X_Y \qquad (S_N1 \text{ type})$$

However, if the molecule to be added (XY) displaces the leaving molecule, AB, in a concerted or stepwise fashion, then a free carbene need never be an intermediate in the reaction:

$$R_2C\big\langle{}^A_B + XY \longrightarrow \left[\begin{matrix} A & R & X \\ & C & \\ B & R & Y \end{matrix}\right]^{\ddagger} \longrightarrow AB + R_2C\big\langle{}^X_Y$$

or

$$R_2C\big\langle{}^A_B + XY \longrightarrow AB\!-\!\overset{R}{\underset{R}{C}}\!-\!XY \longrightarrow AB + R_2C\big\langle{}^X_Y \qquad \Bigg\} \quad (S_N2 \text{ type})$$

These two potential mechanisms are related in the same way as the S_N1 and S_N2 mechanisms for unimolecular and bimolecular nucleophilic displacement. Much of the evidence used to distinguish between them is similar, such as trapping techniques for the intermediate carbene, kinetic analyses, and alternative precursors to the same carbene.

Although the intermediacy of carbenoids raises a difficult mechanistic ambiguity, it allows a much greater versatility in carbene reactions. By complexation with suitable donors, carbenes lose much of their extreme reactivity, and carbenoid reactions can be extremely selective and useful synthetically. Thus, although either singlet or triplet methylene reacts with alkenes to give both C–H insertion and cycloaddition, the Simmons–Smith reaction involving an iodomethyl zinc iodide complex produces only cyclopropanes in excellent yields.[28]

$$CH_2I_2 \xrightarrow[Et_2O]{Zn-Cu} ICH_2ZnI \text{ (or } ZnI_2 \cdot Zn(CH_2I)_2)$$

Simmons–Smith reaction

The Simmons–Smith reagent is an isolable species, very similar to a Grignard reagent, which can be considered to be a ZnI_2 complex of methylene. The absence of free methylene is clearly indicated by the lack of any insertion products. The stereochemistry of the Simmons–Smith cycloaddition, like that of singlet methylene, retains the alkene stereochemistry.

As we now examine the various reactions of carbenes we must necessarily pay close attention to the mechanistic possibilities already alluded to: does the reaction involve free carbene or a carbenoid complex; and if the reaction involves a free carbene, what is its spin state?

7.6 Generation of Carbenes

Carbenes (and carbenoids) are chiefly formed in two types of reactions: (a) decomposition of reactive molecules like ketenes or diazo compounds, and (b) α eliminations. Both methods provide the strong driving force necessary for the generation of molecules as reactive as carbenes. Ketenes eliminate CO (bond strength 256 kcal/mole), diazo compounds eliminate N_2 (bond strength 226 kcal/mole), and an α elimination produces a stable ionic species such as a metal halide:

$$R_2C=C=O \xrightarrow{h\nu} R_2C: + CO$$

$$R_2C=\overset{+}{N}=\overset{-}{N} \xrightarrow[\text{or }\Delta]{h\nu} R_2C: + N_2$$

$$R_2C{\overset{X}{\underset{Y}{\Big\langle}}} \longrightarrow R_2C: + XY$$

For example,

$$CHBr_3 \xrightarrow{KOtBu} Br_2C: + tBuOH + KBr$$

Diazo compounds constitute the most common class of carbene precursors. These compounds are usually prepared *in situ* and are readily available from either an amine or a carbonyl functional group. The very useful Bamford–Stevens reaction involves elimination from the tosylhydrazone derivative of an aldehyde or ketone.

$$R_2C=O \xrightarrow{Ts-NH-NH_2} R_2C=N-NHTs \xrightarrow{n\text{-}BuLi} R_2C=N-\overset{\overset{\displaystyle Li}{|}}{N}-Ts$$

$$R_2C=N-\overset{\overset{\displaystyle Li}{|}}{N}-Ts \xrightarrow{\alpha\text{ elim}} R_2C=N_2 + Ts^-Li^+$$

Bamford–Stevens reaction

From amines, a variety of N-nitroso derivatives provide suitable precursors to diazo compounds. Diazomethane is most commonly prepared by base treatment of N-nitroso-N-methyl-p-toluenesulfonamide, which is commercially available.

$$CH_3-\overset{\overset{\displaystyle N=O}{|}}{N}-Ts \longrightarrow CH_3-N=N-OTs \xrightarrow{OH^-} CH_2=N_2 + TsO^- + H_2O$$

α Eliminations

In 1950, 88 years after Geuther's original suggestion,[1] Hine demonstrated that the basic hydrolysis of chloroform, which ultimately yields formic acid or carbon monoxide plus HCl, proceeds by α elimination to give dichloro-carbene as an intermediate:[29]

$$CHCl_3 + OH^- \rightleftharpoons {}^-:CCl_3 \rightleftharpoons :CCl_2 + Cl^-$$

This mechanism is supported by the following evidence. Proton abstraction had been earlier demonstrated by the rapid incorporation of deuterium when chloroform is mixed with basic D_2O.[30] Furthermore, the relative reactivities of dichloromethane and carbon tetrachloride are much less than the reactivity of chloroform toward hydroxide; carbon tetrachloride has no abstractable proton, and dichloromethane cannot stabilize a negative charge as well and probably undergoes a slow S_N2 reaction. The intermediacy of the dichlorocarbene was demonstrated by several trapping experiments. A concentration of 0.16 M NaCl decreases the hydrolysis rate by about 15%, apparently by reversing the α elimination.[31] This is an example of a common ion rate depression (see page 154), since it was shown that equivalent concentrations of inert salts, such as sodium perchlorate, sodium nitrate, or sodium fluoride, had no detectable effects upon the rate. In the presence of sodium bromide or sodium iodide, an even larger rate depression was noted, as well as the incorporation of bromine or iodine into the starting haloform.

Similar α eliminations may occur from trihaloacetate anions:

$$CF_3-\overset{\overset{\displaystyle O}{\|}}{C}-O^-Na^+ \xrightarrow{\Delta} :CF_2 + CO_2 + Na^+F^-$$

A highly novel method of carbene generation involves α elimination via a carbocation intermediate. Note that a carbene may be considered to be the conjugate base of a carbocation. Thus, a stable carbocation such as di-methoxymethyl fluoroborate may be treated with a nonnucleophilic base such as diisopropylethylamine to yield dimethoxycarbene:[32]

$$CH_3O-\overset{+}{\underset{\underset{\displaystyle H}{|}}{C}}-OCH_3 + (iPr)_2\ddot{N}Et \longrightarrow CH_3O-\overset{\overset{\displaystyle H}{|}}{\underset{}{C}}-OCH_3 + (iPr)_2\overset{+}{N}Et$$

Recently, some examples of photoeliminations leading to carbenes have been shown:[33]

$$\phi-\triangleleft \xrightarrow{h\nu} \phi CH{=}CH_2 + :CH_2$$

$$\overset{\phi}{\underset{\phi}{\Big\rangle}}\!\!O \xrightarrow{h\nu} \phi CHO + \phi CH:$$

Organometallic Precursors

An important preparation of dihalocarbenes pioneered by Seyferth and co-workers[34] involves thermolysis of phenyl trihalomethyl mercury reagents. These are particularly useful since they lead to carbene generation under neutral conditions at moderate temperatures (usually refluxing benzene):

$$\phi HgCCl_2Br \xrightarrow{\Delta} \phi HgBr + :CCl_2$$

Besides photolysis or thermolysis, diazo compounds are readily decomposed by copper or copper salts. There is strong evidence, however, that copper-catalyzed reactions do not proceed through free carbene intermediates. As in the Simmons–Smith reaction, C–H insertion does not occur with these copper carbenoids. However, the strongest evidence that the copper remains an integral part of the carbenoid complex involved the use of an optically active copper complex to produce carboethoxycarbene in the presence of styrene. The cycloaddition products displayed optical activity (asymmetric induction) to an extent of about 6%.[35] This clearly indicates that the copper complex is a significant part of the transition state for cycloaddition.

$$N_2=CHCOOEt + R_2{}^*Cu \longrightarrow [R_2{}^*Cu \ldots CHCOOEt]$$

(both optically active)

Asymmetric induction in the reaction of a copper carbenoid

There is also substantial evidence that free carbenes are not present in many reactions involving lithium derivatives. The use of an alkyllithium to promote α elimination from an alkyl halide may actually proceed in two distinct ways. The alkyllithium may serve as a base and remove a proton, or it may undergo metal–halogen interchange. The interchange becomes the preferred pathway for iodine and bromine derivatives and leads to a carbene (or carbenoid) different from what would be expected for a reaction as a base:

Reaction as a base

$$\phi CHCl_2 + CH_3Li \longrightarrow CH_4 + \left[\begin{array}{c} Li \\ | \\ \phi CCl \\ | \\ Cl \end{array} \right] \longrightarrow$$

Lithium–halogen interchange

$$\phi CHBr_2 + CH_3Li \longrightarrow CH_3Br + \left[\begin{array}{c} Li \\ | \\ \phi CBr \\ | \\ H \end{array} \right] \longrightarrow LiBr +$$

Given these two different pathways, the question still remains: do these reactions involve free carbenes or are the organolithium carbenoids the reactive intermediates? The evidence indicates that free carbenes are not involved. Although both benzal bromide and benzal iodide lead to products attributable to phenylcarbene (and not phenylhalocarbene), the selectivity towards olefins is different for the two precursors.[36] Furthermore, both show a selectivity which differs from that of phenylcarbene generated from phenyldiazomethane.[37] Finally, it has been possible to isolate some α-halolithium compounds at −78° C and observe that their decomposition rate is specifically enhanced in the presence of olefin, thus indicating a direct reaction between the two.[38]

Many other metals have been successfully utilized in carbene or carbenoid reactions. As we have already mentioned, the Simmons–Smith reaction involves an organozinc carbenoid which is very useful as a stereospecific and mild cyclopropanation reagent. Mercurial precursors apparently lead to free carbenes. Other carbenoids have been prepared using magnesium, aluminum and many of the transition metals, including tin, iron, lead, cadmium, chromium, indium, iridium, molybdenum, tungsten, and manganese.

7.7 Carbene Rearrangements

Carbene rearrangements occur quite frequently, and may be considered as intramolecular examples of some of the more typical carbene reactions. Unlike rearrangements discussed for other reactive intermediates, singlet carbene rearrangements may lead directly to stable molecules. For example, an intramolecular C–H insertion of isopropyl carbene leads to isobutylene or to methylcyclopropane:

$$(CH_3)_2CH—CH: \longrightarrow (CH_3)_2C=CH_2 + CH_3—CH\overset{CH_2}{\underset{CH_2}{\diagdown}}$$

Cyclic carbenes undergo a variety of possible rearrangements, depending upon ring size and experimental conditions:

$$\triangleright—CH: \longrightarrow \square \;+\; \diagup\!\diagup \;+\; \overset{CH_2}{\underset{CH_2}{\|}} \;+\; HC\!\equiv\!CH$$

(~80%) (~3%) (~5%) (~5%) (ref. 39)

$$\square \overset{\cdot\cdot}{} \longrightarrow \triangleright\!=\!CH_2 + \square \;+\; \diagup\!\diagup$$

(80%) (19%) (1%) (ref. 40)

(100%)

(6%) (14%) (ref. 41)

(18%) (62%)

Intramolecular cycloadditions have also been obtained in synthetically useful reactions:

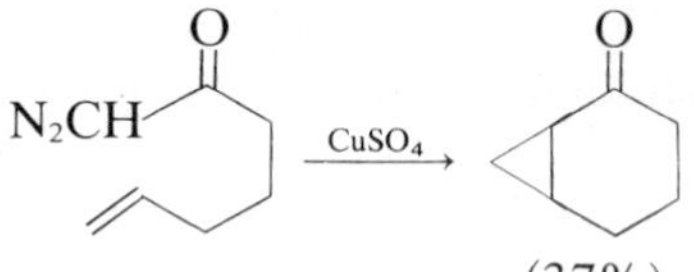

(37%)   (ref. 42)

Undoubtedly, the most synthetically useful carbene rearrangement is the Wolff rearrangement, which is the heart of the Arndt–Eistert synthesis. This series of reactions converts a carboxylic acid to its next higher homolog:[43]

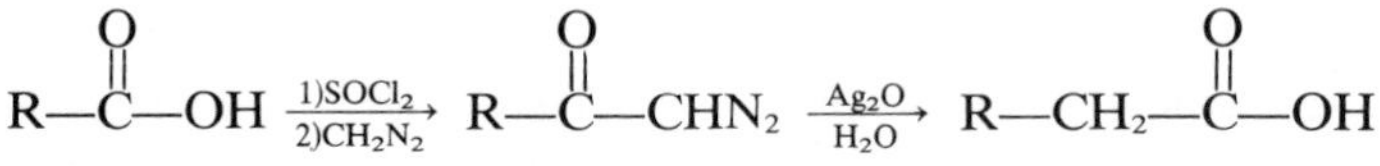

Arndt–Eistert synthesis

The mechanism of the Wolff rearrangement is normally considered to involve a carbene intermediate that rearranges to a ketene. Although ketene intermediates have been conclusively implicated by a variety of trapping experiments, firm evidence for the carbene intermediate is lacking, and a rearrangement that is concerted with nitrogen loss is also quite plausible.[44]

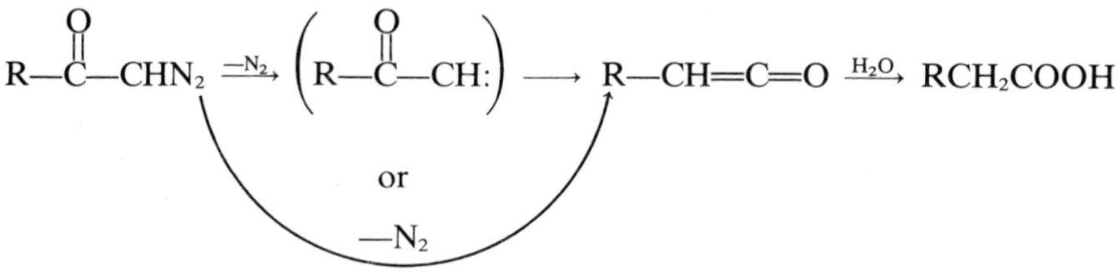

Wolff rearrangement

Cyclopropylidenes undergo ready electrocyclic ring opening to allenes in a synthetically useful reaction:

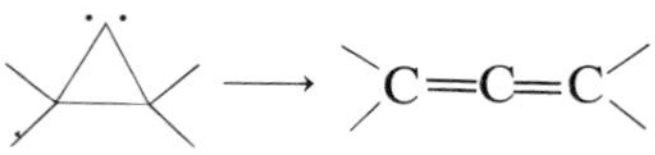

The reaction is especially intriguing in that an optically active cyclopropane precursor gives rise to an optically active allene with high stereoselectivity. Orbital symmetry considerations lead to the prediction of a conrotatory ring opening. However, there are two conrotatory openings that lead to two different allene enantiomers. The preferred opening may be predicted on the basis of minimization of nonbonded interactions.[45]

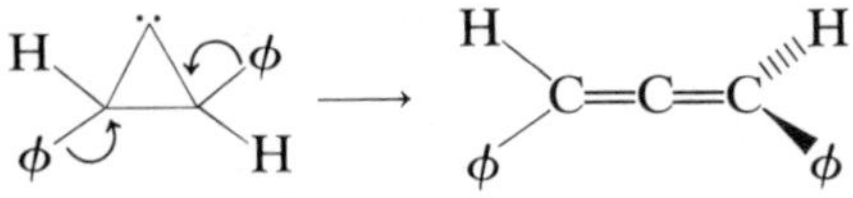

Disfavored–strong ϕ–ϕ interactions

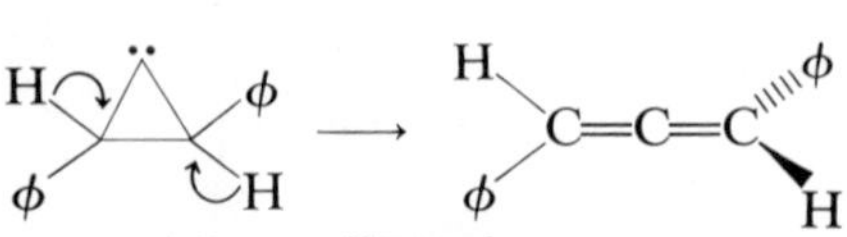

Favored

Although arylcarbenes show no tendency to rearrange in solution, they exhibit a rather surprising rearrangement in vapor phase pyrolysis:[46]

7.8 Other Carbene Reactions

Carbenes are typically strongly electron-deficient species, and their reactions are characteristic of electrophiles. However, if the vacant p orbital can be delocalized, the electrophilic properties should be suppressed and the nucleophilic properties of the lone pair should be enhanced. An appropriate system is cycloheptatrienylidene. Delocalization of the vacant p orbital creates an aromatic cycloheptatrienyl cation (six π electrons), and leaves an exocyclic lone pair of electrons in an sp^2 orbital. The nucleophilicity of this carbene was demonstrated by cycloaddition reactions with styrenes which showed substituent effects correlating with a Hammett ρ value of +1.05.[47]

Nucleophilic addition of cycloheptatrienylidene to styrenes

Besides the most important types of carbene reactions, C–H insertion and olefin cycloaddition, many other carbene reactions have been observed. Carbenes have been shown to insert (or at least lead to insertion-type products) with O—H, N—H, S—H, Si—H, C—Cl, C—Br, and C—O bonds, and to cycloadd to C=O, C=N, C=C, C≡C, and aromatic π bonds. With the tremendous reactivity borne by carbenes, it is perhaps most surprising that some types of reaction have not yet been observed. A C—C insertion has not yet been conclusively demonstrated, for example. A 1,4-cycloaddition to conjugated dienes also has not been conclusively demonstrated. Finally, carbenes are rarely observed to dimerize to olefins, partly because the reactivity is so great and hence concentrations so low that bimolecular reaction of two carbenes would be rare, and partly because the olefin would be formed with an excessive amount of energy and would undergo further "hot molecule" reactions.

7.9 Other Carbenelike Intermediates

In recent years, organic chemists have uncovered increasing numbers of reaction intermediates resembling carbenes. The most frequently studied carbene analogs are the nitrenes, molecules that involve a monovalent nitrogen atom with only a sextet of electrons. Nitrenes are usually produced in much the same way as carbenes, and their reactions are closely analogous. For example, photolytic or pyrolytic expulsion of nitrogen from an azide yields a nitrene that may then undergo an insertion reaction or cycloaddition to a π bond:[48]

$$EtOOCN_3 \xrightarrow[\text{or } \Delta]{h\nu} EtOOC—\ddot{N}: + N_2$$

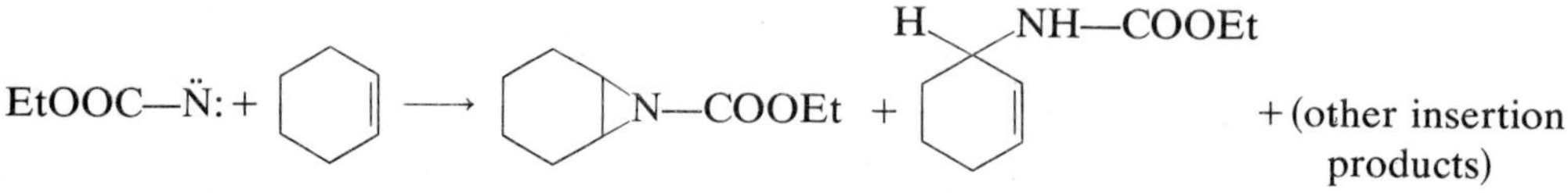

$$EtOOC—\ddot{N}: + \bigcirc\!\!\| \longrightarrow \bigcirc\!\!\triangleright\!N—COOEt + \bigcirc\!\!\text{(H, NH—COOEt)} + \text{(other insertion products)}$$

The nitrene analog of the Wolff rearrangement is called the Curtius rearrangement. Again there is no firm evidence for a free nitrene intermediate, and the rearrangement seems to be synchronous with nitrogen loss.[48]

$$R—\overset{O}{\overset{\|}{C}}—N\!\!=\!\!\overset{+}{N}\!\!=\!\!\overset{-}{N} \longrightarrow R—\overset{O}{\overset{\|}{C}}—\ddot{N}: + N_2$$

$$R—\overset{O}{\overset{\|}{C}}—\ddot{N}: \longrightarrow R—N\!\!=\!\!C\!\!=\!\!O \xrightarrow{H_2O} RNHCOOH \longrightarrow RNH_2 + CO_2$$

Curtius rearrangement

Carbene analogs of the other group IV elements—silicon, germanium, and tin—have also been observed. They show interesting combinations of similarities and differences with carbenes in their chemical and physical properties.[49]

The chemistry of carbon itself involves some other unusual carbenelike intermediates. Monovalent carbon intermediates are formed in the following reaction. They behave as carbene–radicals, with the carbene reactivity apparently satisfied first:[50]

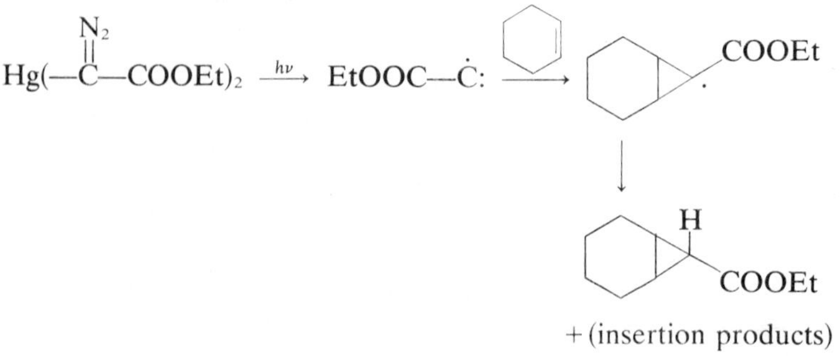

$$Hg(—\overset{N_2}{\overset{\|}{C}}—COOEt)_2 \xrightarrow{h\nu} EtOOC—\dot{\ddot{C}}: \longrightarrow$$

+ (insertion products)

Carbon vapor contains C_3 molecules that may be considered to be 1,3-dicarbenes:[51]

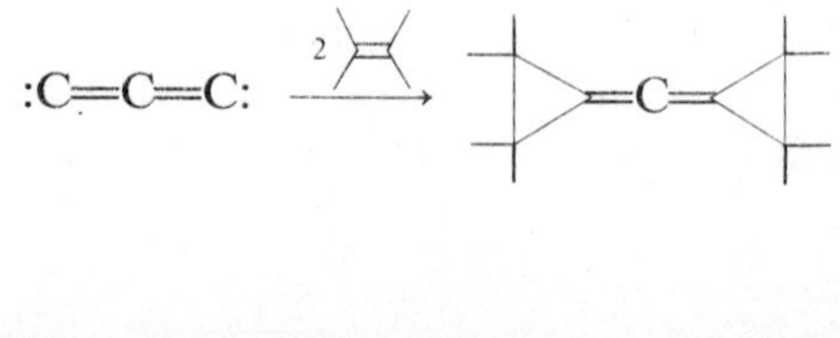

$$:C\!\!=\!\!C\!\!=\!\!C: \xrightarrow{2}$$

Perhaps the ultimate organic reactive intermediate is the carbon atom, another constituent of carbon vapor. It may be considered a 1,1-dicarbene (with triplet ground state), and yields spiro compounds with two cycloadditions. The first cycloaddition has been shown to be a stereospecific singlet addition, and the second a nonstereospecific triplet addition.[52] Carbon atoms are also capable of deoxygenation of carbonyl compounds to give carbenes and carbon monoxide:[53]

References

1. A. Geuther, *Justus Liebig's Ann. Chem.*, **123,** 121 (1862).
2. Walt Disney's Comics and Stories, **4** (8), 2 (1944).
3. G. Herzberg, *Proc. Roy. Soc. London, Ser. A*, **262,** 291 (1961).
4. G. Herzberg and J. W. C. Johns, *J. Chem. Phys.*, **54,** 2276 (1971).
5. J. A. Schellman, *J. Chem. Phys.*, **58,** 2882 (1973).
6. H. M. Frey, *J. Chem. Soc., Chem. Commun.*, 1024 (1972).
7. W. L. Hase, R. J. Phillips, and J. W. Simons, *Chem. Phys. Lett.*, **12,** 161 (1971).
8. M. J. S. Dewar, R. C. Haddon, and P. K. Weiner, *J. Amer. Chem. Soc.*, **96,** 253 (1974).
9. A. M. Trozzolo, R. W. Murray, and E. Wasserman, *J. Amer. Chem. Soc.*, **84,** 4990 (1962).
10. E. Wasserman, R. W. Murray, W. A. Yager, A. M. Trozzolo, and G. Smolinsky, *J. Amer. Chem. Soc.*, **89,** 5076 (1967).
11. A. M. Trozzolo, R. W. Murray, G. Smolinsky, W. A. Yager, and E. Wasserman, *J. Amer. Chem. Soc.*, **85,** 2526 (1963).
12. A. M. Trozzolo, E. Wasserman, and W. A. Yager, *J. Amer. Chem. Soc.*, **87,** 129 (1965).
13. R. C. Woodworth and P. S. Skell, *J. Amer. Chem. Soc.*, **81,** 3383 (1958).
14. R. Hoffman, *J. Amer. Chem. Soc.*, **90,** 1475 (1968).
15. W. Kirmse and M. Buschhof, *Chem. Ber.*, **102,** 1098 (1969).
16. W. von E. Doering and H. Prinzbach, *Tetrahedron*, **6,** 24 (1959).
17. B. M. Herzog and R. W. Carr, Jr., *J. Phys. Chem.*, **71,** 2688 (1967).
18. R. A. Moss in "Carbenes," Vol. 1, M. Jones, Jr., and R. A. Moss, Eds., Wiley, New York, 1973, pp. 153 ff.
19. D. F. Ring and B. S. Rabinovitch, *J. Amer. Chem. Soc.*, **88,** 4285 (1966).
20. D. F. Ring and B. S. Rabinovitch, *Can. J. Chem.*, **46,** 2435 (1968).
21. S.-Y. Ho and W. A. Noyes, Jr., *J. Amer. Chem. Soc.*, **89,** 5091 (1967).
22. D. Seyferth and Y. M. Cheng, *J. Amer. Chem. Soc.*, **95,** 6763 (1973).
23. H. M. Frey and G. B. Kistiakowsky, *J. Amer. Chem. Soc.*, **79,** 6373 (1957).
24. H. M. Frey, *Proc. Roy. Soc. London, Ser. A*, **251,** 575 (1959).
25. T. F. Thomas, C. I. Sutin, and C. Steel, *J. Amer. Chem. Soc.*, **89,** 5107 (1967).
26. R. F. W. Bader and J. I. Generosa, *Can. J. Chem.*, **43,** 1631 (1965).
27. M. Jones, Jr., and K. R. Rettig, *J. Amer. Chem. Soc.*, **87,** 4013, 4015 (1965).
28. E. P. Blanchard and H. E. Simmons, *J. Amer. Chem. Soc.*, **86,** 1337, 1347 (1964).
29. J. Hine, *J. Amer. Chem. Soc.*, **72,** 2438 (1950).
30. Y. Sakamoto, *J. Chem. Soc. Jap.*, **57,** 1169 (1936); *Bull. Chem. Soc. Jap.*, **11,** 627 (1936).

31. J. Hine and A. M. Dowell, Jr., *J. Amer. Chem. Soc.*, **76**, 2688 (1954).
32. R. A. Olofson, S. W. Walinsky, J. P. Marino, and J. L. Jernow, *J. Amer. Chem. Soc.*, **90,** 6554 (1968).
33. G. W. Griffin and N. R. Bertoniere in "Carbenes," Vol. 1, M. Jones, Jr., and R. A. Moss, Eds., Wiley, New York, 1973, pp. 305 ff.
34. D. Seyferth, J. M. Burlitch, R. J. Minasz, J. Y.-P. Mui, H. D. Simmons, Jr., A. J. H. Treiber, and S. R. Dowd, *J. Amer. Chem. Soc.*, **87,** 4259 (1965).
35. H. Nozaki, H. Takaya, S. Moriuti, and R. Noyori, *Tetrahedron*, **24,** 3655 (1968).
36. R. A. Moss, *J. Org. Chem.*, **30,** 3261 (1965).
37. G. L. Closs and R. A. Moss, *J. Amer. Chem. Soc.*, **86,** 4042 (1964).
38. G. Köbrich, *Angew. Chem. Int. Ed. Engl.*, **6,** 41 (1967).
39. J. A. Smith, H. Schechter, J. Bayless, and L. Friedman, *J. Amer. Chem. Soc.*, **87,** 659 (1965).
40. L. Friedman and H. Schechter, *J. Amer. Chem. Soc.*, **82,** 1002 (1960).
41. L. Friedman and H. Schechter, *J. Amer. Chem. Soc.*, **83,** 3159 (1961).
42. W. von E. Doering, *et al.*, *Tetrahedron*, **21,** 25 (1965); **23,** 3943 (1967).
43. W. E. Bachmann and W. S. Struve, *Org. React.*, **1,** 38 (1942).
44. J. Fenwick, G. Frater, K. Ogi, and O. P. Strausz, *J. Amer. Chem. Soc.*, **95,** 124, 133 (1973).
45. J. M. Walbrick, J. W. Wilson, and W. M. Jones, *J. Amer. Chem. Soc.*, **90,** 2895 (1968).
46. R. C. Joines, A. B. Turner, and W. M. Jones, *J. Amer. Chem. Soc.*, **91,** 7754 (1969).
47. L. W. Christensen, E. E. Waali, and W. M. Jones, *J. Amer. Chem. Soc.*, **94,** 2118 (1972).
48. W. Lwowski, *Angew. Chem. Int. Ed. Engl.*, **6,** 897 (1967).
49. P. P. Gaspar and B. J. Herold in "Carbene Chemistry." 2nd ed., W. Kirmse, Ed., Academic, New York, 1971, pp. 504 ff.
50. T. Do Minh, H. E. Gunning, and O. P. Strausz, *J. Amer. Chem. Soc.*, **89,** 6785 (1967); **90,** 1930 (1968).
51. P. S. Skell, L. D. Wescott, Jr., J. P. Golstein, and R. R. Engel, *J. Amer. Chem. Soc.*, **87,** 2829 (1965).
52. P. S. Skell and R. R. Engel, *J. Amer. Chem. Soc.*, **88,** 3749 (1966).
53. P. S. Skell, J. H. Plonka, and R. R. Engel, *J. Amer. Chem. Soc.*, **89,** 1748 (1967).

Bibliography

General References

W. Kirmse, "Carbene Chemistry," 2nd ed., Academic Press, New York, 1971.

R. A. Moss and M. Jones, Jr., Eds., "Carbenes," Vols. 1 and 2, Wiley, New York, 1973, 1975.

T. J. Gilchrist and C. W. Rees, "Carbenes, Nitrenes, and Arynes," Appleton-Century-Crofts, New York, 1969.

J. Hine, "Divalent Carbon," Ronald Press, New York, 1964.

D. Bethell, *Adv. Phys. Org. Chem.*, **7,** 153 (1969).

R. A. Moss, *Chem. Eng. News*, **47,** 60 (June 16, 1969); 50 (June 30, 1969).

C. A. Buehler, *J. Chem. Educ.*, **49,** 239 (1972).

α Eliminations

W. Kirmse, *Angew. Chem. Int. Ed. Engl.*, **4,** 1 (1965).

G. Köbrich, *et al.*, *Angew. Chem. Int. Ed. Engl.*, **6,** 41 (1967).

Differentiation Between Singlet and Triplet Carbenes

M. Jones, Jr., et al., J. Amer. Chem. Soc., **94**, 7469 (1972).

Dihalocarbenes

D. Seyferth, Acc. Chem. Res., **5**, 65 (1972).

Reactions of Carbon Species

P. S. Skell, J. J. Havel, and M. J. McGlinchey, Acc. Chem. Res., **6**, 97 (1973).

Problems

1. Calculate the percentages of each product expected from the reaction of singlet methylene with 3-methylpentane. Assume singlet methylene exhibits no selectivity.

2. Repeat problem 1, considering the reaction of singlet methylene with 3-methylpentane and a selectivity of 1.0:1.2:1.5 for 1°, 2°, 3° hydrogens, respectively. Also list all the products expected in a triplet reaction.

3. Why is CH_3Cl less susceptible to α elimination than $CHCl_3$? What product would be expected from CH_3Cl plus OH^-?

4. Photolysis of diazoethane in cyclohexene leads to a lot of ethylene plus five different C_8 products. Write out the structures of the expected products and explain the formation of ethylene.

5. The gas-phase photolysis of ketene in the presence of propane gives n-butane and isobutane as the major products as well as small amounts of ethane, n-hexane, and 2,3-dimethylbutane. Explain the reactions that lead to these side products.

   If argon is added to the ketene–propane mixture, the side products increase in yield, while if oxygen is added, the side products decrease in yield. Explain these effects.

   What effects will the argon and the oxygen have on the ratio of n-butane to isobutane?

6. Use two successive carbene reactions to synthesize 1,3-diphenylallene from stilbene (1,2-diphenylethylene).

7. In the gas-phase Wolff rearrangement, photolysis of the diazoketone shown below with ^{13}C label at C-2 led to an equal mixture of the products shown:

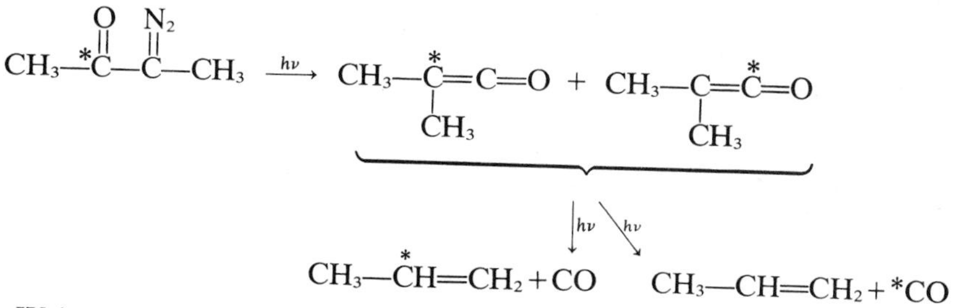

$$CH_3-\overset{*}{\underset{}{C}}-\overset{}{\underset{}{C}}-CH_3 \xrightarrow{h\nu} CH_3-\overset{*}{C}=C=O + CH_3-C=\overset{*}{C}=O$$

$$\begin{array}{ccc} & \underset{CH_3}{|} & \underset{CH_3}{|} \end{array}$$

$$\downarrow h\nu \quad \searrow h\nu$$

$$CH_3-\overset{*}{CH}=CH_2+CO \qquad CH_3-CH=CH_2+{}^*CO$$

Write a mechanism to explain the scrambling.

8. Suggest a mechanism which explains the different pathways followed:

8

Excited States

8.1 Introduction

Photochemistry is the chemistry of electronically excited states, molecules in which the distribution of electrons is temporarily displaced from the most stable arrangement. Since the chemical characteristics of a molecule are mainly determined by the arrangement of electrons in the molecular structure, an electronically excited state usually represents a species very different, both physically and chemically, from the ground state molecule. Most excited states simply return to the ground state in rapid fashion (10^{-13} sec to several seconds covers the normal range of excited state lifetimes), while other excited states undergo reactions that would be thoroughly impossible in the ground state.

Photochemical reactions are distinguished from ground-state reactions by two unique characteristics, both based upon the fact that photochemical reactions are driven by the energy of light while ground state reactions use thermal energy. In a thermal reaction, the molecules in the sample are energized according to the Boltzmann distribution with no possibility of selectively energizing a particular kind of molecule. In a photochemical reaction, however, specific molecules may be excited, depending on which molecules absorb the light of the wavelength chosen for irradiation. For example, a *cis–trans* mixture of the stilbenes may be irradiated at 313 nm where essentially only the *trans*-stilbene absorbs light. This drives the *cis–trans* equilibrium to nearly pure *cis*.[1] Thermally, the equilibrium strongly favors *trans*.

A second unique aspect of photochemistry is that the amount of energy injected into a molecule by the absorption of a photon of light can far exceed the amount of energy available in a normal thermal reaction. When benzene absorbs the ultraviolet light of a mercury resonance lamp (254 nm), it absorbs 113 kcal/mole of energy. This energy corresponds to thermal energy (RT) at about 60,000°C. With this huge injection of energy, benzene is induced to undergo the following unusual rearrangements:[2]

This photoreaction can be performed at room temperature, and it is possible to isolate the rearrangement products as stable species. Thermally, sufficient energy would be available only with difficulty, and at the required temperatures, the rearrangement products would certainly be unstable.

Besides the potential for creating new and unusual chemical reactions, photochemistry is important as a vital factor in life processes. Except for nuclear energy, all the energy on the earth comes from the radiation of the sun, either absorbed directly as heat or converted to chemical energy by photosynthesis in plants. With proper utilization of such photochemical processes, vast new stores of energy could become available.

To understand the nature and the reactivity of electronically excited states, we will consider the following aspects: the formation of excited states by absorption of visible or ultraviolet light; the different kinds of excited states and the possibilities of interconversions among them; and finally, some typical reactions of electronically excited states.

8.2 Formation of Electronically Excited States

Electronic Absorption Spectra

Depending upon the properties being considered, light may be treated either as a particle or as a wave phenomenon. The wave properties are exemplified by the specific speed of propagation ($c = 3.0 \times 10^{10}$ cm/sec in a vacuum), the capability of being diffracted according to wavelength λ, the existence of electromagnetic oscillations which define a frequency ν, and the interrelationship among these properties according to

$$c = \lambda\nu$$

The particle nature of light was not appreciated until the quantum revolution. In fact, the nature of light provided some of the most intriguing mysteries that necessitated the development of a quantum theory. The photoelectric effect indicated that light of a particular frequency carries energy in bundles of specific size, the amount of energy being proportional to the frequency

$$E = h\nu$$

where h is Planck's constant.

Furthermore, the fine structure in the absorption and emission spectra of atoms and molecules suggested that these quanta of light energy could be absorbed or emitted only when the quanta were of certain specific energies.

These bundles of light energy are now called photons, and it is recognized that the absorption of a photon by a molecule occurs only when the energy of the photon is exactly right to raise the molecule from its existing state to a higher state. The photon then ceases to exist and its energy is converted to internal energy in the molecule:

$$\Delta E = E_f - E_i = h\nu$$

where E_f is the energy content of the final state and E_i is the energy content of the initial state.

The absorption spectrum of a molecule is simply a scan of wavelengths to determine which wavelengths are absorbed by the molecule. The wavelengths that are absorbed then indicate the location of the various energy levels that exist for the molecule. An infrared spectrum, for example, typically spans the wavelengths from 2.5 to 15 μ. This covers the energy range from approximately 2 to 10 kcal/mole, which represents the spacings between most vibrational quantum levels.

The transitions among electronic quantum levels—excitation of an electron from an occupied molecular orbital to an unoccupied molecular orbital—fall in the range 40 to 150 kcal/mole. Photons carrying these amounts of energy are in the visible (400 to 700 nm) and near ultraviolet (200 to 400 nm) regions of the electromagnetic spectrum.* The terminology uv-visible spectroscopy is often used interchangeably with electronic absorption spectroscopy.

---

* Today the nanometer is the unit of wavelength preferred over millimicrons or Angstroms: 1 nm = 10^{-9} m = 10 Å.

A useful relationship between wavelength and energy is

$$E = \frac{28,600}{\lambda}$$

where E is the energy measured in kilocalories per einstein (1 einstein = 1 mole of photons) and λ is the wavelength in nanometers. Thus, a molecule that absorbs the ultraviolet wavelength of 286 nm has an electronic energy level available that is 100 kcal/mole higher in energy than the ground state of the molecule.

The efficiency with which a compound absorbs light of a given wavelength is measured by the extinction coefficient ϵ. The absorption of light in a path length l by a sample of concentration C is governed by Beer's law:*

$$\log\left(\frac{I_0}{I}\right) = A = \epsilon C l$$

where I_0 is the incident light intensity, I is the transmitted light intensity, and A is the absorbance or optical density. Extinction coefficient values range from essentially zero for highly "forbidden" transitions to about 10^5 1 mole^{-1} cm^{-1} for highly "allowed" transitions, and a plot of ϵ (or log ϵ) versus λ is usually the most useful form of the absorption spectrum.

As a final bit of background material regarding the nature of light and its interaction with matter, we will state Einstein's laws of photochemical equivalence that, although sounding trivial, provide a solid foundation for understanding of absorption of light.

1. Only light which is absorbed is effective in producing photochemical change.
2. Each absorbed photon activates just one molecule in the initial excitation step.

Orbital Transitions

For an organic molecule containing only σ bonds, such as ethane, the possible transitions are relatively simple. An electronic transition can take place only between the filled σ bonding orbitals and the empty σ^* antibonding orbitals. The transition is called a $\sigma \rightarrow \sigma^*$ transition, and the excited state produced is a $\sigma \rightarrow \sigma^*$ excited state. The amount of energy required to excite ethane to its $\sigma \rightarrow \sigma^*$ excited state is quite large, since σ orbitals are very stable and σ^* orbitals very unstable. The beginning of ethane's uv absorption does not occur until about 150 nm (190 kcal/mole), far into the vacuum uv ("vacuum" because air absorbs wavelengths below about 200 nm and evacuated equipment must be used in this region). As would be expected, $\sigma \rightarrow \sigma^*$ excited states lead to dissociative processes.[3]

---

* Beer's law has been restated in more memorable form:

> The longer the glass,
> The darker the brew,
> The less the light
> That transmits through.

$$\sigma^* \ \underline{\qquad} \qquad\qquad\qquad\qquad\qquad \sigma^* \ \underline{\quad\uparrow\downarrow\quad}$$

$$\xrightarrow[\lambda < 150 \text{ nm}]{h\nu}$$

$$\sigma \ \underline{\quad\uparrow\downarrow\quad} \qquad\qquad\qquad\qquad\qquad \sigma \ \underline{\quad\uparrow\quad}$$

ground state $\qquad\qquad\qquad\qquad \sigma \longrightarrow \sigma^*$ excited state

$$CH_3\!-\!CH_3 \ \xrightarrow{\ 147 \text{ nm}\ } \ H_2 + CH_3CH\!:$$

Electronic absorption process for ethane

Ethylene contains one π bond as well as several σ bonds. As a rule, π bonds are higher in energy than σ bonds, so we expect that the highest filled molecular orbital will be the π bonding orbital, and the lowest unfilled molecular orbital will be the π^* antibonding orbital. The first absorption by ethylene at 190 nm is a $\pi \rightarrow \pi^*$ absorption leading to a $\pi \rightarrow \pi^*$ excited state.

There are four π molecular orbitals of 1,3-butadiene, as derived by Hückel theory (see page 8). The lowest two π orbitals are filled; the higher energy two are unfilled. The lowest energy electronic transition in butadiene occurs between $\pi_2 \rightarrow \pi_3^*$ at 220 nm (designated a $\pi \rightarrow \pi^*$ transition). More specifically, the electronic situation of the molecule goes from $(\pi_1)^2(\pi_2)^2 \rightarrow (\pi_1)^2(\pi_2)^1(\pi_3^*)^1$. Other $\pi \rightarrow \pi^*$ transitions of higher energy are possible $(\pi_1 \rightarrow \pi_3^*,\ \pi_2 \rightarrow \pi_4^*,$ and so on) and are observed at shorter wavelengths.

With increasing conjugation of the π system, the energy spacing between the highest occupied molecular orbital and the lowest unoccupied molecular orbital gradually decreases. Table 8.1 shows that the larger conjugated molecules absorb at increasingly longer wavelengths until anthracene is perceptibly yellow, absorbing visible violet light to a slight degree. Note that absorption of violet gives a yellow appearance, since violet and yellow are complementary colors. This kind of observation has been called "eyeball spectroscopy" and can lead to fairly accurate estimations of absorption spectra.

In dealing with the electronic absorption properties of molecules containing heteroatoms, special considerations arise due to the presence of nonbonding electrons. Filled nonbonding orbitals virtually always have energy levels between those of the bonding orbitals and the antibonding orbitals.

Table 8.1 *Approximate Lowest-Energy Transitions for Some Simple Molecules*

Molecule	Orbital Transition	Absorption Onset[a] (nm)	ΔE (kcal/mole)
Ethane	$\sigma \rightarrow \sigma^*$	150	190
Ethylene	$\pi \rightarrow \pi^*$	190	150
1,3-Butadiene	$\pi_2 \rightarrow \pi_3^*$	220	130
Benzene	$\pi_3 \rightarrow \pi_4^*$	280	100
Naphthalene	$\pi_5 \rightarrow \pi_6^*$	320	90
Anthracene	$\pi_7 \rightarrow \pi_8^*$	380	75

[a] Approximate longest wavelength of appreciable absorption (dependent upon solvent).

Thus, the lowest energy transition would be represented by excitation of a nonbonding electron to the lowest antibonding orbital (designated an $n \rightarrow \pi^*$ transition).

The orbital energy levels of formaldehyde are shown below (excluding the C–H bonding and antibonding orbitals). The lowest energy absorption appears at about 300 nm (95 kcal/mole) and has been assigned to an $n \rightarrow \pi^*$ transition. The next absorption does not occur until 180 nm (160 kcal/mole) and is the $\pi \rightarrow \pi^*$ transition very similar to that observed for ethylene. The expected $n \rightarrow \sigma^*$ absorption also appears in this range.

Formaldehyde molecular orbitals

The assignment of a particular absorption to either a $\pi \rightarrow \pi^*$ or an $n \rightarrow \pi^*$ transition is usually made on the basis of two factors. First, $n \rightarrow \pi^*$ absorptions are generally much lower in intensity than $\pi \rightarrow \pi^*$ absorptions. The equation describing absorption intensity includes the extent of overlap between the initial and final orbitals. For truly nonbonding orbitals that are perpendicular to the π system, this overlap is very small and the absorption intensity is weak. The extinction coefficient for the formaldehyde $n \rightarrow \pi^*$ absorption is only 18 liter mole^{-1} cm^{-1}, while for the $\pi \rightarrow \pi^*$ absorption it is 18,000 liter mole^{-1} cm^{-1}.

Second, $n \rightarrow \pi^*$ absorptions shift to shorter wavelengths when solvents of increasing polarity are used. The nonbonding electrons are normally strongly solvated, and more polar solvents will stabilize the n orbital to a greater degree. Thus, the energy separation for an $n \rightarrow \pi^*$ transition will increase in solvents of greater polarity. It is significant to note that the absorption process is too fast (about 10^{-15} sec) for any reorientation of solvent or other molecular motion to take place during the absorption (the Franck-Condon Principle). The $n \rightarrow \pi^*$ absorption of acetone in the gas phase (no solvation) occurs at 280 nm, while in aqueous solution, the absorption is at 265 nm.

Chemical Generation of Excited States

Besides the "traditional" method of generating electronically excited states by absorption of uv-visible radiation, there are a growing number of cases in which chemical reactions yield electronically excited molecules as products. This is commonly referred to as chemiluminescence when the product excitation manifests itself as luminescence, or more generally, these have been called chemiexcitation reactions.[4] A major requirement for a chemiexcitation process is that there be sufficient exothermicity to allow formation of the excited product. Furthermore, most of the unimolecular chemiexcitation reactions that have been discovered are concerted reactions for which orbital symmetry rules predict forbiddenness. Since symmetry forbiddenness means thermal forbiddenness relative to ground state formation, the formation of an electronically excited state would then be symmetry allowed (see page 75). One of the most ubiquitous systems that produces excited products is the 1,2-dioxetane system, which decomposes to give carbonyl products and chemiluminescence.[4] Tetramethyl-1,2-dioxetane decomposes just above room temperature to give one ground state acetone molecule and one. electronically excited (triplet) acetone molecule. This reaction could be described as a $(_\sigma 2_s + _\sigma 2_s)$ cycloreversion. A dioxetane is also thought to be the key intermediate in the firefly's luminescence:[5]

$$CH_3-\underset{\underset{CH_3}{|}}{\overset{\overset{O-O}{|\quad|}}{C}}-\underset{\underset{CH_3}{|}}{C}-CH_3 \quad \xrightarrow{50°} \quad \underset{CH_3 \quad CH_3}{\overset{O}{\overset{||}{C}}} \quad + \quad \left[\underset{CH_3 \quad CH_3}{\overset{O}{\overset{||}{C}}}\right]^{*3}$$

Bimolecular chemiexcitation occurs frequently upon recombination of oppositely-charged ions. In electrogenerated chemiluminescence, both cations and anions are generated electrochemically in the same vicinity, usually by an alternating current. The neutralization of these two species then provides sufficient energy for an electronically excited state (see pages 292–3 for a complete discussion of this process).

Oxidation

$$A \xrightarrow{-e^-} A^+$$

Reduction

$$A \xrightarrow{+e^-} A^-$$

Neutralization

$$A^+ + A^- \longrightarrow A + A^*$$

8.3 Types and Interconversions of Excited States

Primary Photophysical Processes

Once excitation energy is put into a molecule by the absorption of light, the return of the molecule to its ground state may take a variety of pathways. At this point, it is convenient to begin considering state diagrams, rather than orbital diagrams, to describe the possible states of the molecule.

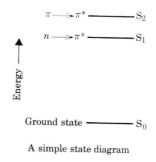

A simple state diagram

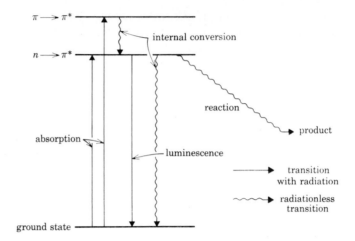

Fig. 8.1   *Possible transitions among singlet states.*

For a molecule with both $n \to \pi^*$ and $\pi \to \pi^*$ excited states, the initial absorption may be either an $n \to \pi^*$ or $\pi \to \pi^*$ transition, depending upon the wavelength of irradiation. Since each of these excited states will have superimposed vibrational and rotational energy levels, absorption spectra for all but quite small molecules show broad absorption peaks. If an excited state is formed with excess vibrational energy, however, it is usually dissipated to the surroundings within a few vibrations (of the order of 10^{-13} sec)—except in the case of small molecules in the gas phase at low pressure, where there are simply no surroundings to accept the excess energy so quickly (see p. 307–9 on pressure effects). In addition, conversion of a higher energy excited state to a lower energy excited state (such as $\pi \to \pi^*$ to $n \to \pi^*$ above) is usually also exceedingly rapid. Thus, the vibrationally equilibrated, lowest-energy excited state is the only excited state with relatively long lifetime and is generally considered the most likely starting place for any further processes that might take place. From this point, the excess electronic energy may be dissipated by emission of a photon (luminescence), by conversion of the energy to vibrational energy (internal conversion), or by chemical reaction (Figure 8.1).

Triplet States

The state diagram just constructed for primary photoprocesses (Figure 8.1) has not yet taken into account one very important aspect of excited states—triplet states. Once an electron is excited into an unoccupied orbital, the two unpaired electrons are no longer bound by the Pauli exclusion principle, which dictates that two electrons must have opposite spins while occupying the same orbital. The two unpaired electrons may have the electronic spins opposite (a singlet state) or parallel (a triplet state*). Generally, an electronic transition will initially produce a singlet state, since an absorption with simultaneous spin change is strongly forbidden. Subsequent to the absorption, however, the singlet excited state may be converted

---

* "Triplet" is derived from the three degenerate spin wavefunctions comprising the triplet state (spin quantum numbers -1, 0, $+1$). They are distinguishable only in the presence of a strong magnetic field. A singlet state has only one spin wavefunction (spin quantum number zero).

into a triplet state by an inversion of one of the electronic spins:

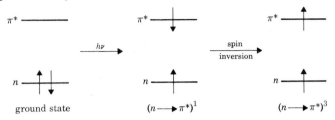

One of the most important properties of a triplet state is that its energy is lower than that of the corresponding singlet state. This has been attributed to the Pauli exclusion principle, which forbids the two electrons with parallel spins from occupying the same area. The triplet thereby minimizes electron–electron repulsions.

Any crossing from a singlet to a triplet state is called intersystem crossing and formally is a forbidden process. The extent of intersystem crossing is extremely variable from molecule to molecule, depending upon a large variety of factors. Some compounds, such as simple alkenes, show no evidence of intersystem crossing subsequent to their absorption of light. Other compounds, such as benzophenone and its derivatives, undergo intersystem crossing so rapidly and completely that the singlet state is too short-lived to be observed by standard techniques. The major factor that breaks down this spin forbiddenness is the coupling of spin and orbital angular momenta of electrons, which is maximized in the presence of atoms of high atomic number (the heavy atom effect). For example, the rate of intersystem crossing is calculated to increase from $1 \times 10^6 \, \text{sec}^{-1}$ in naphthalene to $3 \times 10^{10} \, \text{sec}^{-1}$ in 1-iodonaphthalene.[6] A similar effect can be observed with a heavy-atom solvent, such as methyl iodide. In the case of benzophenone, which does not have a heavy atom, the rapid intersystem crossing seems to be caused by the relatively small energy difference (5 kcal/mole) between the lowest $n \rightarrow \pi^*$ singlet state and the lowest $n \rightarrow \pi^*$ triplet state.

Thus, a complete state diagram for a molecule (often called a Jablonski diagram) should include intersystem crossings to triplet states and deactivation of the triplet by emission or by radiationless processes. Emission from a triplet state is called phosphorescence, while fluorescence refers to emission from a singlet state (Figure 8.2).

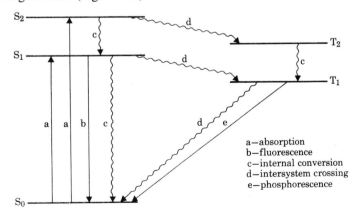

Fig. 8.2   *Jablonski diagram.*

Luminescence

For many compounds, the fluorescence spectrum bears a strong mirror-image relationship to the absorption spectrum. Both spectra have one wavelength in common; absorption and emission between the lowest vibrational level of the ground state and the lowest vibrational level of the singlet state occur at the same wavelength, called the (0,0) band. This will be the only point in common, since absorption virtually always takes place from a vibrationally equilibrated ground state and fluorescence virtually always takes place starting from the lowest vibrational level of the lowest singlet state. The absorption spectrum will include shorter wavelengths and the fluorescence spectrum will include longer wavelengths (Figure 8.3).

Most frequently, the (0,0) band in both absorption and emission spectra has a lesser intensity than other parts of the spectra. This is because the geometry of the excited state may not be the same as that of the ground state. Since absorption of a photon takes place too quickly for any motion of nuclei (the Franck-Condon principle), the excited state, which is the immediate product of absorption, will be in a nonequilibrium geometry and will contain excess vibrational energy. This can be viewed as a displacement of the minimum of the excited state potential well relative to the minimum of the ground state potential well. Biphenyl is an extreme case; the ground state is most stable with a dihedral angle of about 22° between the two phenyl groups whereas the excited state geometry is completely planar. Thus, the absorption maximum in biphenyl occurs at a much higher energy than the emission maximum (Figure 8.4).[7]

Besides the emission spectrum, the emission lifetime offers useful information about excited states. Experimentally, emission lifetimes are determined in a straightforward manner. The sample is irradiated with a very brief pulse of light and the luminescence is monitored by a photomultiplier tube connected to an oscilloscope to determine how long the emission lasts.

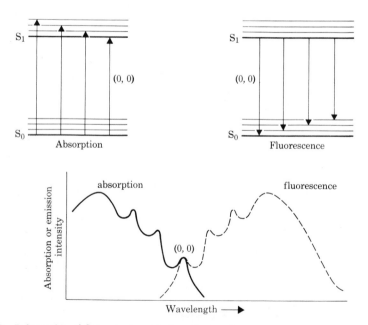

Fig. 8.3 *Relationship of fluorescence and absorption spectra.*

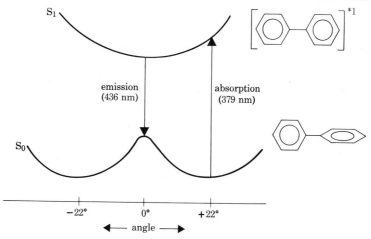

Fig. 8.4 *Potential energy surfaces for biphenyl.* [7]

For fluorescent lifetimes, which directly represent the lifetime of the singlet states from which the fluorescence arises, typical lifetimes range from about 10^{-10} to 10^{-6} sec (0.1 to 1000 nsec). Phosphorescent lifetimes are substantially longer, due to the forbiddenness of the transition from triplet to singlet ground state and range from about 10^{-3} to 10 sec. Normally, phosphorescence is only observed in rigid media such as crystals or frozen glasses. The lack of significant phosphorescence in solution is because of the reactivity of the triplet excited state, which normally will find some other way to dissipate its excess energy rather than its slow phosphorescence, provided it is free to diffuse about in solution.[8]

The Quantum Yield

The quantum yield is defined as the fraction of excitation that leads to a particular photoprocess, or

$$\Phi = \frac{\text{no. of molecules undergoing a particular process}}{\text{no. of photons absorbed}}$$

The sum of the quantum yields for all photoprocesses equals unity if the fate of all absorbed photons is accounted for. For example, a hypothetical molecule might show the following quantum yields: Φ(fluorescence) = 0.45, Φ(phosphorescence) = 0.50, Φ(isomerization) = 0.05. It is important to note that a quantum yield need not correspond to a chemical reaction yield. Thus, the preceding hypothetical molecule could be made to isomerize in a 100% chemical yield by continued irradiation, since all those excited states that do not isomerize return to the ground state via luminescence, where they may be recycled by continued irradiation. The only exception to this description would occur if the isomerization product were unstable to the reaction conditions or absorbed some of the exciting light and went on to do photochemistry of its own.

Quantum yields are determined by the relative rates of all competing processes that deactivate a given excited state. These rates determine the proportions of the various processes by their relative magnitudes. Thus, if an excited singlet state is deactivated by fluorescence ($k_f = 9 \times 10^7$ sec^{-1}), radiationless internal conversion ($k_{ic} = 1 \times 10^5$ sec^{-1}), intersystem crossing ($k_{isc} = 1 \times 10^8$ sec^{-1}), and a photochemical reaction such as isomerization ($k_x = 1 \times 10^7$ sec^{-1}), the various quantum yields from the singlet state are calculated as the ratio of the rate of the specific process over the sum of all competing processes (processes which deactivate the singlet state).

$$\Phi_j^S = \frac{k_j}{\sum_S k}$$

Quantum yields from a triplet state must additionally include Φ_{isc} in order to take into account the fraction of molecules that reach the triplet state per photon absorbed:

$$\Phi_j^T = \Phi_{isc} \frac{k_j}{\sum_T k}$$

A complete set of rate constants and calculated quantum yields for our hypothetical molecule is shown in Figure 8.5.

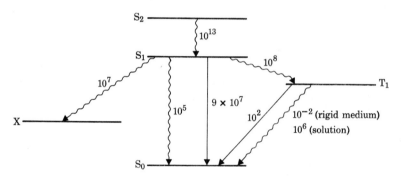

Fig. 8.5 *Typical rate constants and quantum yields for a hypothetical molecule (rates in sec^{-1}).*

$k_f = 9 \times 10^7$ sec^{-1}	$\Phi_f = 0.45$	Fluorescence
$k_x = 1 \times 10^7$ sec^{-1}	$\Phi_x = 0.05$	Isomerization
$k_{isc} = 1 \times 10^8$ sec^{-1}	$\Phi_{isc} = 0.50$	Intersystem crossing
$k_{ic} = 1 \times 10^5$ sec^{-1}	$\Phi_{ic} = 0.00$	Internal conversion
$\left.\begin{array}{l} k_p = 1 \times 10^2 \text{ sec}^{-1} \\ k_d = 1 \times 10^{-2} \text{ sec}^{-1} \end{array}\right\}$	$\Phi_p = 0.50$	Phosphorescence in a rigid medium
$k_d' = 1 \times 10^6$ sec^{-1}	$\Phi_p' = 0.00$	Phosphorescence in a solution

The previous diagram shows that the relative rates of radiationless deactivation of the triplet state vary substantially depending upon the medium. It is these rates that determine whether or not phosphorescence is observed. In a rigid medium, radiationless deactivation is so slow that phosphorescence is virtually the only means of returning the triplet to the ground state. In solution, radiationless deactivation is so rapid that it accounts for virtually all of the triplet decay.

As in most kinetic analyses, it is useful to refer to a half-life, $t_{1/2}$, or lifetime, τ. In photochemistry, it is more common to consider the lifetime of a state, which is the reciprocal of the total rates of disappearance of the state.

$$\tau = \frac{1}{\sum k} \qquad t_{1/2} = \frac{0.693}{\sum k}$$

In the hypothetical molecule described above, the following triplet and singlet lifetimes may be calculated: $\tau_T = 10^{-2}$ sec in rigid media, $\tau_T = 10^{-6}$ sec in solution, and $\tau_S = 5 \times 10^{-9}$ sec. Although these lifetimes are normally observed directly by monitoring emission lifetimes, it is important to note that these measured lifetimes do not relate to the rate of emission but to the total rate of deactivation of the emitting state. The fluorescence lifetime in combination with a quantum yield for fluorescence, however, will allow the absolute fluorescence rate to be calculated, since

$$\Phi_f = \frac{k_f}{\sum\limits_s k} = k_f \cdot \tau_S$$

Electronic Energy Transfer

Electronic energy transfer may be defined as a process in which electronic excitation is transferred from an excited donor molecule to an acceptor molecule:

$$D^* + A \rightarrow D + A^*$$

where the asterisk denotes electronic excitation, either singlet or triplet.

The classic demonstration of energy transfer is the spectroscopic evidence obtained from benzophenone and naphthalene mixtures.[9] Benzophenone and naphthalene each display characteristic phosphorescence spectra when irradiated in a rigid glass of ether–ethanol at $-190°C$. Upon irradiation of a mixture of naphthalene and benzophenone with 366 nm radiation, where only benzophenone absorbs, the phosphorescence spectrum of naphthalene was clearly observed, even though naphthalene could not have absorbed any of the excitation. The excitation initially absorbed by the benzophenone must have been transferred to the naphthalene in some manner. The triplet energies of benzophenone and naphthalene may be obtained from the (0,0) bands of the phosphorescence spectra and the singlet energies are available from the absorption spectra.

Benzophenone	Naphthalene
$E_S = 74$ kcal/mole	$E_S = 91$ kcal/mole
$E_T = 69$ kcal/mole	$E_T = 61$ kcal/mole

From these energy data, it is clear that the energy transfer takes place as triplet energy transfer, which is exothermic by 8 kcal/mole.

$$(\phi_2CO)^{*3} + C_{10}H_8 \rightarrow \phi_2CO + (C_{10}H_8)^{*3} \quad \Delta H = -8 \text{ kcal/mole}$$

By adjusting the concentration of the naphthalene in the glass, it was determined that the energy transfer requires relatively close approach of donor and acceptor. The efficiency of energy transfer drops abruptly when the average separation of benzophenone and naphthalene exceeds 10 Å.[9] This close approach is predicted by a theoretical treatment of triplet energy transfer.[10] Since each of the individual transitions involved is spin-forbidden $(D^{*3} \rightarrow D$ and $A \rightarrow A^{*3})$, an electron exchange mechanism must be invoked. In its simplest form, it may be visualized as an exchange of electrons between donor and acceptor, thereby changing the spin state of each. As long as the total spin state of the system (donor and acceptor) remains constant, energy transfer by an electron exchange mechanism is allowed (the Wigner spin rule):

$$(D^{*3})^{\uparrow\uparrow} + A^{\uparrow\downarrow} \longrightarrow D^{\uparrow\downarrow} + (A^{*3})^{\uparrow\uparrow}$$

Transfer of singlet electronic energy has been shown to take place by at least three different mechanisms.[10] The simplest mechanism is fluorescence of the donor singlet state and absorption of the radiation by the acceptor to form a new singlet excited state. A second mechanism operates at relatively close distances and is equivalent to the electron exchange mechanism for triplet energy transfer. Finally, there exists a relatively long-distance mechanism (up to 50 Å), called Förster resonance excitation transfer. The theoretical treatment of this mechanism considers the donor transition and acceptor transition as coupled dipole–dipole interactions.

PHOTOSENSITIZATION

The process of electronic energy transfer is exceptionally useful, since in many cases the energy transfer method is the only way to form certain excited states. For example, most aliphatic olefins and dienes do not undergo intersystem crossing from their excited singlet states and their triplet states would be inaccessible if it were not for energy transfer. Irradiation of benzophenone in the presence of cis-1,3-pentadiene leads to rapid formation of an equilibrium mixture of cis and trans isomers.[1] Direct irradiation of the diene would require light of much shorter wavelengths, and does not lead to efficient cis–trans isomerization. Since only the benzophenone absorbs the incident 366 nm irradiation, energy transfer from benzophenone to the diene must activate the diene for the isomerization:

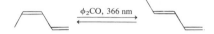

The foregoing reaction is called a photosensitized reaction with benzophenone as the photosensitizer. Benzophenone is a very good triplet photosensitizer because it has an exceptionally high intersystem crossing yield ($\Phi_{isc} = 1.00$) and its triplet energy is sufficiently high to allow it to photosensitize a wide variety of other molecules. A complete range of photosensitizers and their triplet energy levels are shown in Table 8.2.

Table 8.2 *Triplet Energy Levels of Some Common Photosensitizers*

Photosensitizer	Triplet Energy (kcal/mole)
Benzene	84.0
p-Xylene	80.1
Propiophenone	74.6
Acetophenone	73.6
Benzophenone	68.5
Triphenylene	66.6
Phenanthrene	62.2
Naphthalene	60.9
β-Acetonaphthone	59.3
α-Acetonaphthone	56.4
Biacetyl	55.6
Benzil	53.0
Pyrene	48.2
Anthracene	42.5
Perylene	36.0
Tetracene	29.3
Oxygen	22.5 (lowest singlet energy)

RATES OF ENERGY TRANSFER

The rates of photochemical processes open up new dimensions for chemists accustomed to kinetics measurable in minutes or hours. Unimolecular rate processes, such as internal conversion or intersystem crossing, may exhibit rates up to 10^{13} sec^{-1} (a rather arbitrary limit selected because it is the rate of a typical molecular vibration). Some bimolecular rate processes occur upon every collision and, therefore, are limited only by the rate at which the two species can diffuse together. Rates approaching the diffusion-controlled limit are commonplace in photochemistry. For typical solvents, maximum rates of about $10^{10}\, M^{-1}$ sec^{-1} are observed. The magnitudes of diffusion rates are strongly dependent on solvent viscosity and temperature and have been observed to follow the Debye equation:[11]

$$k = \frac{8RT}{3000\eta}$$

where k is the bimolecular rate constant in M^{-1} sec^{-1}, η is the solvent viscosity in poise, R is the gas constant in ergs mole^{-1} deg^{-1}, and T is absolute temperature.

The measurement of rates as rapid as these can be handled by the method of flash photolysis. Development of flash photolysis by R. G. W. Norrish (of the University of Cambridge) and G. Porter (of the Royal Institution, London) was recognized by the 1967 Nobel Prize in Chemistry.[12] Excited states are produced by a brief flash of light and their disappearance is monitored by direct observation of decay of the absorption spectrum of the excited states. The kinds of reactions observable by this method are limited by the duration of the excitation flash, since the formation of excited states continues until the flash terminates. "Conventional" flash photolysis works in the range of microseconds, but the use of pulsed lasers has brought the scale down to nanoseconds and even picoseconds (10^{-12} sec).[13] For a typical triplet state in solution, which decays either by radiationless deactivation or by energy transfer to an added quencher Q, the following rate equations apply:

$$S^{*3} \xrightarrow{k_0} S$$

$$S^{*3} + Q \xrightarrow{k_Q} S + Q^{*3}$$

$$-\frac{d[S^{*3}]}{dt} = k_0[S^{*3}] + k_Q[S^{*3}][Q]$$

$$\ln \frac{[S^{*3}]_0}{[S^{*3}]} = kt \qquad \text{(First-order decay)}$$

where $k = k_0 + k_Q[Q]$

A plot of the observed first-order decay rate constant k as a function of quencher concentration, therefore, yields the energy transfer rate k_Q as the slope.

Study of a large variety of energy transfer processes with different donor and acceptor molecules has led to the conclusion that triplet energy transfer approaches the diffusion-controlled rate whenever the energy transfer reaction is exothermic by 3 kcal/mole or more. Endothermic energy transfer is much less efficient and requires thermal activation. Thus, for a given quencher molecule and a variety of sensitizers of different triplet energy, the relationship in Figure 8.6 is typical.[14,15]

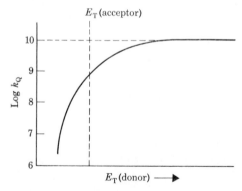

Fig. 8.6 *Energy transfer rates as a function of triplet energy (see also Figure 8.10).*

The importance of energy transfer in photochemical processes may be illustrated by photosynthesis in green plants. The vast majority of leaf pigments that absorb visible light act simply to harvest the excitation energy and transfer it to an active site where the photosynthetic reactions begin. Even among the chlorophyll molecules, only one in 300 is capable of performing the initial photoreduction, with the others simply passing on the excitation energy to the active molecule.

OTHER QUENCHING MECHANISMS

There is a rapidly growing number of examples of quenching of excited states that do not proceed by energy transfer. The best-studied cases involve charge-transfer quenching,[16] as in the case of the quenching of naphthalene fluorescence by amines:

$$(C_{10}H_8)^{*1} + R\ddot{N}H_2 \rightarrow (C_{10}H_8^{\bar{\cdot}} \cdots\cdots R\overset{+}{N}H_2) \rightarrow C_{10}H_8 + R\ddot{N}H_2$$

Another example is the unusual quenching of naphthalene fluorescence by quadricyclane, which is isomerized to norbornadiene.[17] It is not yet known whether charge-transfer or other interactions lead to the quenching and the isomerization:

$$(C_{10}H_8)^{*1} \quad + \quad \text{[image]} \quad \longrightarrow \quad C_{10}H_8 \quad + \quad \text{[image]}$$

8.4 Reactions of Excited States

The study of photochemical reactions will be broken down according to typical functional group reactions. This is especially appropriate for photochemical reactions, since the electronic energy can usually be considered to be localized at the chromophore portion of the molecule. Alkanes and other molecules with only σ bonds will not be specifically considered. It will suffice to say that a $\sigma \rightarrow \sigma^*$ transition is usually of very high energy and most often results in cleavage of the σ bond. For example, all of the halogens undergo photodissociation upon irradiation with visible light:[18]

$$Cl_2 \xrightarrow{h\nu} 2Cl\cdot$$

8.5 Olefin Photoreactions

Cis–Trans Isomerizations

The lowest energy transition in alkenes is a $\pi \rightarrow \pi^*$ transition that removes the overall bonding character of the π bond. The most stable excited state structure for both the singlet and the triplet is a perpendicular geometry.

This structure minimizes electron–electron repulsions between the p orbitals:

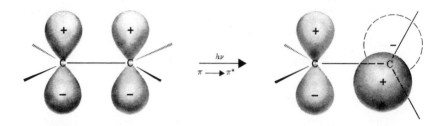

The photochemical reaction that is facilitated by such a structure is obviously *cis–trans* isomerization. The *cis–trans* isomerization of olefins is efficiently performed with triplet photosensitization, or less efficiently by direct irradiation. The different reactivities of the triplet state and the singlet state are apparently caused by the differences in their energy content and lifetimes. The short-lived singlet state of alkenes has a greater tendency to return directly to ground state without isomerization, while the longer-lived triplet state probably attains its equilibrium perpendicular geometry before returning to ground state. In addition, the singlet state has substantially more energy that may make other reactions feasible. The potential energy diagram for the electronic states of ethylene is shown in Figure 8.7.[19] There remains some question as to whether the thermal isomerization of olefins involves the triplet state as an intermediate, or whether it proceeds entirely on the singlet surface.

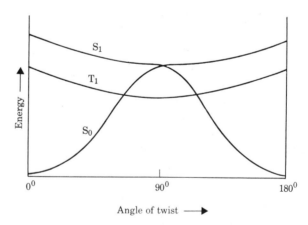

Fig. 8.7   *Potential energy surfaces for ethylene.*[19]

Dimerizations

The direct irradiation of olefins at high concentrations gives predominantly dimerization to cyclobutane products. This $(2_s + 2_s)$ cycloaddition is photochemically allowed by orbital symmetry rules, and the dimerization has been shown to proceed with stereospecificity:[20]

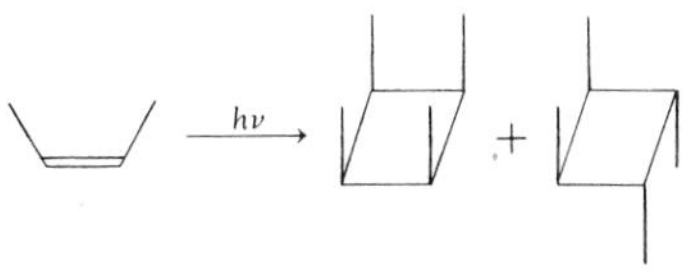

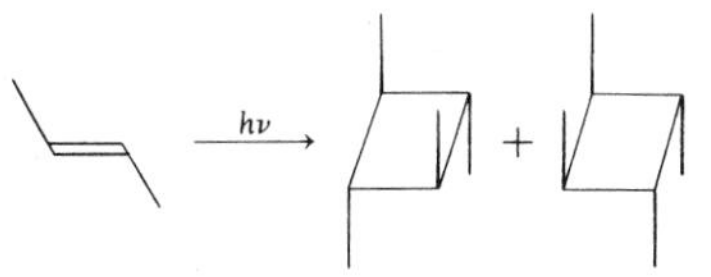

Ionic Additions to Alkenes

Alkenes, which normally are susceptible to electrophilic attack in their ground states, can be more reactive in the excited states. In acidic methanol, 1-methylcyclohexene undergoes acid-catalyzed addition of methanol upon either direct or sensitized irradiation.[21] This ionic addition is a specific photoreaction of medium-ring cycloalkenes and is not observed for acyclic alkenes:

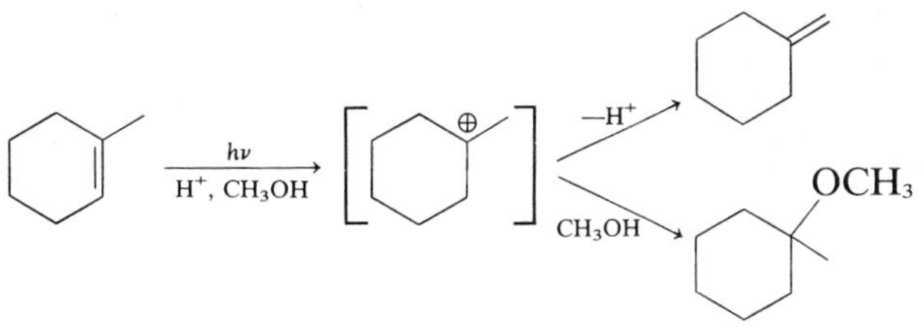

The Stilbenes

PHOTOSTATIONARY STATES

The *cis–trans* photoisomerization of the stilbenes has been studied in great detail; the depth of this work has led to a variety of interesting observations. Sensitized isomerization of either of the stilbene isomers leads ultimately to an equilibrium mixture of *cis*- and *trans*-stilbene.[1] An equilibrium situation attained photochemically rather than thermally is called a *photostationary state*. The attainment of such an equilibrium may be explained by considering that both *cis*- and *trans*-stilbene may be sensitized to a common triplet state, through which they equilibrate:

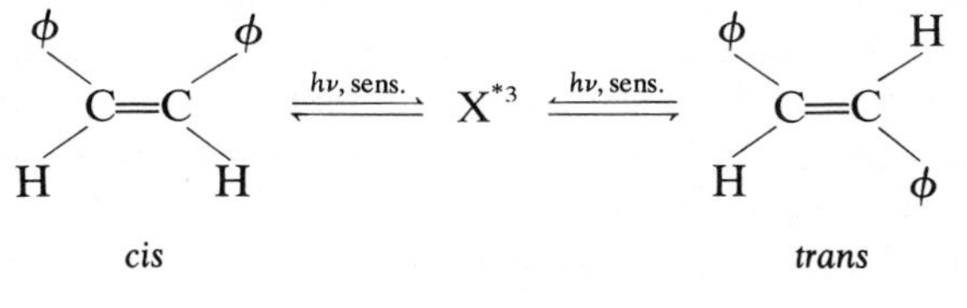

Using this assumption, the following kinetic arguments apply:

$$S^{*3} + cis \xrightarrow{k_{QC}} S + X^{*3}$$

$$S^{*3} + trans \xrightarrow{k_{QT}} S + X^{*3}$$

$$X^{*3} \xrightarrow{k_{DC}} cis$$

$$X^{*3} \xrightarrow{k_{DT}} trans$$

At the photostationary state, the concentrations of the *cis* and *trans* isomers are no longer changing, so their rates of formation and disappearance must be equal:

$$k_{QC}[S^{*3}][cis]_0 = k_{DC}[X^{*3}]$$
$$k_{QT}[S^{*3}][trans]_0 = k_{DT}[X^{*3}]$$

Dividing and rearranging, we obtain

$$\frac{[cis]_0}{[trans]_0} = \frac{k_{DC}}{k_{DT}} \cdot \frac{k_{QT}}{k_{QC}}$$

The ratio of *cis*- and *trans*-stilbene present at the photostationary state depends only upon the ratio of their formation rates from X^{*3} and their ratio of sensitization rate constants. In the situation in which very high energy triplet sensitizers are used, both *cis*- and *trans*-stilbene are sensitized equally efficiently at the diffusion-controlled limit:

$$k_{QC} = k_{QT}$$

$$\frac{[cis]_0}{[trans]_0} = \frac{k_{DC}}{k_{DT}}$$

In this high-energy region, the observed photostationary state reflects only the partitioning ratio of the stilbene triplet state. Experimentally, this ratio is found to be 1.4—that is, about 60% *cis* and 40% *trans* at the photostationary state. This high-energy region is found to occur with sensitizers of triplet energy above 60 kcal/mole. The observed photostationary states can be correlated rather nicely with the observed triplet energy levels for *cis*-stilbene ($E_T = 57$ kcal/mole) and *trans*-stilbene ($E_T = 48$ kcal/mole). With sensitizers of energy 60 or more kcal/mole, energy transfer to both stilbene isomers is expected to be at maximum efficiency. With sensitizers of lesser energy, the energy transfer to *cis*-stilbene should drop off quickly, while the energy transfer to *trans*-stilbene remains efficient. This will drive the photo-equilibrium farther to the *cis* side, since the *cis* isomer is no longer being converted to triplet so rapidly. The photostationary state observed for sensitizers with triplet energy values between 50 and 60 kcal/mole becomes increasingly rich in *cis*-stilbene. Finally, with sensitizers of triplet energy below about 50 kcal/mole, neither isomer is sensitized efficiently. The photostationary state ratios can be graphed as a function of triplet energy to give the following plot (Figure 8.8).[22] These ratios may also be successfully calculated from measured energy transfer rates to *cis*- and *trans*-stilbene.[22]

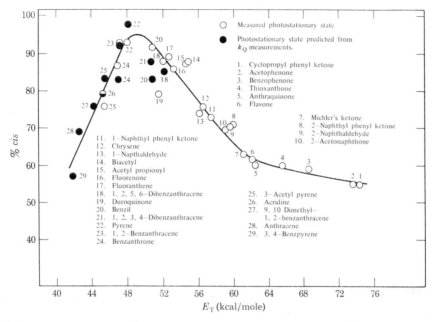

Fig. 8.8   *Photostationary states in the sensitized isomerization of the stilbenes.*[22] *$E_T > 60$, sensitization of cis and trans equal; $48 < E_T < 60$, sensitization of trans favored by energetics; and $E_T < 48$, sensitization of neither cis nor trans efficient.*

NONVERTICAL ENERGY TRANSFER

For a long time the triplet-sensitized isomerization of the stilbenes was interpreted in terms of a potential energy diagram that included two distinct triplet states—a planar *trans* triplet, which can take full advantage of resonance, as well as a "phantom" twisted triplet, similar to the perpendicular triplet state of simple alkenes.[23] The most recent evidence now indicates only one minimum in the triplet potential energy surface, at or near the *trans* configuration (Figure 8.9).[24]

A potential energy diagram of this type has been used to explain an interesting phenomenon referred to as "nonvertical" excitation.[23,24] The observation was made that *cis*-stilbene is a reasonably effective acceptor of triplet sensitization from sensitizers with energy insufficient to raise *cis*-stilbene to its "spectroscopic" triplet state of 57 kcal/mole. A flash photolysis study of absolute energy transfer rates showed that there was indeed a

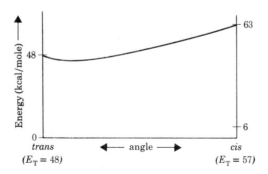

Fig. 8.9   *Potential energy surface for triplet stilbene.*[24]

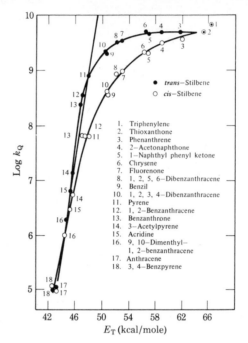

Fig. 8.10   *Triplet energy transfer rates to the stilbenes.*[22] *(The straight line indicates the calculated rate dropoff for endothermic energy transfer to trans-stilbene.)*

significant difference between *trans*-stilbene, which behaved as a normal acceptor, and *cis*-stilbene, which acted as an unusually good acceptor for low energy sensitizers (Figure 8.10).[22]

This phenomenon has been interpreted in terms of the stilbene triplet potential energy diagram as a "nonvertical" transition. In other words, sensitizers with energy less than 57 kcal/mole may still excite *cis*-stilbene onto the triplet energy surface, if the *cis*-stilbene is excited nonvertically—that is, with a change in geometry (Figure 8.11).

This explanation would seem to be in violation of the Franck–Condon principle (see p. 328) that had always restricted photochemists to writing only vertical excitation arrows. However, the Franck–Condon principle is intended to apply to processes that take exceptionally short time periods, such as the absorption of a photon. Under these circumstances, the requirement that there be no molecular motion is reasonable. However, in an

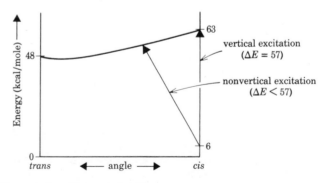

Fig. 8.11   *Nonvertical excitation of cis-stilbene.*

energy transfer process, particularly in solution, the donor and acceptor molecules remain in contact for a relatively substantial period of time, probably 10 to 100 collisions per encounter. Under these circumstances, it would be reasonable to expect that the molecules may be able to align themselves in an orientation which facilitates reaction. Because nonvertical energy transfer has a substantial steric requirement, the efficiency is not as high as it is with vertical energy transfer.

8.6   Diene Photoreactions

Cis–Trans Isomerizations

Like alkenes, dienes also undergo *cis-trans* photoisomerization. As with the alkenes, the isomerization is most efficient with triplet photosensitization, but dienes will also undergo some isomerization upon direct irradiation. The major side products observed during direct irradiation of conjugated and nonconjugated dienes are electrocyclic reactions that may be viewed as internal dimerizations:

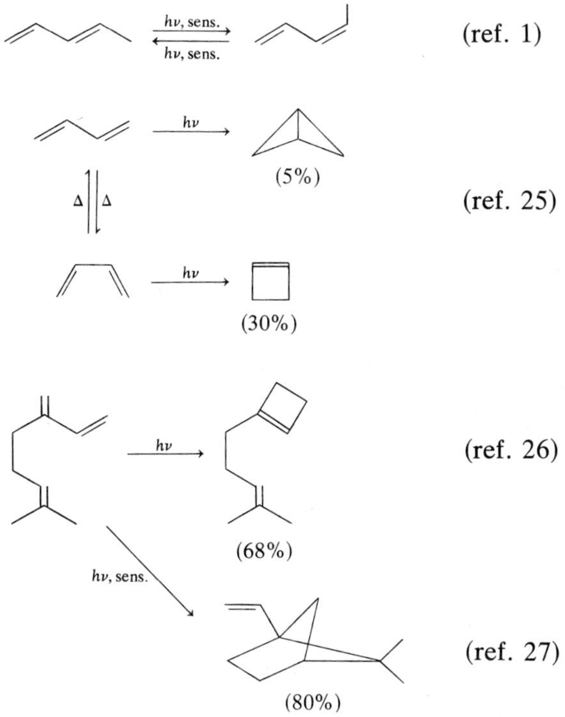

(ref. 1)

(5%)   (ref. 25)

(30%)

(68%)   (ref. 26)

(80%)   (ref. 27)

The *cis–trans* photoisomerization of dienes that contain two different sites of geometric isomerism leads to the interesting question of whether a single photon can lead to isomerization of both sites of isomerism or just one of them. This question has been resolved for the conjugated diene 2,4-hexadiene. In the case of triplet photosensitization, analysis of the quantum yields indicated that all of the geometric isomers were interconvertible, indicating that there could be isomerization of both ends with a single sensitization:[28,29]

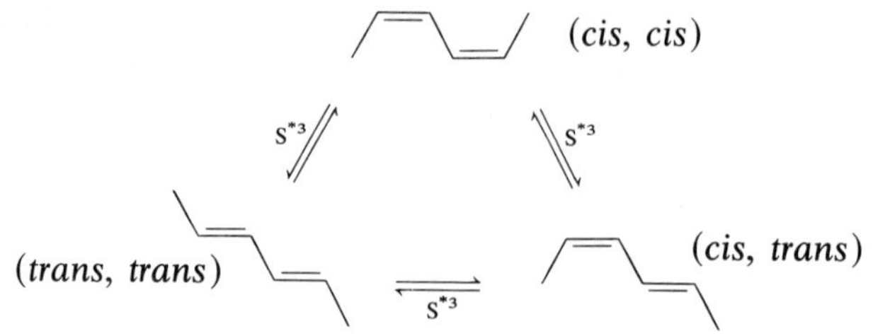

This phenomenon can be explained by either of two triplet state config-urations: a biradical structure in which both termini are perpendicular to a central π bond or an equilibrating system in which one p orbital perpen-dicular to an allyl system alternates from end to end:

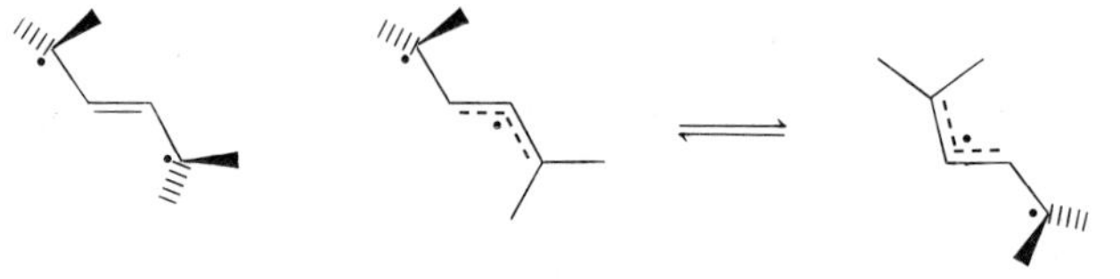

"Biradical" Equilibrating "allyl-methylene"

Interestingly, *direct* irradiation of the 2,4-hexadienes leads to a different photostationary state mixture and other isomers.[30] This points out that the singlet states of the 2,4-hexadienes do not cross to triplets and emphasizes the different reactivities and, presumably, different structures for the singlet and triplet states of these dienes.

Dimerizations

The photosensitized cyclodimerization of 1,3-butadiene yields $(2+2)$ dimers, *cis*- and *trans*-1,2-divinylcyclobutane, and a $(4+2)$ dimer,* 4-vinyl-cyclohexene:[31,32]

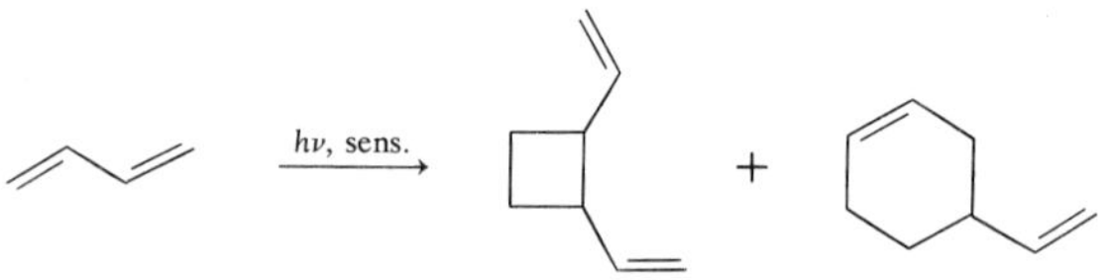

$(2+2)$ $(4+2)$

As was observed with the stilbenes, there was a substantial effect upon the products depending on the sensitizer used. High-energy sensitizers above 60 kcal/mole always gave products corresponding to 96% $(2+2)$ products and 4% of the $(4+2)$ product. At lower sensitizer energies, however, the proportion of vinylcyclohexene product rose sharply to a maximum of 60% for triplet sensitizers of energy about 50 kcal/mole.

As in the case of the stilbenes, this effect is caused by two isomeric acceptors that are competing for the sensitization. In this case, there may be energy transfer to either the s-*trans* conformer or the s-*cis* conformer of the butadiene. The triplet energy level of the s-*trans* can be placed at

---

* We will use the $(2+2)$ and $(4+2)$ designations simply to classify these product types, even though they are not concerted cycloadditions.

60 kcal/mole and the triplet energy of an *s-cis* diene, using 1,3-cyclohexadiene as a model, is about 53 kcal/mole. Sensitization of these two conformers gives two distinct excited states, which do not equilibrate during their lifetime and which lead to different products.

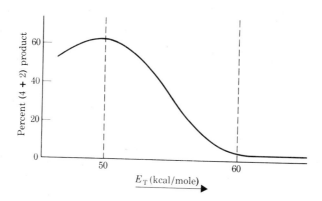

Mechanism of the photosensitized dimerization of butadiene

At room temperature, butadiene is known to exist about 97% in its *s-trans* form. Therefore, high-energy sensitizers that would sensitize either conformer equally well will predominantly transfer triplet energy to the *s-trans* conformer simply because of its abundance. The 4% of the $(4+2)$ product observed may come from s-cis triplets plus s-*trans* diene or from s-*trans* triplets plus s-cis diene. Note that one cis component is necessary for the $(4+2)$ product formation. With sensitizers of energy less than 60 kcal/mole, the energy transfer to s-*trans* becomes decreasingly efficient, while energy transfer to s-cis remains completely efficient. Thus, these sensitizers become selective and predominantly sensitize s-cis butadiene for energetic reasons. Sensitizers below about 50 kcal/mole in triplet energy sensitize neither conformer well (Figure 8.12).

Fig. 8.12 *Product ratio from the photosensitized dimerization of butadiene.*

Di-π-methane Rearrangements

A general photoreaction of 1,4-dienes is the so-called di-π-methane rearrangement. The reaction occurs with great generality upon photolysis of molecules having two π bonds attached to a single saturated carbon (hence the name di-π-methane).[33]

Di–π–methane rearrangements

Note that the general form of the di-π-methane rearrangement suggests many mechanistic and stereochemical possibilities. If there are two different π systems on the central carbon, which will migrate and which will become part of the cyclopropane ring? Furthermore, how is the stereochemistry about each of the five atoms affected during the rearrangement? And finally, does the rearrangement occur by a singlet or triplet excited state?

Detailed product studies with a large number of substrates have answered the first two questions. With two different π systems, there is a strong preference for the less conjugated π system to migrate to the more conjugated π system:

The stereochemistry about the migrating π bond has been shown to be retained:

The stereochemistry of formation of the new cyclopropane ring is quite complex and is the subject of continuing study. For simple systems under direct photolysis, the stereochemistry may be described in terms of a disrotatory C_3—C_5 ring closure, specifically on the opposite side of the ring from the migrating group, which undergoes a 1,2-sigmatropic shift.

Stereochemistry of the singlet di–π–methane rearrangement

All the foregoing data pertain to the di-π-methane rearrangement under direct photolysis conditions. Under conditions of triplet photosensitization, different results are obtained indicating that the direct photolysis proceeds through the singlet state. Occasionally, triplet photosensitization leads to no products at all, and in some instances a di-π-methane rearrangement of differing stereochemistry results.[33]

8.7 Carbonyl Photoreactions

Excited States

For most carbonyl compounds, the lowest singlet and triplet excited states are $n \rightarrow \pi^*$ in nature. An $n \rightarrow \pi^*$ excited state involves decreased electron density at the oxygen, since an oxygen n electron has been removed and distributed in the π^* orbital.

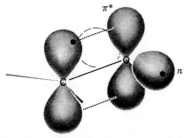

Orbital picture of a carbonyl $n \rightarrow \pi^*$ excited state

The carbonyl oxygen in an $n \to \pi^*$ excited state behaves very much like an alkoxy radical, capable of ready hydrogen abstraction from a suitable hydrogen donor. A comparison of abstraction of different kinds of hydrogen atoms by *tert*-butoxy radical and by benzophenone triplet indicates that these two species have similar selectivities.[34] This suggests that the picture of an $n \to \pi^*$ state as an alkoxy free radical is appropriate.

Oxetane Formation

One reaction that might be expected of an electrophilic species would be an addition to an alkene π bond. Carbonyl $n \to \pi^*$ excited states do undergo a (2+2) cycloaddition to alkenes (nonconcerted), so long as the alkenes do not quench the carbonyl excited state by energy transfer. For simple alkenes, energy transfer is usually inefficient and the formation of an oxetane is a facile reaction. The preferred orientation of addition can be predicted on the basis of the formation of the most stable biradical intermediate:[35]

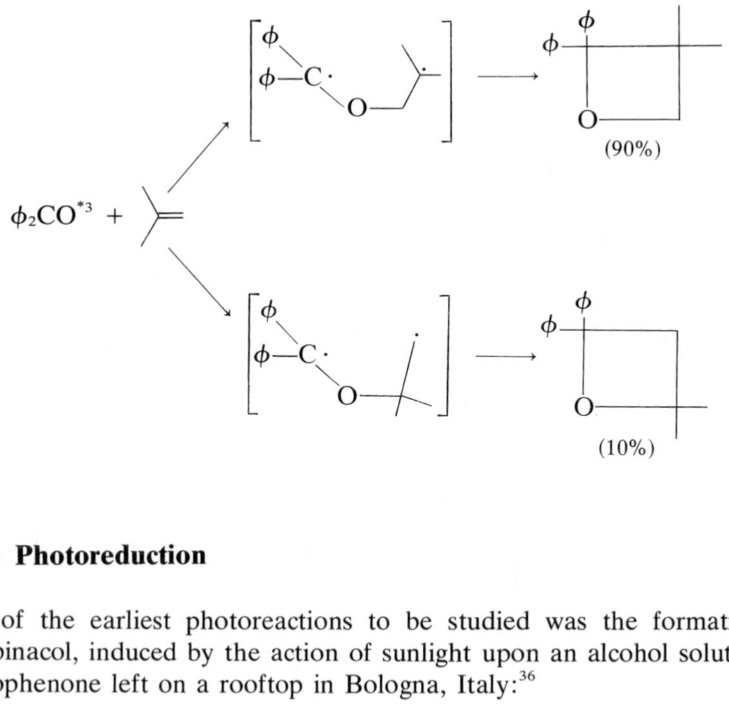

(90%)

(10%)

Benzophenone Photoreduction

One of the earliest photoreactions to be studied was the formation of benzpinacol, induced by the action of sunlight upon an alcohol solution of benzophenone left on a rooftop in Bologna, Italy:[36]

This reaction has now become one of the most thoroughly studied of all photoreactions. The reducing agent may be virtually any molecule with an abstractable hydrogen atom, such as alkanes, alkyl aromatics, and alcohols with α-hydrogens. With benzhydrol as the reducing agent, the following mechanism has been demonstrated.[37,38]

$$B \longrightarrow B^{*1} \longrightarrow B^{*3} \xrightarrow{\text{BH}_2} 2BH\cdot \longrightarrow BHBH$$

$(B = \phi_2CO, BH_2 = \phi_2CHOH, BH\cdot = \phi_2\dot{C}OH, BHBH = \phi_2C(OH)C(OH)\phi_2)$

Reaction	Rate
$B \to B^{*1}$	I
$B^{*1} \to B$	$k_1[B^{*1}]$
$B^{*1} \to B^{*3}$	$k_2[B^{*1}]$
$B^{*3} \to B$	$k_3[B^{*3}]$
$B^{*3} + BH_2 \to 2BH\cdot$	$k_4[B^{*3}][BH_2]$
$2BH\cdot \to BHBH$	$k_5[BH\cdot]^2$

The above kinetic analysis describes all of the important rate processes that take place during the photoreaction and leads to an expression for the quantum yield of the photoreduction. The quantum yield of a photoprocess that proceeds through several stages will be the product of the quantum yields for attaining each of the intermediate stages. That is, the quantum yield for benzpinacol formation will be the fraction of excited molecules that become triplets times the fraction of triplets that abstract hydrogen to become ketyl radicals times the fraction of ketyl radicals that dimerize to become benzpinacol:

$$\Phi(B \to \tfrac{1}{2}BHBH) = \Phi(B^{*1} \to B^{*3}) \times \Phi(B^{*3} \to BH\cdot) \times \Phi(BH\cdot \to \tfrac{1}{2}BHBH)$$

$$\Phi = \left(\frac{k_2[B^{*1}]}{k_1[B^{*1}] + k_2[B^{*1}]}\right) \cdot \left(\frac{k_4[B^{*3}][BH_2]}{k_3[B^{*3}] + k_4[B^{*3}][BH_2]}\right) \cdot \left(\frac{k_5[BH\cdot]^2}{k_5[BH\cdot]^2}\right)$$

$\Phi(B^{*1} \to B^{*3}) = \Phi_{\text{isc}} = 1.00$ (determined experimentally)

$\Phi(BH \to \tfrac{1}{2}BHBH) = 1.00$ (assume no competing pathways)

$$\Phi = \frac{k_4[BH_2]}{k_3 + k_4[BH_2]}$$

$$\frac{1}{\Phi} = 1 + \frac{k_3}{k_4[BH_2]} = 1 + \frac{1}{\tau_0 k_4[BH_2]}$$

$$\text{where } \tau_0 = \frac{1}{k_3} \text{ (natural triplet lifetime)}$$

This kinetic analysis (Figure 8.13) yields only $\tau_0 \cdot k_4$, the product of the natural triplet lifetime and the abstraction rate constant. By adding an additional step of known rate constant, these two constants may be resolved. Addition of naphthalene as a triplet quencher includes an energy transfer step for deactivation of triplet benzophenone. This rate constant may be measured by flash photolysis techniques or may be assumed to be the diffusion-controlled rate constant, since the triplet energy transfer from ben-zophenone to naphthalene is exothermic by 8 kcal/mole. This quenching

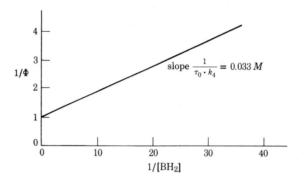

Fig. 8.13 *Quantum yield for benzophenone photoreduction by benzhydrol.*

step establishes a competition between naphthalene and benzhydrol for the benzophenone triplets and leads to the following revised kinetic analysis:

$$B^{*3} + N \rightarrow B + N^{*3} \qquad \text{Rate} = k_Q[B^{*3}][N]$$

At $[N] = 0$,

$$\Phi_0 = \frac{k_4[BH_2]}{k_3 + k_4[BH_2]}$$

At variable $[N]$, constant $[BH_2]_0$:

$$\Phi = \frac{k_4[BH_2]_0}{k_3 + k_4[BH_2]_0 + k_Q[N]}$$

$$\frac{\Phi_0}{\Phi} = 1 + \frac{k_Q[N]}{k_3 + k_4[BH_2]_0}$$

$$\tau = \frac{1}{k_3 + k_4[BH_2]_0} \text{ (triplet lifetime with } [BH_2]_0)$$

$$\frac{\Phi_0}{\Phi} = 1 + k_Q\tau[N]$$

This final expression is the common Stern–Volmer equation that quantitatively describes the quenching of a photoreaction by an added quencher. The plot of Φ_0/Φ should be linear with $[N]$, with slope $k_Q\tau$ (Figure 8.14). With a quencher having a known k_Q, τ may be determined. This provides a

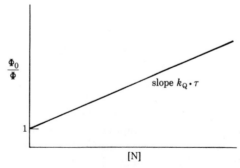

Fig. 8.14 *A typical Stern–Volmer quenching plot.*

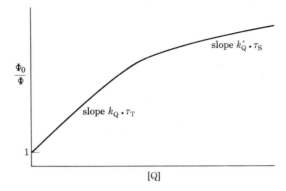

Fig. 8.15 *Curved Stern–Volmer plot indicating two quenchable species.*

second equation relating k_3 and k_4, so each may be determined. In the particular case of benzophenone triplet abstraction from benzhydrol, $k_4 \sim 10^7 M^{-1} sec^{-1}$.[38]

Curvature of a Stern–Volmer plot, occasionally observed at very high concentrations of quencher, indicates that another species is capable of being quenched by high quencher concentrations. Most often, this is the singlet state, and if a new linear region were observed, its slope should be $k'_Q \cdot \tau_S$ reflecting the quenching constant and lifetime of the singlet state (Figure 8.15).

Synthetically, the photoreduction of benzophenone in 2-propanol provides an excellent yield of benzpinacol, which precipitates during the irradiation. The quantum yield of benzophenone disappearance increases as the benzophenone concentration increases, and at very high concentrations approaches two.[39] The disappearance of two benzophenone molecules per photon has been explained by the following mechanism, in which the first (excited) benzophenone abstracts hydrogen from the alcohol and the second (ground state) benzophenone accepts the other hydrogen from the solvent ketyl radical:

$$\phi_2CO \xrightarrow{h\nu} (\phi_2CO)^{*1} \longrightarrow (\phi_2CO)^{*3}$$

$$(\phi_2CO)^{*3} + Me_2CHOH \longrightarrow \phi_2\dot{C}OH + Me_2\dot{C}OH$$

$$Me_2\dot{C}OH + \phi_2CO \longrightarrow Me_2CO + \phi_2\dot{C}OH$$

$$2\phi_2\dot{C}OH \longrightarrow \phi\underset{\underset{\phi}{|}}{\overset{\overset{OH}{|}}{C}}\underset{\underset{\phi}{|}}{\overset{\overset{OH}{|}}{C}}\phi$$

The key step that increases the quantum yield above unity is the hydrogen transfer step that becomes increasingly important at higher benzophenone concentrations. At lower benzophenone concentrations, direct evidence for the solvent ketyl radical appears in the products; that is, the mixed pinacol is also observed in small amounts.[40]

Norrish Type II Cleavage

A special case of photoreduction is the intramolecular abstraction of hydrogen by an $n \to \pi^*$ excited state. This abstraction is typically from a γ-position, since this provides a favorable six-membered ring transition state. The resulting 1,4-biradical may then undergo cyclization to a cyclobutanol product or undergo cleavage to olefin plus enol. This cleavage is called the Norrish type II reaction[41] (see p. 356 for the Norrish type I reaction).

Carbonyl Compounds with Lowest $\pi \to \pi^*$ States

The two main photoreactions typical of carbonyl compounds, photoreduction and oxetane formation, are dependent upon the electrophilic nature of the oxygen in an $n \to \pi^*$ state. Carbonyl compounds for which the lowest excited states are $\pi \to \pi^*$, rather than $n \to \pi^*$, undergo neither of these photoreactions efficiently.

Michler's ketone

Michler's ketone (p,p'-bis(dimethylamino)benzophenone) represents the classic case of an unreactive benzophenone derivative. Like benzophenone, intersystem crossing in Michler's ketone is highly efficient ($\Phi_{isc} = 1.00$) but in sharp contrast to benzophenone, Michler's ketone undergoes no detectable photoreduction in alcohol solvents ($\Phi < 0.001$).[42] The lack of reactivity may be explained in terms of the molecular orbitals for Michler's ketone. There are now four nonbonding orbitals, two on the carbonyl oxygen and one on each of the nitrogen atoms. Typically, nitrogen nonbonding orbitals will be higher in energy than oxygen nonbonding orbitals (this could be estimated simply by their relative basicity). Thus, the lowest energy transition is from a nitrogen nonbonding orbital to the π^* system, and is not the typical carbonyl $n \to \pi^*$ absorption.

The nonbonding nitrogen orbital differs in several important ways from the carbonyl oxygen nonbonding orbitals. The nitrogen lone pair will be in an unhybridized p orbital, and the orientation of the p orbital will be parallel to the aromatic π system, rather than perpendicular to the π system, as the oxygen lone pairs would be. Thus, the nitrogen p orbital can be considered part of the π system, and in fact, the absorption process is fully allowed. The extinction coefficient at the long wavelength maximum for Michler's ketone is 36,000 liter mole^{-1} cm^{-1}, while for benzophenone it is only 120 liter mole^{-1} cm^{-1}. This type of $\pi \rightarrow \pi^*$ excited state is often referred to as a charge-transfer excited state, since one electron is removed from nitrogen and distributed into the overall π^* system. In fact, this type of excited state involves an increase in electron density on the carbonyl oxygen, rather than the decrease usually observed for an $n \rightarrow \pi^*$ state. In this light, Michler's ketone would not be expected to undergo those photoreactions that are based upon an electrophilic $n \rightarrow \pi^*$ state.

Michler's ketone charge-transfer excited state

The importance of the nitrogen lone pair can be dramatically demonstrated by neutralization of the amine functions with excess acid. Doubly protonated Michler's ketone has a weak absorption spectrum and is reactive in photoreduction processes.[42]

Carbonyl compounds that have lowest $\pi \rightarrow \pi^*$ states are usually aromatic ketones that have either nitrogen or oxygen substituents or ketones with a lower energy chromophore elsewhere in the molecule. For example, α-acetonaphthone has lowest $\pi \rightarrow \pi^*$ excited states that are naphthalene-like rather than carbonyl-like. These molecules are often more useful as sensitizers than the $n \rightarrow \pi^*$ carbonyl compounds, since hydrogen abstraction can occasionally be a troublesome side reaction.

Cyclohexadienone Photorearrangements

Conjugated enones and dienones undergo a variety of interesting photoreactions, many of which can be considered to be characteristic carbonyl photoreactions or characteristic π-bond photoreactions or combinations of both. One such system that has been subjected to detailed mechanistic study is the crossconjugated cyclohexadienone system.[43] The photorearrangement

Mechanism of the 2,5-cyclohexadienone photorearrangement

can be considered an example of a di-π-methane rearrangement (see p. 347). However, the photorearrangement of cyclohexadienones has been shown to proceed through the triplet state, and, therefore, it must follow a mechanism other than the concerted mechanism we considered for the singlet di-π-methane rearrangement.

The proposed mechanism is supported by the following evidence. Either direct irradiation or triplet photosensitization (with acetophenone, $E_T = 73$ kcal/mole) leads to rearrangement with a quantum yield of 0.85. This indicates that the dienone undergoes efficient intersystem crossing and the reaction occurs from the triplet state. Although quenching of the photoreaction by standard triplet quenchers was not observed, this was taken to be an indication of a very rapid reaction from the triplet state, calculated to have a rate constant greater than 2×10^{10} sec^{-1}. The proposed zwitterion intermediate was synthesized by a chemical route and shown to give the expected photoproduct.[44] The availability of the zwitterion, furthermore, allowed the stereochemistry of the final step, the rearrangement itself, to be ascertained. The final rearrangement is a 1,4-sigmatropic shift with inversion:

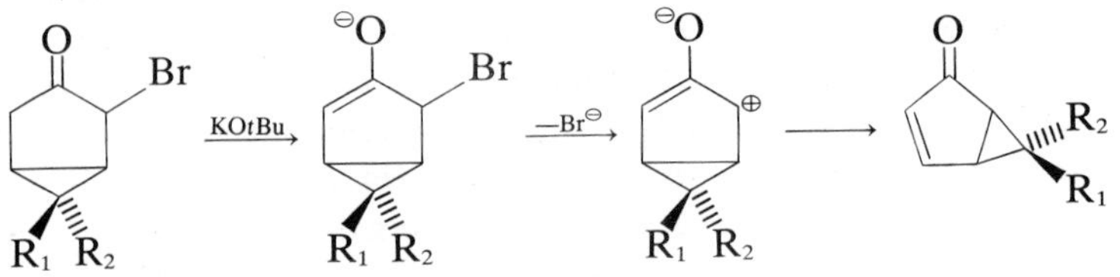

Vapor Phase Photochemistry

In most cases, the photochemistry observed for a given molecule in the vapor phase and in condensed phases is fairly similar. However, one class of reactions, photofragmentations, is substantially enhanced in the vapor phase. This enhancement may be due to two factors: the lack of a solvent cage so the fragments may escape from one another very efficiently and the lack of rapid vibrational deactivation so that vibrationally excited molecules retain their excess vibrational energy longer. An example of a photofragmentation that is observed almost exclusively in the vapor phase, and rarely in solution, is the Norrish type I cleavage of ketones:

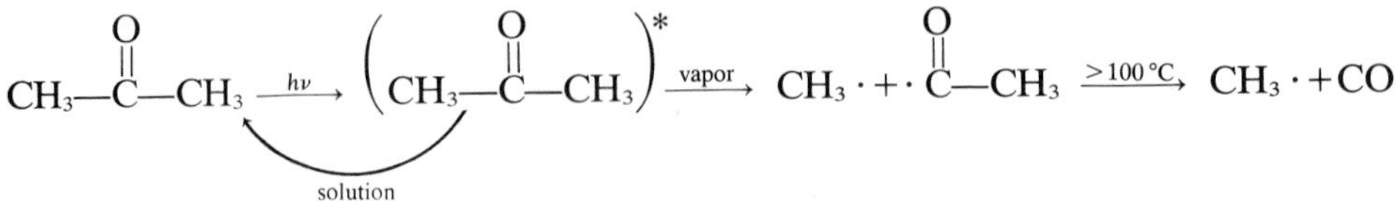

$$\Phi_{(-CO)} = 0.001 \quad \text{in solution}$$

$$\Phi_{(-CO)} = 0.1 \quad \text{in vapor at 25°C}$$

$$\Phi_{(-CO)} = 1.0 \quad \text{in vapor at 120°C}$$

In solution, type I cleavage is mainly observed in cyclic systems, especially when the cleavage leads to particularly stable free radicals:

(ref. 46)

(18% *cis*, 75% *trans*)

(ref. 47)

Thus far, we have considered the photochemistry of unsaturated compounds and carbonyl compounds. While these two classes of compounds provide a rich variety of photoreactions, and in fact cover the largest portion of known organic photochemistry, there are many other functional groups with interesting photochemistry of their own. For example, nitrobenzene undergoes photoreduction to aniline,[48] anthracene photodimerizes across the 9,10-positions,[49] and aryl esters rearrange to hydroxyphenones.[50] Azo and diazo compounds are commonly studied photochemically, but the major interest in these compounds is as precursors to the radical pairs or carbenes that are subsequently generated (see Chapters 6 and 7).

8.8 Singlet Molecular Oxygen

To conclude this survey of photochemistry, we will examine one excited state that is substantially different in nature from those we have considered thus far. This is the lowest singlet excited state derived from the triplet ground state of oxygen. The molecular orbitals for a diatomic molecule like oxygen are rather simply derived from the corresponding atomic orbitals (see page 6). The 16 electrons of the O_2 molecule would fit into molecular orbitals in the following way: $(1\sigma_s)^2 (1\sigma_s^*)^2 (2\sigma_s)^2 (2\sigma_s^*)^2 (2\sigma_z)^2 (2\pi_x)^2 (2\pi_y)^2 (2\pi_x^*)^1 (2\pi_y^*)^1$.

Molecular orbital theory provided the first explanation for the observed paramagnetism of oxygen. That is, the two electrons in the two degenerate π^* orbitals each occupy an orbital singly (Hund's rule) and prefer parallel spins, giving a triplet state. In fact, there are three different ways to locate two electrons in two degenerate molecular orbitals. There is only one triplet state, which is the most stable, and two singlet arrangements. Since these states have equal orbital energy, and differ only in electron–electron interaction terms, the three states all lie quite close in energy (see p. 358).

Once singlet oxygen is formed, the transition back to the triplet ground state is a forbidden process, and this forbiddenness leads to a very long lifetime. For the $^1\Delta g$ state, the "natural" lifetime (without collisions or interactions from other molecules) would be 45 min![51,52] In practice, the

Electronic States of Molecular Oxygen

Spectroscopic State Designation	π^* Orbital Arrangement	Relative Energy
$^1\Sigma_g^+$		38.5 kcal/mole
$^1\Delta_g$		22.5 kcal/mole
$^3\Sigma_g^-$		0 kcal/mole

solution lifetime is substantially shorter because of the numerous collisions and interactions and depends critically upon the solvent.[53] A typical solution lifetime would be about 10^{-3} sec. The lifetime of the $^1\Sigma_g^+$ state is extremely short because it has a spin-allowed transition down to $^1\Delta g$. Since $^1\Sigma_g^+$ deactivates so rapidly to $^1\Delta g$, most of the known reactions of singlet oxygen are reactions of the $^1\Delta g$ state.

Since the transition from triplet ground state to singlet excited state is also a spin-forbidden process, the direct absorption of light is not normally an effective means of generating singlet oxygen. Using very high pressures of oxygen and a long path length, the forbidden transitions may be located, however. They correspond to 762 nm (in the red) and 1269 nm (in the infrared).[54]

Since singlet oxygen cannot be readily formed by direct absorption, indirect methods of generation are usually employed. The most common of these is energy transfer from a triplet sensitizer, usually a dye that conveniently absorbs visible light. Virtually any sensitizer has a triplet energy in excess of the required 22.5 kcal/mole. The interaction of two triplet states to give two singlet states may be a spin-allowed process, as viewed by a simple electron-exchange mechanism:

$$D^{*3(\uparrow\uparrow)} + {}^3O_2^{(\uparrow\uparrow)} \rightarrow D^{(\uparrow\downarrow)} + {}^1O_2^{(\uparrow\downarrow)}$$

$$(E_\text{T} > 22.5 \text{ kcal/mole})$$

Singlet oxygen may also be generated in chemical reactions. The chemiluminescent reaction of basic hydrogen peroxide with chlorine had been observed for a long time, and it is now recognized that it yields singlet oxygen.[55] Since all reactants are in singlet states with all electrons paired, the products must arise initially in singlet states including singlet oxygen. More recently, phosphite adducts with ozone have been shown to decompose to yield singlet oxygen.[56] These chemical methods of generation of singlet oxygen provide an often useful alternative method to the usual dye sensitization:

$$H_2O_2 + Cl_2 + 2OH^- \rightarrow 2Cl^- + 2H_2O + {}^1O_2$$

$$(\phi O)_3 P + O_3 \xrightarrow{-80°C} (\phi O)_3 PO_3 \xrightarrow{-20°C} (\phi O)_3 PO + {}^1O_2$$

The major reactions of singlet oxygen can be conveniently divided into three classes: 1,4-cycloadditions, 1,2-cycloadditions, and hydroperoxide formation. In describing the orbital symmetry requirements for singlet oxygen, it can be considered that singlet oxygen obeys the same rules as ethylene. Thus, the 1,4-cycloadditions are analogous to Diels–Alder reactions.

1,4-Cycloadditions of singlet oxygen

Singlet oxygen undergoes 1,2-cycloaddition with electron-rich olefins. If concerted, the cycloaddition should be a $[2_s + 2_a]$ addition by orbital symmetry rules. The observation of stereospecificity in the cycloaddition has been explained in terms of a concerted reaction with the singlet oxygen taking the antarafacial role.[57] The observed product, a 1,2-dioxetane, decomposes at or above room temperature to yield carbonyl products and chemiluminescence (see p. 329).

1,2-Cycloadditions of singlet oxygen

Singlet oxygen also undergoes reactions with allylic hydrogens giving rearranged allylic hydroperoxides. This reaction had been considered a concerted reaction analogous to the "ene" synthesis, but there is growing evidence that indicates a nonconcerted reaction, with a "perepoxide" intermediate.[58–60]

Singlet oxygen "ene" reaction

The sudden appreciation of the chemistry of singlet oxygen has led to a reevaluation of most of the photoreactions that deal with oxygen. In many of these cases, singlet oxygen rather than ground-state oxygen may be implicated. There is now reason to believe that singlet oxygen may be important in photosynthesis, photocarcinogenicity, chemiluminescence and bioluminescence, photochemical smog, and phototherapy.

References

1. G. S. Hammond, J. Saltiel, A. A. Lamola, N. J. Turro, J. S. Bradshaw, D. O. Cowan, R. C. Counsell, V. Vogt, and C. Dalton, *J. Amer. Chem. Soc.*, **86,** 3197 (1964).
2. K. E. Wilzbach, J. S. Ritschen, and L. Kaplan, *J. Amer. Chem. Soc.*, **89,** 1031 (1967).
3. J. G. Calvert and J. N. Pitts, Jr., "Photochemistry," Wiley, New York, 1966, p. 499.
4. N. J. Turro, P. Lechtken, N. E. Schore, G. Schuster, H.-C. Steinmetzer, and A. Yekta, *Acc. Chem. Res.*, **7,** 97 (1974).
5. F. McCapra, *Quart. Rev.*, **20,** 485 (1966).
6. N. J. Turro, "Molecular Photochemistry," Benjamin, New York, 1965, p. 75.
7. P. J. Wagner, *J. Amer. Chem. Soc.*, **89,** 2820 (1967).
8. S. C. Tsai and G. W. Robinson, *J. Chem. Phys.*, **49,** 3184 (1968).
9. A. Terenin and V. Ermolaev, *Trans. Faraday Soc.*, **52,** 1042 (1956).
10. A. A. Lamola, in "Energy Transfer and Organic Photochemistry," A. A. Lamola and N. J. Turro, Eds., Wiley-Interscience, New York, 1969, pp. 17ff.
11. P. J. Debye, *Trans. Electrochem. Soc.*, **82,** 265 (1942).
12. G. Porter, *Science*, **160,** 1299 (1968). Nobel Prize address.
13. R. R. Alfano and S. L. Shapiro, *Scientific American*, **228,** 42 (June 1973).
14. H. L. J. Bäckström and K. Sandros, *Acta Chem. Scand.*, **12,** 823 (1958); **14,** 48 (1960).
15. G. Porter and F. Wilkinson, *Proc. Roy. Soc. London Ser. A*, **264,** 1 (1961).
16. M. R. J. Dack, *J. Chem. Educ.*, **50,** 169 (1973).
17. S. L. Murov, R. S. Cole, and G. S. Hammond, *J. Amer. Chem. Soc.*, **90,** 2957 (1968).
18. J. G. Calvert and J. N. Pitts, Jr., "Photochemistry," Wiley, New York, 1966, p. 226.
19. R. S. Mullikan and C. C. J. Roothaan, *Chem. Rev.*, **41,** 219 (1947).
20. H. Yamazaki and R. J. Cvetanovic, *J. Amer. Chem. Soc.*, **91,** 520 (1969).
21. P. J. Kropp, *J. Amer. Chem. Soc.*, **91,** 5783 (1969).
22. W. G. Herkstroeter and G. S. Hammond, *J. Amer. Chem. Soc.*, **88,** 4769 (1966).
23. G. S. Hammond and J. Saltiel, *J. Amer. Chem. Soc.*, **85,** 2516 (1963).
24. S. Yamauchi and T. Azumi, *J. Amer. Chem. Soc.*, **95,** 2709 (1973).
25. R. Srinivasan and F. I. Sonntag, *J. Amer. Chem. Soc.*, **87,** 3778 (1965).
26. K. J. Crowley, *Tetrahedron*, **21,** 1001 (1965).
27. R. S. H. Liu and G. S. Hammond, *J. Amer. Chem. Soc.*, **86,** 1892 (1964).
28. J. Saltiel, L. Metts, and M. Wrighton, *J. Amer. Chem. Soc.*, **91,** 5784 (1969).
29. J. Saltiel, D. Townsend, and A. Sykes, *J. Amer. Chem. Soc.*, **95,** 5968 (1973).
30. R. Srinivasan, *J. Amer. Chem. Soc.*, **90,** 4498 (1968).
31. R. S. H. Liu, N. J. Turro, and G. S. Hammond, *J. Amer. Chem. Soc.*, **87,** 3406 (1965).
32. W. L. Dilling, R. D. Kroening, and J. C. Little, *J. Amer. Chem. Soc.*, **92,** 928 (1970).
33. S. S. Hixson, P. S. Mariano, and H. E. Zimmerman, *Chem. Rev.*, **73,** 531 (1973).
34. C. Walling and M. J. Gibian, *J. Amer. Chem. Soc.*, **87,** 3361 (1965).
35. D. R. Arnold, *Adv. Photochem.*, **6,** 301 (1969).
36. G. Ciamician and P. Silber, *Chem. Ber.*, **33,** 2911 (1900).
37. W. M. Moore, G. S. Hammond, and R. P. Foss, *J. Amer. Chem. Soc.*, **83,** 2789 (1961).

38. W. M. Moore and M. Ketchum, *J. Amer. Chem. Soc.*, **84,** 1368 (1962).
39. A. Beckett and G. Porter, *Trans. Faraday Soc.*, **59,** 2038 (1963).
40. S. A. Weiner, *J. Amer. Chem. Soc.*, **93,** 425 (1971).
41. P. J. Wagner, *Acc. Chem. Res.*, **4,** 168 (1971).
42. G. Porter and P. Suppan, *Trans. Faraday Soc.*, **61,** 1664 (1965).
43. H. E. Zimmerman and J. S. Swenton, *J. Amer. Chem. Soc.*, **89,** 906 (1967).
44. H. E. Zimmerman and D. S. Crumrine, *J. Amer. Chem. Soc.*, **90,** 5612 (1968).
45. W. A. Noyes, Jr., G. B. Porter, and J. E. Jolley, *Chem. Rev.*, **56,** 49 (1956).
46. R. Srinivasan, *Adv. Photochem.*, **1,** 83 (1963).
47. G. Quinkert, K. Opitz, W. Wiersdorf, and M. Finke, *Ann. Chem.*, **693,** 44 (1966).
48. J. A. Barltrop and N. J. Bunce, *J. Chem. Soc. Ser. C,* 1467 (1968).
49. E. J. Bowen, *Adv. Photochem.*, **1,** 23 (1963).
50. D. Bellus, *Adv. Photochem.*, **8,** 109 (1971).
51. R. M. Badger, A. C. Wright, and R. F. Whitlock, *J. Chem. Phys.*, **43,** 4345 (1965).
52. R. W. Nicholls, *Can. J. Chem.*, **47,** 1847 (1969).
53. P. B. Merkel and D. R. Kearns, *J. Amer. Chem. Soc.*, **94,** 7244 (1973).
54. D. R. Kearns, *Chem. Rev.*, **71,** 397 (1971).
55. C. S. Foote, *Acc. Chem. Res.*, **1,** 104 (1967).
56. P. D. Bartlett and G. D. Mendenhall, *J. Amer. Chem. Soc.*, **92,** 210 (1970).
57. P. D. Bartlett and A. P. Schaap, *J. Amer. Chem. Soc.*, **92,** 3223 (1970).
58. A. P. Schaap and G. R. Faler, *J. Amer. Chem. Soc.*, **95,** 3381 (1973).
59. W. Fenical, D. R. Kearns, and P. Radlick, *J. Amer. Chem. Soc.*, **91,** 7771 (1969).
60. N. Hasty, P. B. Merkel, P. Radlick, and D. R. Kearns, *Tetrahedron Lett.*, 49 (1972).

Bibliography

General References

N. J. Turro, "Molecular Photochemistry," Benjamin, New York, 1965.

J. G. Calvert and J. N. Pitts, Jr., "Photochemistry," Wiley, New York, 1966.

D. C. Neckers, "Mechanistic Organic Photochemistry," Reinhold, New York, 1967.

A. A. Lamola and N. J. Turro, "Energy Transfer and Organic Photochemistry," Wiley-Interscience, New York, 1969.

W. A. Noyes, Jr., G. S. Hammond, and J. N. Pitts, Jr., Eds., "Advances in Photochemistry," Wiley-Interscience, New York, vols. 1– , 1963– .

O. L. Chapman, Ed., "Organic Photochemistry," Dekker, New York. vols. 1– , 1967– .

Spectroscopy

H. H. Jaffe and M. Orchin, "Theory and Applications of Ultraviolet Spectroscopy," Wiley, New York, 1962.

S. P. McGlynn, T. Azumi, and M. Kinoshita, "Molecular Spectroscopy of the Triplet State," Prentice-Hall, Englewood Cliffs, N.J., 1969.

Energy Transfer

F. Wilkinson, *Adv. Photochem.*, **3,** 241 (1964).

A. A. Lamola, in "Energy Transfer and Organic Photochemistry," A. A. Lamola and N. J. Turro, Eds., Wiley-Interscience, New York, 1969, pp. 17ff.

Chemiluminescence

"Chemiluminescence and Bioluminescence," M. J. Cormier, D. M. Hercules, and J. Lee, Eds., Plenum, New York, 1973.

Carbonyl Photoreactions

J. S. Swenton, *J. Chem. Educ.*, **46**, 217 (1969).

Singlet Oxygen

D. R. Kearns, *Chem. Rev.*, **71**, 395 (1971).

C. S. Foote, *Acc. Chem. Res.*, **1**, 104 (1968).

Photobiology

R. K. Clayton, "Light and Living Matter," Vols. 1 and 2. McGraw-Hill, New York, 1971.

Problems

1. The $n \to \pi^*$ absorption of benzophenone shows a $(0, 0)$ band at 380 nm.

   What is the lowest singlet energy for benzophenone?

   If the lowest triplet energy of benzophenone is 68.5 kcal/mole, at what wavelength would the $(0, 0)$ band for $S_0 \to T_1$ absorption be expected? Why is this absorption not normally observed? Under what conditions could it be observed?

2. For naphthalene, $\tau_S = 100$ nsec, $\Phi_f = 0.55$, and $\Phi_{isc} = 0.40$. Describe how each of these three numbers might have been determined experimentally.

   Calculate the absolute rates for fluorescence, internal conversion, and intersystem crossing.

3. Explain why singlet oxygen is rarely quenched by energy transfer but is very susceptible to quenching by basic molecules. Show the two types of quenching mechanisms.

4. Chlorophyll is the pigment chiefly responsible for the green color of leaves and plants. Dissolved in nonpolar solvents like anhydrous diethyl ether, there is weak fluorescent emission in the near infrared, with a lifetime of about 100 nsec.

   Dissolved in polar solvents, or even with the addition of small amounts of water to the ether solvent, a new strong fluorescence is observed in the visible red region with a lifetime of about 1 nsec.

   Explain these phenomena in terms of the excited states involved.

5. Show the mechanism and all the products expected from the following photochemical reactions. When more than one compound is present, assume that only the first absorbs the radiation.

$$(a)\ \phi\!-\!\overset{\displaystyle O}{\overset{\displaystyle \|}{C}}\!-\!\phi\ +\ \overset{\displaystyle CH_3}{\underset{\displaystyle H}{}}C\!=\!C\overset{\displaystyle H}{\underset{\displaystyle CH_3}{}}\ \xrightarrow{\ h\nu\ }$$

(b) 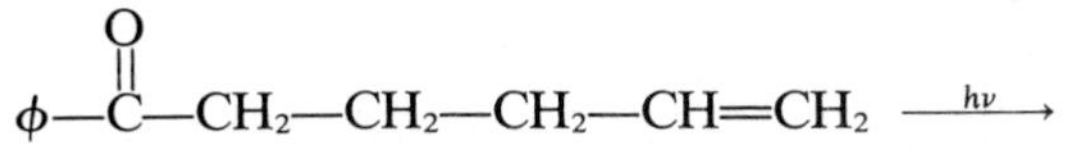 ϕ—$\overset{O}{\overset{\|}{C}}$—CH$_2$—CH$_2$—CH$_2$—CH=CH$_2$ $\xrightarrow{h\nu}$

(c) ϕ—$\overset{O}{\overset{\|}{C}}$—O—O—$\overset{O}{\overset{\|}{C}}$—$\phi$ $\xrightarrow{h\nu}$

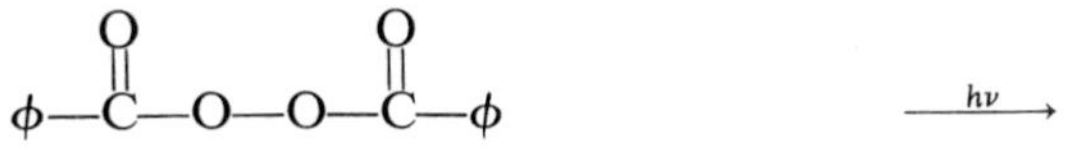

(d)  ϕ—$\overset{O}{\overset{\|}{C}}$—CH$_3$ + $\xrightarrow{h\nu}$

(e) $\xrightarrow{h\nu}$

(f) $\xrightarrow{h\nu, \text{ vapor}}$

(g) $\xrightarrow{h\nu}$

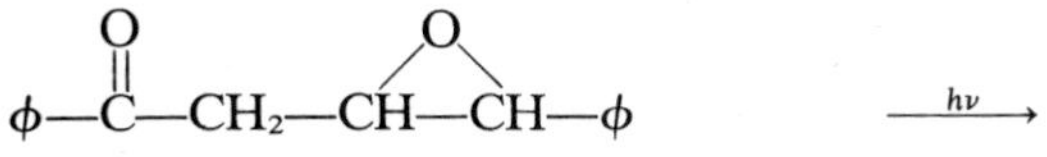

(h) ϕ—$\overset{O}{\overset{\|}{C}}$—CH$_2$—CH—CH—$\phi$ $\xrightarrow{h\nu}$

(i) 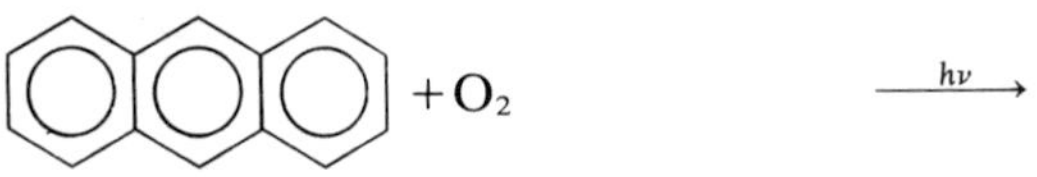 +O$_2$ $\xrightarrow{h\nu}$

(j) 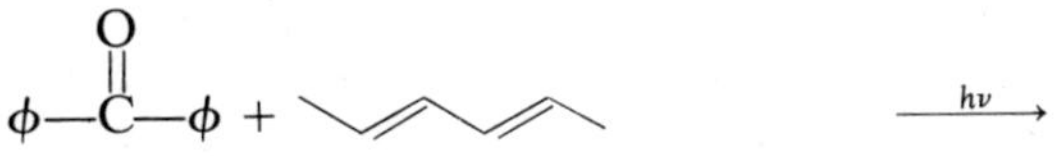 ϕ—$\overset{O}{\overset{\|}{C}}$—$\phi$ + $\xrightarrow{h\nu}$

6. Devise reasonable mechanisms for the following photochemical transformations:

(a) 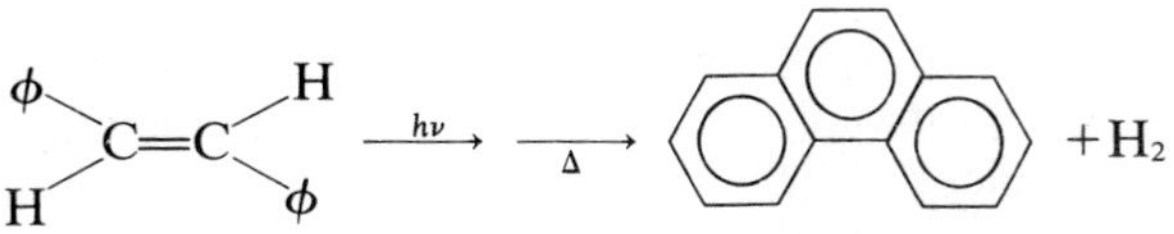 $\xrightarrow{h\nu}$ $\xrightarrow{\Delta}$ +H$_2$

(b) 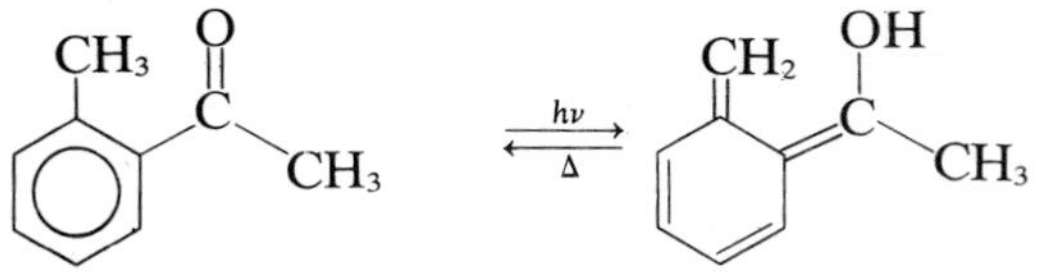 $\underset{\Delta}{\overset{h\nu}{\rightleftharpoons}}$

(c) 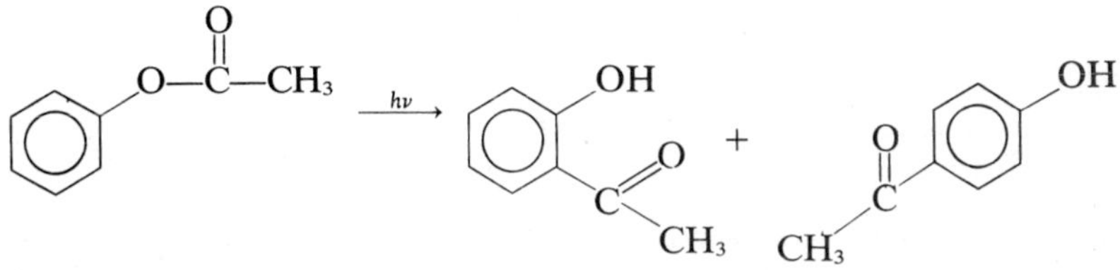 $\xrightarrow{h\nu}$ +

(d) 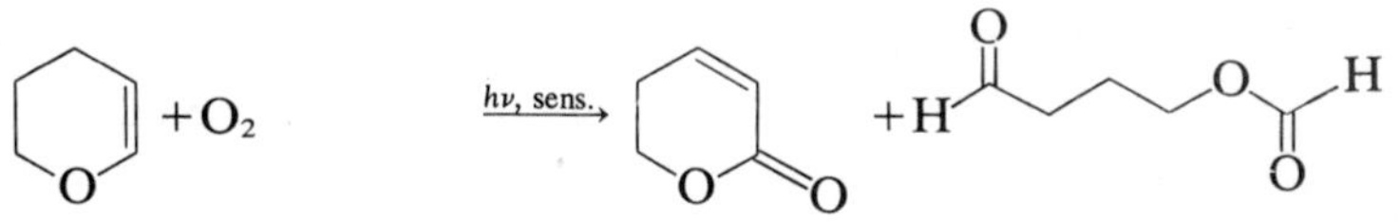 +O$_2$ $\xrightarrow{h\nu, \text{ sens.}}$ +H

(e) $\xrightarrow{hv, \text{ sens.,}}$

7. Cyclopentadiene undergoes triplet sensitized photodimerization. Predict all of the dimer structures expected. Predict how the ratio of these dimers will vary with the triplet energy of the sensitizer.

$\xrightarrow{hv, \text{ sens.}}$ dimers

8. Santonin **I** is a light-sensitive natural product that undergoes photorearrangement to lumisantonin **II**. Show the complete mechanism for this rearrangement.

$\xrightarrow{hv}$

I **II**

9. Explain the following observations:

(a) The phosphorescence spectrum of benzophenone shows a progression of peaks separated by the energy of the carbonyl stretching frequency (peaks spaced about 1700 cm^{-1} apart).

(b) Anthracene $(E_T = 43 \text{ kcal/mole})$ rather efficiently photosensitizes the decomposition of azomethane $(E_T = \text{about } 55 \text{ kcal/mole})$. Azomethane is a yellow gas.

(c) p-Aminoacetophenone does not photoreduce in 2-propanol, but in acidic 2-propanol it does undergo photoreduction.

(d) Naphthalene $(E_T = 61 \text{ kcal/mole})$ and anthracene $(E_T = 43 \text{ kcal/mole})$ both quench the photoreduction of benzophenone equally effectively.

(e) The quantum yield for photoinitiated chlorination of an alkane can approach 10,000.

10. 2-Hexanone undergoes type II cleavage from both the singlet and triplet states.

$$CH_3-\overset{\overset{\displaystyle O}{\|}}{C}-CH_2-CH_2-CH_2-CH_3 \xrightarrow{hv} CH_3-\overset{\overset{\displaystyle O}{\|}}{C}-CH_3 + CH_2=CH-CH_3$$

Quantum yields were measured for the following: intersystem crossing $(\Phi_{isc} = 0.27)$, reaction from the singlet state, $(\Phi_s = 0.10)$, and reaction from the triplet state $(\Phi_t = 0.27)$.

The singlet and triplet lifetimes may be determined by quenching of the reaction by 1,3-pentadiene, which will quench both singlets and triplets at the diffusion-controlled rate $(k_Q = 5 \times 10^9 \ M^{-1} \sec^{-1})$. The Stern–Volmer plot gives a curve with two linear regions of slopes 50 and $3.5 \ M^{-1}$.

Calculate the singlet and triplet lifetimes and then the absolute rates of reaction of the singlet and triplet states.

Brief Answers to Selected Problems

Chapter 1

3. Orbitals: 4,4,4,5,8,3; Electrons: 4,4,4,4,8,2.

4. See Figure 1.2: π_x^* and π_y^* will each contain one electron.

5. Join two allyls.

6. See p. 24.

7. (a) $\psi_1 = 0.5\phi_1 + 1/\sqrt{2}\phi_2 + 0.5\phi_3$; $\psi_2 = 1/\sqrt{2}\phi_1 - 1/\sqrt{2}\phi_3$;
$\psi_3 = 0.5\phi_1 - 1/\sqrt{2}\phi_2 + 0.5\phi_3$; $x = 0, \pm\sqrt{2}$.

   (b) $\psi_1 = 1/\sqrt{3}\phi_1 + 1/\sqrt{3}\phi_2 + 1/\sqrt{3}\phi_3$; $\psi_3 = 1/\sqrt{2}\phi_1 - 1/\sqrt{2}\phi_2$;
$\psi_2 = 1/\sqrt{6}\phi_1 + 1/\sqrt{6}\phi_2 - 2/\sqrt{6}\phi_3$; $x = 1, 1, -2$.

   (c) ψ_1 (all coefficients 0.5); $\psi_2 = 0.5\phi_1 + 0.5\phi_2 - 0.5\phi_3 - 0.5\phi_4$
$\psi_3 = 0.5\phi_1 - 0.5\phi_2 - 0.5\phi_3 + 0.5\phi_4$;

$$\psi_4 = 0.5\phi_1 - 0.5\phi_2 + 0.5\phi_3 - 0.5\phi_4;$$

   $x = 0, 0, \pm2$.

   (d) $\psi_1 = 0.56\phi_1 + 0.44\phi_2 + 0.56\phi_3 + 0.44\phi_4$;
$\psi_2 = 0.71\phi_2 - 0.71\phi_4$; $\psi_3 = 0.71\phi_1 - 0.71\phi_3$;
$\psi_4 = 0.45\phi_1 - 0.59\phi_2 + 0.45\phi_3 - 0.59\phi_4$.
$x = -2.56, 0, +1, +1.56$.

   (e) $\psi_1 = 0.28\phi_1 + 0.61\phi_2 + 0.52\phi_3 + 0.52\phi_4$;
$$\psi_2 = -0.81\phi_1 - 0.25\phi_2 + 0.37\phi_3 + 0.37\phi_4;$$
$\psi_3 = 0.71\phi_3 - 0.71\phi_4$; $\psi_4 = 0.51\phi_1 - 0.75\phi_2 + 0.30\phi_3 + 0.30\phi_4$;
$x = -2.17, -0.31, +1.00, +1.48$.

8. (a) 0.83β. (b) 0, β, 2β. (c) 0. (d) 1.12β. (e) 0.96β.

10. (a) Cation: 0.5, 1, 0.5; 0.71. Radical: 1, 1, 1; 0.71. Anion: 1.5, 1, 1.5;
0.71. (b) Cation: $q_1 = q_2 = q_3 = 0.677$; 0.677. Radical (assuming
one-half electron in ψ_2 and ψ_3—see ref. 4, p. 53): 1; 0.5. Anion:
1.33; 0.33. (c) $q_n = 1$; $p_{12} = 1$, $p_{23} = 0.5$. (d) $q_n = 1$; $p_{12} = 0.62$, $p_{23} = 0$, $p_{13} = 0.80$.

   (e) 1.48, 0.88, 0.82, 0.82; $p_{12} = 0.75$, $p_{23} = 0.45$, $p_{34} = 0.82$.

12. (a)  (b)

13. (a) exocyclic carbon, -0.57; o, p, -0.143; others zero. (b) Exocyclic
carbon, -0.40; o, p, -0.10; others zero. (c) exocyclic carbon,
-0.31; o, p, -0.08; others zero.

14. 0.089β.

15. Assume coefficients are equal; this will lead to self-consistent first solution. $E_1 = -0.570$ eV, $E_2 = -0.061$ eV.

Chapter 2

1. (a) Use

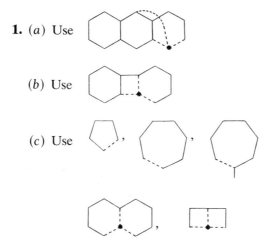

(b) Use

(c) Use

2. (a) Nonaromatic from PMO intramolecular union.
   (b) Antiaromatic, 4π electrons.
   (c) Aromatic, but no more so than benzene, PMO union even HCs.
   (d) Aromatic, but no more so than the annulene; the transannular single bond is essential.
   (e) Aromatic.
   (f) Nonaromatic; a classical molecule.
   (g) Aromatic; 6π electrons.
   (h) Aromatic; zwitterionic form important.
   (i) Aromatic.
   (j) Aromatic.
   (k) Nonaromatic; a classical molecule.
3. (c) -1.08β. (d) -0.42β. (e) 0.
4. The bicyclics have stabilities equal to the annulenes.
5. Nuclear magnetic resonance ring current deshields "outside" protons.
6. g, e, f, b, c, a = d.
7. Anthracene-9, anthracene-1, biphenylene-2, naphthalene-1, anthracene-2, biphenylene-1, naphthalene-2, benzene.
8. Yes.
9. (a) One *trans* and one *cis* double bond.
   (b) Electrocyclic to give new six-membered ring with methyl and hydrogen *cis*; two possibilities.
   (c) Four-membered ring, hydrogens *cis*.
   (d) $[4_s + 2_s]$ cycloaddition

10. The [1,3] shift will place deuteriums at every position but the [1,5] will not.

11. A = 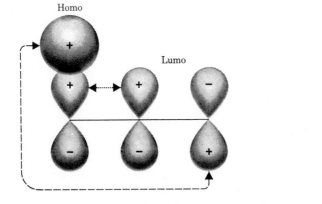 ⟶ product

12. **II** and **IV** produce *cis* double bonds in the rings, but **I** and **III** each produce a *trans* double bond in one ring.

13. (a)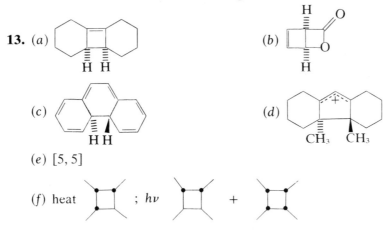

(b)

H H

(c)

H H

(d)

CH₃ CH₃

(e) [5, 5]

(f) heat ; hν +

14. [3,3] shifts rapidly interconvert all carbons; at −25°C bridgehead and vinylic hydrogens are interconverted slowly on the nmr time scale.

15. No. Yes.

16. 21.18, 21.26, 32.99; 0.53β.

17. Möbius–Hückel: The [1, 3] shift is a $4n$ system and must involve one sign inversion or antarafacial union; the electrocyclic opening involves two electrons and must involve no sign inversion—therefore disrotatory.

Frontier orbital:

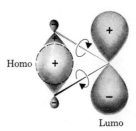

Chapter 3

1. All unreasonable.

2. Bromoethyl cation intermediate.

3. Rate-determining step is formation of the nitrating agent.

4. For *a*, use excess B and C; for *b* and *c*, vary the excess B and C.

5. Different rate-determining step at high and low Cl_2 concentration.

6. Slow ring opening, followed by Diels–Alder reaction.

7. Intermolecular: compare rate of acetone with acetone-d_6. Intramolecular: analyze products from acetone-d_n ($n = 1$ to 5).

8. 7.4.

9. Cyclopropanone gains greater relief of ring strain.

10. Goes through phthalic anhydride (M. L. Bender, Y.-L. Chow, and F. Chloupek, *J. Amer. Chem. Soc.*, **80**, 5380 (1958)).

11. Bimolecular *trans–cis* isomerization is rate-determining step (P. C. Huang and E. M. Kosower, *J. Amer. Chem. Soc.*, **90**, 2367 (1968)).

12. Break in Hammett plot indicates change in rate-determining step (G. Ostrogovich, G. Csunderlik, and R. Bacaloglu, *J. Chem. Soc.*, Ser. B, 18 (1971)).

Chapter 4

1. (*a*) A = NaCl, B = NaOH. (*b*) A = CH_3OH, B = TsOH. (*c*) A = NaN_3, B = NaCl. (*d*) The diol.

2. Inverted product is formed that cancels the rotation of a reactant molecule.

3.

$k_a > k_t$ because of ion pair return; differing amounts of ^{18}O scrambling possibly from

—see ref. 37, p 279.

4. Use mY equation and Table 4.5.

5. Less steric hindrance to nucleophile approach for quinuclidine.

6. Cyclopentyl. Yes, same eclipsing interactions relieved.

7. Larger R groups could interfere with approach of nucleophile. Also, if elimination is kinetically important, increasing the number of hydrogens on the R groups should increase reaction rate.

8. Adjacent orbitals will interact to give two new orbitals, one of higher and one of lower energy. The presence of a higher energy orbital increases nucleophilicity.

9. Use Hammett equation.

10. Numbering from left and most reactive first: 1, 4, 3, 2, 5.

11. No, not if return is not occurring.

12. Intermediate

13. Increasing the degree of nucleophilic involvement in the transition state, lowers the degree of positive charge development in the transition state and lowers the response to changes in solvent polarity.

14.
$$\frac{\% RN_3}{100} = \frac{k_N[N_3^-]}{k_N[N_3^-] + k_w[H_2O]} = \frac{k_N[N_3^-]}{X}$$

$$\frac{1}{r.a.} = \frac{k_w[H_2O]}{X}$$

$$1 - \frac{1}{r.a.} = \frac{X}{X} - \frac{k_w[H_2O]}{X} = \frac{k_N[N_3^-]}{X}$$

$$1 - \frac{1}{r.a.} = \frac{\% RN_3}{100}$$

Using this equation, rates can be used to predict $\% RN_3$ for 2-propyl but not for 2-adamantyl; therefore, only the reaction of 2-propyl is second-order.

15.

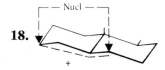

16. Scrambled to every position by repeated 1,2-alkyl shifts.

17.

carbonyl
participation

18.

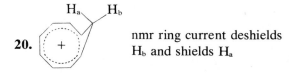

19. At 0°C, combination of 6,2 and 2,3 hydride shifts and 1,2 carbon shifts leads to equivalence of all carbons.

20.   nmr ring current deshields H_b and shields H_a

21. (a) $\longrightarrow$ R—O—$\overset{\overset{\displaystyle O}{\|}}{S}$—Cl $\longrightarrow$ R$^+$Cl$^-$ + SO$_2$

(b) CH$_3$—$\overset{\overset{\displaystyle CH_3}{|}}{\underset{\underset{\displaystyle H—O}{|}}{C}}$—$\overset{}{\underset{\underset{\displaystyle Cl}{|}}{C}}$(CH$_3$)$_2$

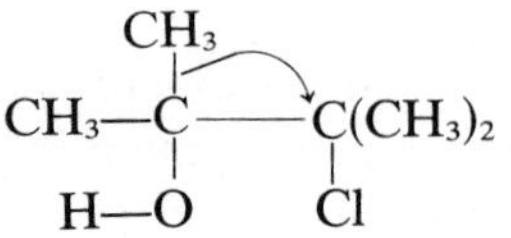

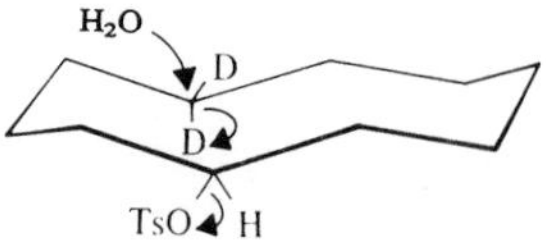

(c)

(d) R—$\overset{\overset{\displaystyle N—OH_2}{+}}{\underset{\underset{\displaystyle}{}}{C}}$—R' $\longrightarrow$ R—N=$\overset{+}{C}$—R' $\xrightarrow{H_2O}$ R—N=$\overset{}{\underset{\underset{\displaystyle OH}{|}}{C}}$—R' $\searrow$

$\overset{\overset{\displaystyle O}{\|}}{}$
RNHCR'

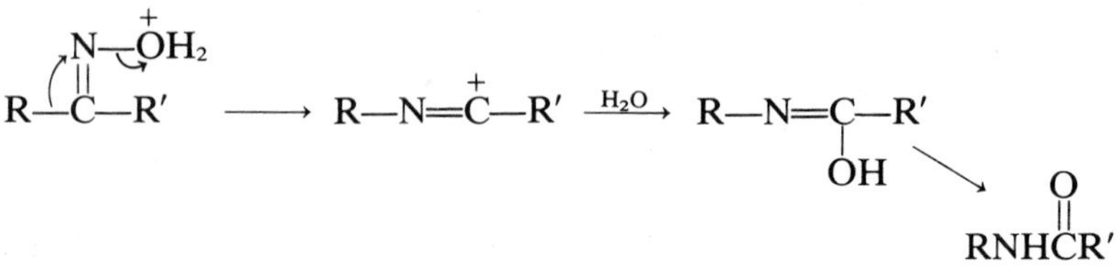

22. 2-Norbornyl derivative.

23. log $k_t = -3.50$

log $k_s = -4.0$

$k_t = k_s + k_\Delta$

$\dfrac{k_\Delta}{k_s + k_\Delta} \times 100 = $ per cent retention

24. (a) 1-Alkene. (b) 2-Alkene.

(c) CH$_2$=CH—CH$_2$—CH$_2$—CH=CH$_2$ (d) CH$_2$=CH—$\overset{}{\underset{\underset{\displaystyle CH_3}{|}}{C}}$=CH$_2$

25. Both the large base and the sulfonium leaving group promote removal of terminal hydrogen.

26. tBu $> i$Pr $>$ Et.

27. Charge is dispersed in both E2 and S$_N$2 transition states, but more so for E2. Therefore, elim:sub ratio will decrease in more polar solvent.

28.

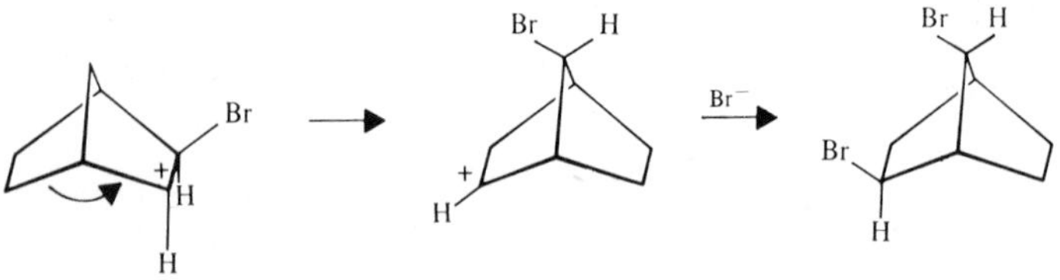

29.

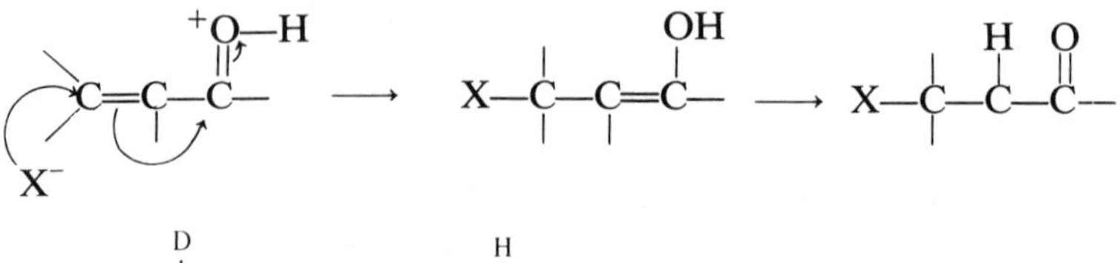

30. (a)

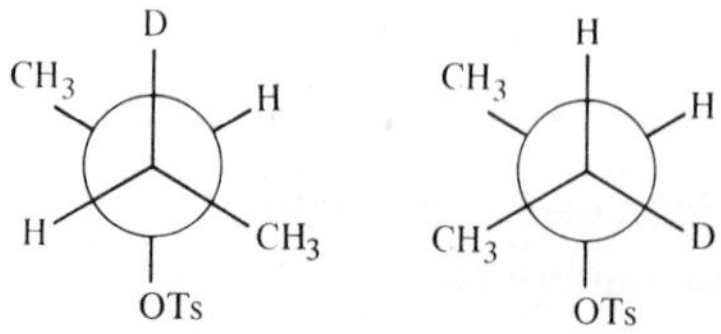

(*b*) In order to get *trans* orientation of leaving groups, it is necessary to place all six chlorines in axial positions.

(*c*) $\text{Cl}\text{—}\text{CH}_2\text{CH}_2\text{CH}\text{—}\text{CN}$
$\quad\quad\quad\quad\quad\quad\mid$
$\quad\quad\quad\quad\quad\quad\text{H}$
$\quad\quad\quad\quad\quad\text{:NH}_2^-$

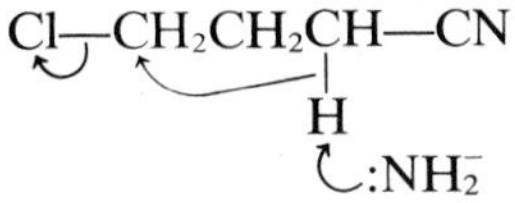

(*d*) Resonance.

31. Hydroboration involves a cyclic, four-center transition state subject to steric effects. The HCl addition is two step and less subject to steric effects.

32. −4.4; 14.2.

33. (*a*)

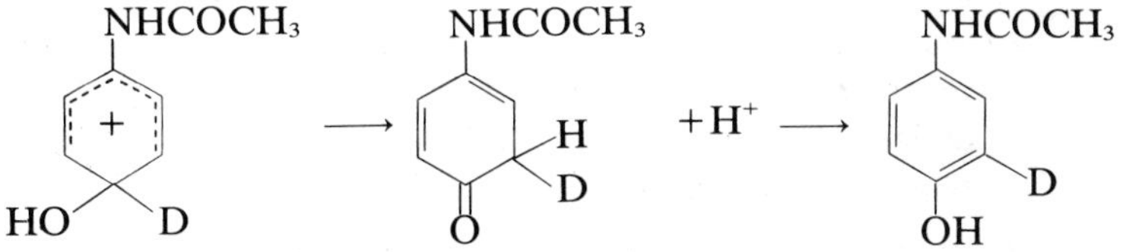

(*b*)

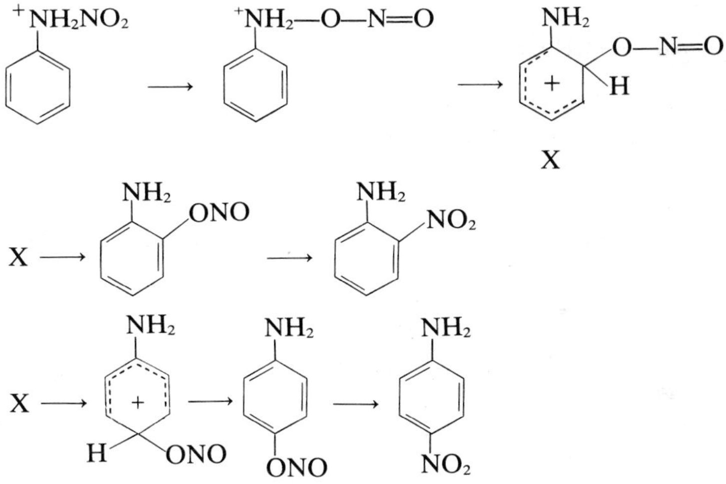

X →

X →

(*c*)

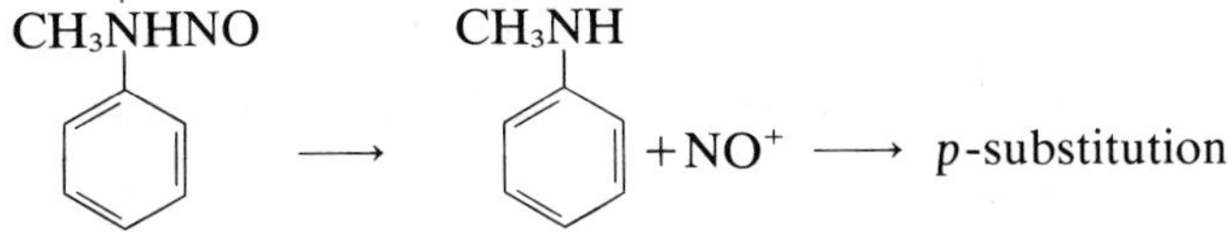

$+\text{NO}^+ \longrightarrow$ *p*-substitution

(*d*)

$\longrightarrow$ product

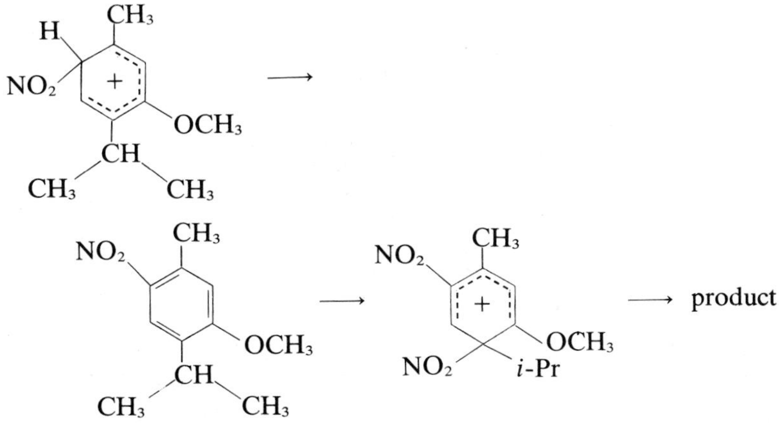

(e) ⟶ ⟶ product

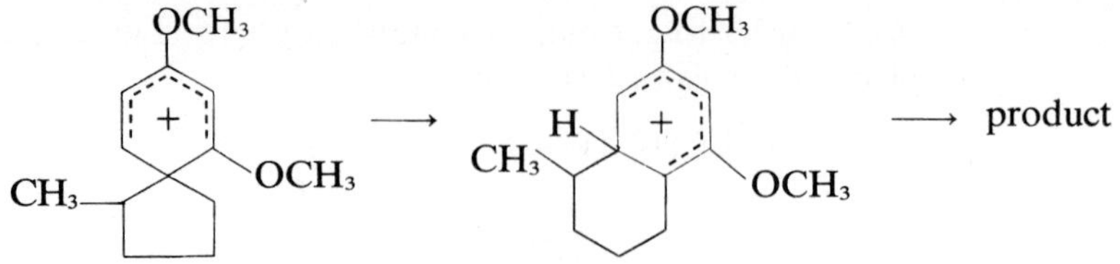

Chapter 5

1. Percent s character and degree of substitution: ethylene, cyclopropane, methane, and cyclohexane.

2. Acetylacetone, two resonance forms for enolate; ethyl malonate, two resonance forms for enolate but -OEt makes carbonyl less electron deficient; enolization of acetone involves loss of freedom of rotation relative to cyclohexanone.

3. $^{\ominus}CH(CO_2Et)_2 + CH_3Br \rightarrow CH_3CH(CO_2Et)_2 \xrightarrow[\text{hydrolysis}]{\text{ester}} CH_3CH(CO_2H)_2$

   $\xrightarrow[-CO_2]{\Delta} CH_3CH_2CO_2H$

4. (a) Enamine + cyclohexanone.
   (b) Base + ethylene bromide.
   (c) Wittig.

5.

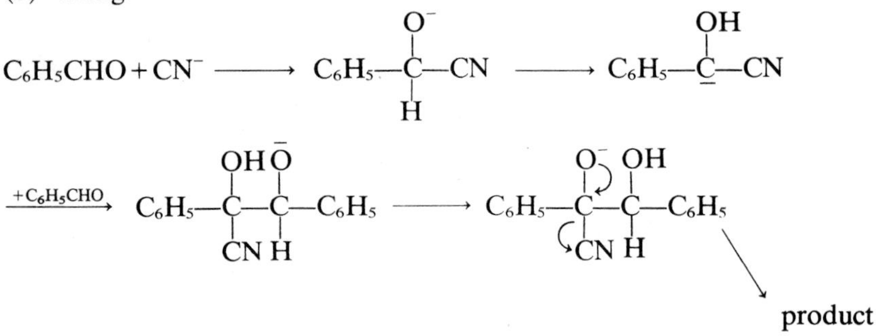

6. (a)

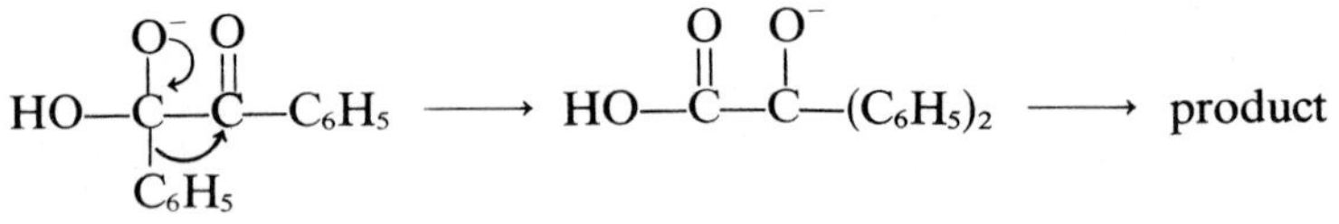

   (b)

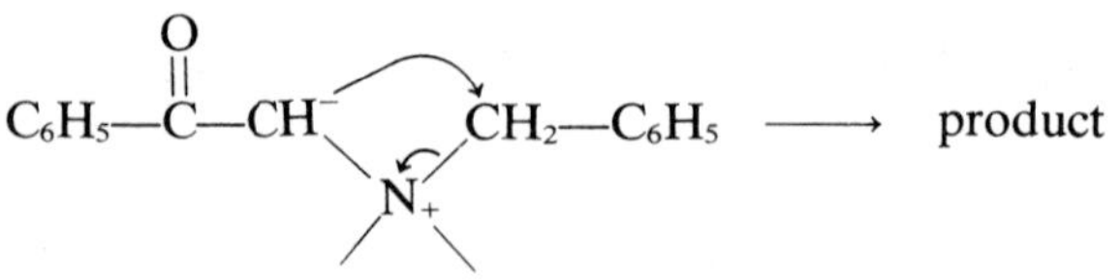

   (c)

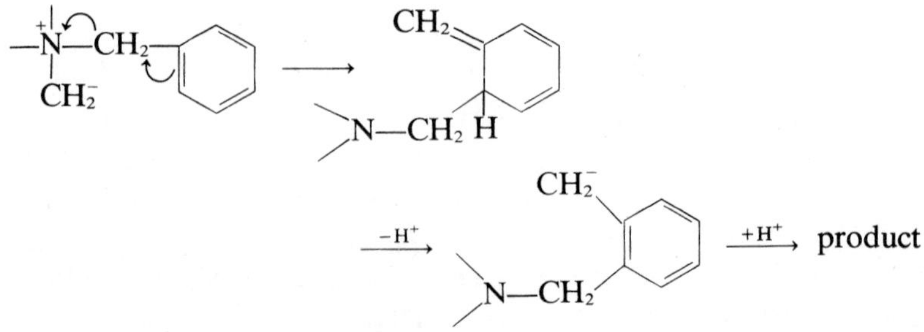

(d)

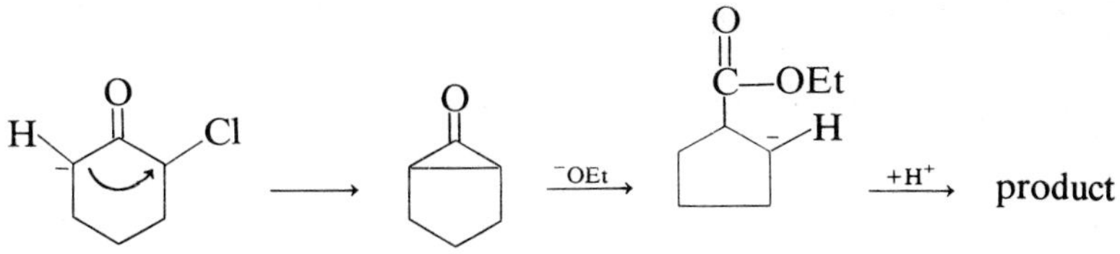

(e) CH_3—$\overset{OH}{\underset{\underset{CH_3}{|}}{\overset{|}{C}}}$—$\overset{OH}{\overset{|}{C}}$—R $\longrightarrow$ CH_3—$\overset{H—O^+}{\overset{||}{C}}$—$\overset{OH}{\underset{CH_3}{\overset{|}{C}}}$—R $\xrightarrow{-H^+}$ product

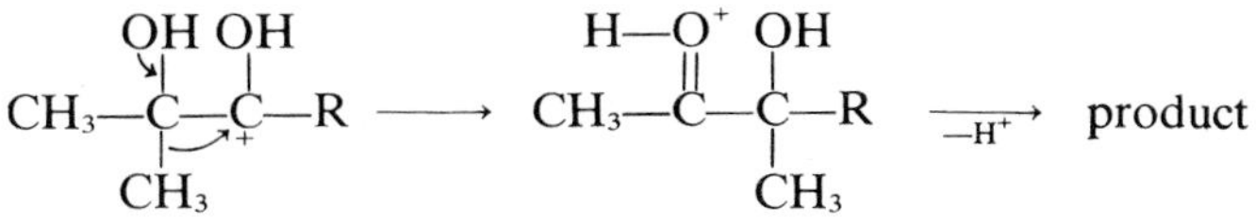

7. 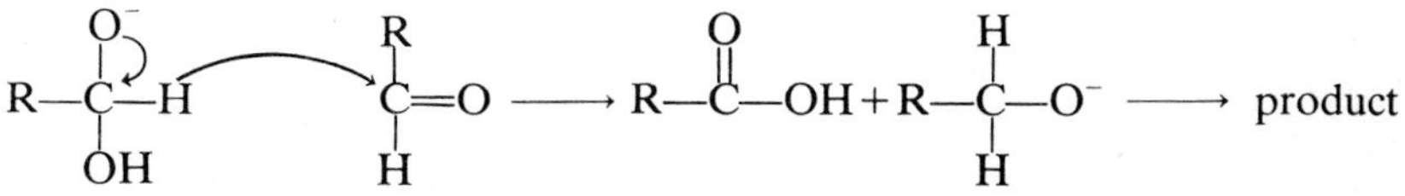 R—$\overset{O^-}{\underset{OH}{\overset{|}{\underset{|}{C}}}}$—H $\quad$ $\overset{R}{\underset{H}{C}}$=O $\longrightarrow$ R—$\overset{O}{\overset{||}{C}}$—OH + R—$\overset{H}{\underset{H}{\overset{|}{\underset{|}{C}}}}$—O$^-$ $\longrightarrow$ product

8. The rate changes despite a constant buffer ratio. Therefore, it is general base catalysis.

9. There will be steric interactions between the *ortho* hydrogens and the tetrahedral carbon formed upon alcohol attack.

10. The reaction is aided by electron withdrawal, and probably involves a negative intermediate.

11. Approach of the AlH_4^- ion is less sterically hindered from the H side.

12.

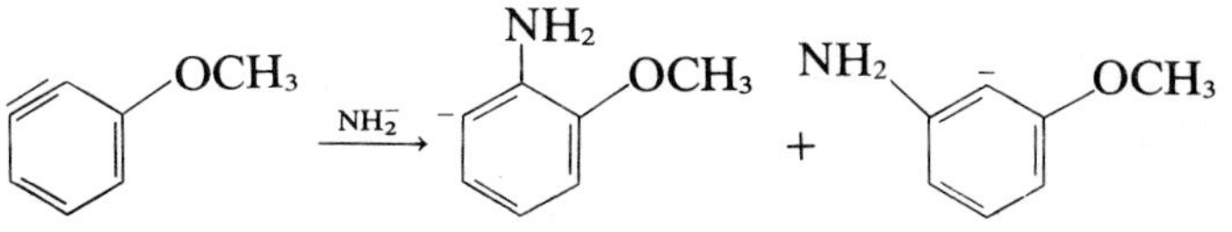

more stable;
– near OCH_3.

Chapter 6

1. All answers in kilocalories per mole; the more negative value indicates the more favorable reaction.
(a) +8; −19.

X =	F	Cl	Br	I	
(b)	−32	+1	+16	+33	
	−4	+20	+34	+48	
(c)	−40	−26	−5	+7	(see p.279)
	−32	+1	+16	+33	
(d)	−98	−65	−50	−33	(π bond ~60)
	−108	−84	−70	−56	

(e) ~−60; ~−85 (π bond versus σ bond).

2. First, the absorption is split to a septet by H_β ($a_\beta = 25$ gauss), then each of the seven lines is split to two by H_α ($a_\alpha = 22$ gauss).

3. Delocalization to *ortho* and *para* positions and each nitro oxygen.

4. Autoxidation at α–C–H bond.

5. (*a*) Decarbonylation.
   (*b*) Bromination at benzylic position.
   (*c*) Autoxidation at benzylic position.
   (*d*) Addition of aldehyde to double bond.
   (*e*) Allylic bromination.
   (*f*) Addition to double bond.
   (*g*) Polymerization.
6. (*a*) Greater cage effect in more viscous solvent (H. Kiefer and T. G. Traylor, *J. Amer. Chem. Soc.*, **89**, 6667 (1967)).
   (*b*) Cage return of a radical pair (A. Tsolis, S. G. Mylonakis, M. T. Nieh, and S. Seltzer, *J. Amer. Chem. Soc.*, **94**, 829 (1972); see also N. A. Porter, M. E. Landis, and L. J. Marnett. *J. Amer. Chem. Soc.*, **93**, 795 (1971)).
   (*c*) More intervening molecules minimize cage recombination (same reference as 6(*a*)).
7. Radical abstraction from methoxide to give formaldehyde radical anion, a good electron transfer agent (J. F. Bunnett and C. C. Wamser, *J. Amer. Chem. Soc.*, **89**, 6712 (1967)).
8. Abstraction of phenolic hydrogen, leading to quinone.
9. Polar stabilization of transition states makes the greatest difference with an A terminus.
10. Competitive abstraction of allylic H is minimized by deuteration.
11. Termination of unlike radicals gives best possibility for polar stabilization of transition state.
12. The *t*-BuOH comes from H abstraction and the acetone from β-scission. Cumene is a better H donor, hence expect more alcohol.
13. Of the two allylic resonance forms, the tertiary cation makes the greater contribution.
14. Normal abstraction involves removal of electron density. This case might involve a diaryliodine intermediate (W. C. Danen and D. G. Saunders, *J. Amer. Chem. Soc.*, **91**, 5924 (1969)).

Chapter 7

1. 3-Methylhexane (43%); 2,3-dimethylpentane (29%); 3,3-dimethyl-pentane (7%); 3-ethylpentane (21%).
2. Same products as Problem 1: 39, 31, 10, 20%. Also products from $CH_3\cdot$ and $C_6H_{13}\cdot$ radicals in the triplet reaction.
3. Carbanion stabilized by three Cl substituents. CH_3Cl would undergo S_N2 reaction.
4. Ethylene formed by H shift; three different C—H insertion products (two of them racemic mixtures) and two different cycloaddition products (*syn* and *anti*).
5. Side products arise from the triplet (radical) reactions. Argon enhances singlet–triplet intersystem crossing, while oxygen scavenges triplets and radicals. The *n*-butane:isobutane ratio is lower for triplet reaction.
6. (*a*) φHgCBr₃, heat: (*b*) CH_3Li (L. Skattebøl, *Acta Chem. Scand.*, **17**, 1683 (1963)).
7. Symmetrical dimethyloxirene intermediate (I. G. Csizmadia, J. Font, and O. P. Strausz, *J. Amer. Chem. Soc.*, **90**, 7360 (1968)).

8. Singlet → Wolff rearrangement; triplet → cycloaddition (M. Jones, Jr., and W. Ando, *J. Amer. Chem. Soc.*, **90,** 2200 (1968)).

Chapter 8

1. 75 kcal/mole; 418 nm. The spin-forbidden transition might be observed in a heavy-atom solvent, such as CH_3I.

2. $k_f = 5.5 \times 10^6 \text{ sec}^{-1}$; $k_{ic} = 5 \times 10^5 \text{ sec}^{-1}$; $k_{isc} = 4.0 \times 10^6 \text{ sec}^{-1}$.

3. Energy transfer requires the quencher to have $E_T < 22$ kcal/mole. Electron transfer is a common alternative quenching mechanism (E. A. Ogryzlo and C. W. Tang, *J. Amer. Chem. Soc.*, **92,** 5034 (1970)).

4. In nonpolar solvents, lowest $n \to \pi^*$ singlet; in polar solvents, lowest $\pi \to \pi^*$ singlet.

5. (*a*) Oxetanes.
  (*b*) Type II cleavage, vinylcyclobutanol, and cyclohexenol formation.
  (*c*) Homolytic cleavage and decarboxylation.
  (*d*) Sensitized dimerizations: [2+2], *cis* and *trans*; [4+2], *exo* and *endo*.
  (*e*) Conrotatory ring closure.
  (*f*) Type I cleavage, followed by decarbonylation or α or γ H shift.
  (*g*) Expulsion of N_2.
  (*h*) Dibenzoylethane.
  (*i*) Singlet oxygen cycloaddition at the 9,10-positions.
  (*j*) Sensitized *cis–trans* isomerizations.

6. (*a*) (F. B. Mallory, C. S. Wood, and J. T. Gordon, *J. Amer. Chem. Soc.*, **86,** 3094 (1964)).
  (*b*) (G. Wetterman, *Photochem. Photobiol.*, **4,** 621 (1965)).
  (*c*) (J. W. Meyer and G. S. Hammond, *J. Amer. Chem. Soc.*, **94,** 2219 (1972)).
  (*d*) [2+2] cycloaddition and allylic H abstraction with dehydration.
  (*e*) (J. R. Edman, *J. Amer. Chem. Soc.*, **91,** 7103 (1969)).

7. See 5 (*d*). No change with sensitizers (N. J. Turro and G. S. Hammond, *J. Amer. Chem. Soc.*, **84,** 2841 (1962)).

8. (M. H. Fisch and J. H. Richards, *J. Amer. Chem. Soc.*, **85,** 3029 (1963)).

9. (*a*) Emission produces ground state with different vibrational energies.
  (*b*) Singlet sensitized decomposition.
  (*c*) Lowest $\pi \to \pi^*$ triplet state, except when the amine is protonated.
  (*d*) All quenchers with $E_T < 66$ kcal/mole quench equally at the diffusion-controlled rate.
  (*e*) Radical chain reaction.

10. $\tau_S = 0.7$ nsec; $\tau_T = 10$ nsec; $k_S = 1.4 \times 10^8 \text{ sec}^{-1}$; $k_T = 1.0 \times 10^8 \text{ sec}^{-1}$ (P. J. Wagner, *Acc. Chem. Res.*, **4,** 168 (1971)).

Index

Page numbers *in italics* refer to tables.

DATE DUE			
GAYLORD			PRINTED IN U.S.A.